SHAPE CASTING:
2nd International Symposium

Shape Casting: 2nd International Symposium
TMS Member Price: $47   TMS Student Member Price: $37   List Price: $67

Related Titles
- *Shape Casting: The John Campbell Symposium*
  Edited by M. Tiryakioğlu and P.N. Crepeau
- *Modeling of Casting, Welding, and Advanced Solidification Processes XI*
  Edited by C.A. Gandin and M. Bellet
- *Solidification Processes and Microstructures:*
  A Symposium in Honor of Professor W. Kurz
  Edited by M. Rappaz, C. Beckerman and R. Trivedi

HOW TO ORDER PUBLICATIONS

For a complete listing of TMS publications, contact TMS at (800) 759-4TMS or visit the TMS Document Center at http://doc.tms.org:

- Purchase publications conveniently online.
- View complete descriptions, tables of contents and sample pages.
- Find award-winning landmark papers and reissued out-of-print titles.
- Compile customized publications that meet your unique needs.

MEMBER DISCOUNTS

TMS members receive a 30% discount on TMS publications. In addition, members receive a free subscription to the monthly technical journal *JOM* (both in print and online), discounts on meeting registrations, and additional online resources to name a few of the benefits. To begin saving immediately on TMS publications, complete a membership application when placing your order in the TMS Document Center at http://doc.tms.org or contact TMS.

Telephone: (724) 776-9000 / (800) 759-4TMS
E-mail: membership@tms.org or publications@tms.org
Web: www.tms.org

SHAPE CASTING:
2nd International Symposium

Proceedings of a symposium sponsored by
the Aluminum Committee of
the Light Metals Division (LMD) and
the Solidification Committee of
the Materials Processing & Manufacturing Division (MPMD) of
TMS (The Minerals, Metals & Materials Society)

Held at the 2007 TMS Annual Meeting & Exhibition
Orlando, Florida, USA
February 25 - March 1, 2007

Edited by

**Paul N. Crepeau
Murat Tiryakioğlu
John Campbell**

A Publication of

A Publication of **The Minerals, Metals & Materials Society (TMS)**
184 Thorn Hill Road
Warrendale, Pennsylvania 15086-7528
(724) 776-9000

Visit the TMS Web site at
http://www.tms.org

Library of Congress Catalog Number 2006939990
ISBN Number 978-0-87339-660-8

If you are interested in purchasing a copy of this book, or if you would like to receive the latest TMS publications catalog, please telephone (724) 776-9000, ext. 270, or (800) 759-4TMS.

SHAPE CASTING: THE 2ND INTERNATIONAL SYMPOSIUM TABLE OF CONTENTS

Shape Casting: The 2nd International Symposium

Liquid Metal/Solidification

Process Design/Analysis

Structure/Property

Modeling

Applications/Novel Processes

Foreword

Significant progress has been made in the quality and reliability of shape castings in the past several decades through a better understanding of the nature of structural defects, how they form, and how to avoid them. Light alloy castings are now replacing complex assemblies by offering weight savings and significant reduction in tooling, assembly, and quality costs. These advances have been made possible by research on the physics behind simulation codes, the quality of molten metal, the hydraulics of mold filling, and the nature of bifilms and their effect on mechanical properties.

The 1st Shape Casting Symposium held in 2005 initiated a forum where researchers and foundry engineers could exchange their latest findings to improve the quality and reliability of shape castings. The 2nd Shape Casting Symposium has followed this trend. Leading edge technologies and the latest innovations in casting process design and quality improvements relative to shape casting were explored through presentations by researchers from around the world, and are documented in the articles in this book.

We would like to express our appreciation to the Organizing Committee, authors, presenters, participants, and of course, Cheryl Moore and Christina Raabe at TMS for making this symposium a very successful event.

Paul N. Crepeau
John Campbell
Murat Tiryakioğlu
22-Nov-06

Editors' Biographies

Paul N. Crepeau is a Technical Specialist in Advanced Materials Engineering at General Motors Powertrain Group in Pontiac, MI USA. He supports aluminum intensive engine programs and leads a multidisciplinary team merging CAE and Materials Engineers. Dr. Crepeau received his B.S. in metallurgical Engineering at the University of Alabama (1978) and, after a 5-year respite at an iron foundry, both M.S. in Metallurgy (1985) and Ph.D. in Metallurgical Engineering (1989) from the Georgia Institute of Technology. He has published in the areas of fracture mechanics, molten metal processing, quantitative metallography and image analysis, aluminum heat treatment, and more recently Monte Carlo simulation of fatigue test methods. Dr. Crepeau is a registered professional engineer and former chairman of both the AFS Aluminum Division and the TMS Aluminum Committee. Dr. Crepeau was editor of *Light Metals 2003*.

Murat Tiryakioğlu is Professor in the Department of Engineering at Robert Morris University. He received his B.Sc. in Mechanical Engineering from Boğaziçi University (1990), M.S. (1991) and Ph.D. (1993) in Engineering Management from the University of Missouri-Rolla, and another Ph.D in Metallurgy and Materials. (2002) from the University of Birmingham, England.

Dr. Tiryakioğlu grew up in his family's foundry, which continues to thrive in Istanbul, Turkey. This has led to research interests in process design for high quality castings, aluminum heat treatment modeling and optimization, process-structure-property relationships in metals, statistical modeling and quality and reliability improvement, on which he has written a chapter, published over 40 technical papers, and edited 5 books.

Dr. Tiryakioğlu has received a *Certificate of Outstanding Performance* from The Boeing Company and a *Certificate of Appreciation* from the American Society for Quality (ASQ). He was selected a TMS *Young Leader* in Light Metals and was awarded the SME *Eugene Merchant Outstanding Young Manufacturing Engineer*. He is a member of the ASM International, TMS, and a senior member of. ASQ, and is an ASQ *Certified Quality Engineer.*

John Campbell has retired from his post at The University of Birmingham and his editorship of the International Journal of Cast Metals Research. He keeps in touch, retaining an Emeritus Status at the University.

Since these moves he has mainly occupied himself with practical work in foundries around the world, applying, testing, and extending the latest technology to upgrade quality and reduce costs. Despite some casting successes in which the author is grateful and proud, he has also accumulated a few scrapped castings that confirm the technology of filling castings is still not fully developed. Good, targeted research into the most valuable means to reduce scrap, increase quality and improve profit margins remains systematically neglected for lack of vision and funds. In China he has recently met foundries with the courage and entrepreneurial spirit to tackle huge castings using revolutionary technology. In Mexico the application of correct filling systems promises to revolutionize profitability. These foundries are maneuvering themselves into world leading positions. There are salutary lessons here.

Thus the globetrotting activity constitutes valuable education. However, longer-term updating of existing books and the writing of further books have been indefinitely delayed. In the meantime, *Castings, 2ⁿᵈ ed* (2003) and *Castings Practice* (2004), in addition to the proceedings of the first two Shape Casting Symposia, all remain bargains!

Session Chairs and Review Committee

Liquid Metal/Solidification

Sumanth Shankar is Associate Professor, and Braley-Orlick Chair in Advanced Manufacturing in the Department of Materials Science and Engineering at McMaster University and also serves as Adjunct Professor in the Department of Materials Science and Engineering at Worcester Polytechnic Institute. He received his B.Tech in Metallurgical Engineering at the Institute of Technology – Banaras Hindu University and Ph.D. in Materials Science and Engineering at Worcester Polytechnic Institute. He is currently developing a research program at the newly commissioned CRC in Light Metals at McMaster and teaches courses in transport phenomena, thermodynamics, solidification processing, and metal casting. He has published in the areas of solidification processing, silicon modification, die soldering and alloy design.

Process Design/Analysis

Makhlouf M. Makhlouf is professor of Mechanical Engineering and Director of the Advanced Casting Research Center at Worcester Polytechnic Institute (WPI). He received his BS with high honors from the American University in Cairo (1978), MS from New Mexico State University (1980), and Ph.D. from WPI (1990). Dr. Makhlouf's teaching and research focus on solidification of metals, and the application of thermodynamics, kinetics, and heat and mass transfer to modeling. He has authored over a hundred articles on topics ranging from metal matrix composites to hydrogen embrittlement in steel. Makhlouf has received the AFS Scientific Merit Award, the NACE Uhlig Award, and the ACRC Flemings Award. Dr.. Makhlouf is a member of TMS International, ASM International, AFS, NADCA, NACE, ASEE, and Sigma XI - the Scientific Research Society.

Structure/Property

Glenn Byczynski is Manager of the Nemak Engineering Centre, the site for new product and process development for Nemak Canada. He received his BS in Mechanical Engineering (1994) MS in Materials Science and Engineering (1997) from the University of Windsor, and, on a Ford Fellowship, his doctoral degree at the University of Birmingham, England (2002). Dr. Byczynski is an active member AFS and is currently Chair of the Detroit/Windsor chapter. He has served on the board of directors of the Foundry Educational Foundation. He is also a registered Professional Engineer with the Province of Ontario.

Modeling

Mark Jolly is Manager in the Castings Centre in the IRC in Materials and heads the Process Modeling Group at University of Birmingham, England. He has a degree in Metallurgy from University of Sheffield and a Ph.D. from the University of Cambridge. He has worked in the areas of rapid solidification technology, powder metallurgy, squeeze casting, metal matrix composites, foundry technology, process simulation software, and manufacturing problem solving. Dr. Jolly has run a number of UK Government funded research programs in investment casting, extrusion, and metal quality. He has published some 75 articles and presented over 35 invited lectures. His current research interests are in process modeling and promotion of latest best foundry practice.

Modeling (continued)

Jacob W. Zindel is currently a Technical Expert in the Materials Research and Engineering Department at the Ford Research and Innovation Center, Dearborn, MI. He has been working in the areas aluminum casting technology and solidification research for the past 15 years. As an experimentalist, Dr. Zindel has led studies on topics ranging from melt cleanliness assessments in a foundry to fundamental studies on microstructural evolution during solidification.

Applications/Novel Processes

David StJohn is CEO of the Cooperative Research Centre for Cast Metals Manufacturing in Australia (CAST). He is a graduate of the University of Queensland and since 1994 has held a Chair in Solidification Technology at the University of Queensland. His major research interests are in casting and heat treatment of aluminium and magnesium alloys with particular focus on how interactions between solidification, casting and forming affect microstructure, defects and properties. Professor StJohn has a long track record of working with industry on collaborative research programs. He has published widely and is currently on the editorial committee of the International Journal of Cast Metals Research.

SHAPE CASTING:
2nd International Symposium

Liquid Metal/Solidification

Session Chair:
Sumanth Shankar

PREVENTATIVE METAL TREATMENT THROUGH ADVANCED MELTING SYSTEM DESIGN

Mark Osborne[1], C. Edward Eckert[2] Thomas Meyer[3], Mike Kinosz[2],

[1] GM Powertrain, 893 Joslyn Ave, Pontiac, MI 48340 USA
[2] Apogee Technology, Inc.; 1600 Hulton Road P.O. Box 101; Verona, PA 15147; USA,
[3] Thomas Meyer Engineering; 3987 Murry Highlands; Murrysville, PA 15668; USA,

Keywords: Aluminum, melting, metal quality, Isothermal Melting

Abstract

Recent technology advances in aluminum melting developed under a U.S. Department of Energy program are changing how melt shops are designed and operated. These developments avoid temperature and turbulence during the melting, transfer, and distribution operations that degraded metal quality in the past. First, aluminum is isothermally melted under high flux subsurface conduction, with a maximum metal temperature less than 22°C above mean holding temperature. No products of combustion (i.e. water) are produced. Second, metal is quiescently transferred into integrally heated molten metal delivery ladles detained at the casting unit for controlled metal dispensation. These vessels functionally serve as portable holding furnaces. Finally, molten aluminum is delivered to the casting unit by a new high heat flux launder/trough system design based on conduction. Surface heating of metal does not occur and temperature can be closed loop controlled. This paper will describe the system and expected metal quality benefits.

Introduction

Dissolved hydrogen and suspended solids (typically oxides) are the contaminants of interest in engineered castings. Wrought alloy metal quality considerations add the requirement of trace element control; typically the alkali (Na) and alkali earth (Ca) elements. Except for trace element control, hydrogen dissolution and melt oxidation are ongoing environmentally dependent processes that usually result in a temporal deterioration of melt quality. Since both of these processes are thermally activated, melt temperature significantly impacts rate. Concentration of dissolved hydrogen, for example, increases exponentially with temperature. Surface renewal also affects molten metal quality by enhancement of transport processes on nascent surfaces. In alloys where aluminum oxide (typically the γ polymorph) is predominant, oxide film thickness is a function of the square root of time, **provided** that the growing oxide film does not rupture. If the oxide isothermally grows from a nascent surface to a thickness of 0.005 cm (50 μm in the first 60 s, it will increase to 71 μm during the next 60 s and to a total thickness of 87 μm after 180 s. In one day (86,400 s) the oxide thickness will only be 0.19 cm and grow to 0.50 cm following one week. The parabolic law oxidation reaction is kinetically self-limiting if the oxide remains contiguous. If ruptured, however, nascent surface renewal results in an oxidation rate characteristic of the initial phase of oxidation. Metal turbulence experienced during metal transfer and movement creates surface renewal.

Traditional Approach to Metal Quality

Past industrial aluminum melt preparation practices relied on remedial metal treatment to attain acceptable molten metal quality.

The need for high quality engineered castings and wrought alloy products (sheet, plate, and forging stock) for aerospace applications in the 1940 period heralded the use of effective remedial metal treatment technology. Such technology continued to improve to the extent that dissolved hydrogen levels below 0.10 ppm could be consistently achieved on a commercial scale. Hydrogen removal is a first order process wherein functionality exists between post-treatment and pre-treatment concentrations. The conversion of modern degassing processes is so great, however, that satisfactory results can be achieved with any reasonable pre-treatment hydrogen concentration. Obviously, such performance increases cost and treatment time.

Suspended oxides, however, create a more difficult separation challenge. Mechanical filtration notwithstanding, body force (gravity) separation is the underlying mechanism responsible for the bulk of inclusion removal in the processing of molten aluminum. Body forces are responsible for suspended oxides to float to the surface of a melt. At low Reynolds numbers and in the Stoke's Law regime, the terminal velocity of a suspended spherical oxide particle occurs is given by:

$$V_t = D_p^2 (\rho_o - \rho_M)g/18\eta \qquad (1)$$

Where: V_t = terminal velocity $\qquad D_p$ = particle diameter
ρ_o = oxide density $\qquad \rho_M$ = melt density
η = melt viscosity

Equation (1) is a force balance between buoyancy and drag forces and represents a velocity where equilibrium is achieved. Inclusion density, melt density, and melt viscosity are important material properties that determine the effectiveness of body force driven separation.

Matrix	T_M, °C	ρ, gm/cm³	ρ_{Oxide}	η, gm/cm-s
Al	660	2.38	3.94	1.79
Cu	1080	8.00	6.40	4.34
Fe	1540	7.03	5.34	6.92
Mg	650	1.59	3.65	1.25
N	1450	7.90	7.45	(5.50)
Zn	420	6.58	5.47	3.50

Figure 1: Key Properties in Body Force-Gravity Separation

It is a common practice in metals processing to allow oxides to rise to the melt surface, form a supernatant layer, and be removed as "skim". The degree to which oxide separation can occur in a metal system is determined by the independent variables in Equation (1). The melt density, oxide density, and melt viscosity are tabulated in Figure 1 for metals of commercial importance.

Matrix	$\rho_o - \rho_M$	$(\rho_o-\rho_M)/\eta$ *	Difficulty (5-most)
Al	+1.56	+0.87	5
Cu	-1.60	-0.37	4
Fe	-1.69	-0.24	2
Mg	+2.06	+1.65	6
Ni	-0.49	-0.09	4
Zn	-1.11	-0.32	3

Figure 2: Difficulty of Gravity Separation for Oxides in Various Metal Systems

The ratio of the body force ($\rho_o - \rho_M$) to melt viscosity (η) can be used as an index of difficulty for separation of a low shape factor or nearly spherical oxide of a particular size[1]. High shape factor oxides, such as films, decrease terminal velocity and render separation even more difficult. This ratio has been tabulated in Figure 2. A positive value of the ratio indicates sedimentation, while negative values imply oxide transport to the melt surface. In aluminum and magnesium, oxide densities exceed melt density with the implication that sedimentation will occur rather than the surface transport of oxides. The economical and well established practice of "sit and skim" used for most commercial metals is not applicable to aluminum. Alternatively, a quiescent aluminum melt *will clarify* through sedimentation; however, such practices require extremely careful sequestration of the oxide sediment during melt dispensation to avoid re-entrainment.

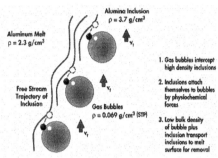

Figure 3: Oxide Flotation by Peripheral Interception

Gas bubble assisted flotation is used in aluminum melt processing to facilitate oxide transport to the melt surface. Efficient flotation requires relatively high metal residence time, high gas loading, a population of bubbles typically under 0.05 cm, and the presence of a surface-active agent to facilitate bubble/oxide attachment. Although high separation performance can be accomplished by flotation, system cost and melt loss due to parasitic skim formation may be high. Further, oxide separation by flotation is practically limited to oxide equivalent diameters greater than 4×10^{-3} cm (40 µ) due to the need for sparging bubble diameters consistently under 0.04 cm. Inclusion flotation by the dominant mechanism of peripheral interception is illustrated in Figure 3.

Melt Contamination

Conventional gas fired reverberatory melters can introduce significant metal contamination through characteristics inherent to the process, namely: surface heating, high p_{O2}, high P_{H2} equivalent (i.e.: water vapor), and surface agitation. The authors previously reviewed the causal mechanisms for such contamination[2]. In most engineered castings applications, molten aluminum is transferred from the melter to some form of delivery means (typically 400 kg – 14,000 kg capacity ladles) where it is ultimately delivered and dispensed to holding furnaces on individual casting units. Alternatively, off-site secondary aluminum processors transfer molten aluminum into large over-the-road ladles to be delivered and used at casting facilities. In all of these cases, molten aluminum is transferred and held, with or without the addition of supplemental heat.

Metal transfer turbulence is a significant source of contamination and affects both dissolved hydrogen and oxide content. Metal transfer ladles are typically filled by transfer (dumping) of molten aluminum through the top of the ladle from some source. The metal stream then falls to the bottom of the ladle through a total height distance of 150 cm to 300 cm. During this time, metal velocity can reach 350 cm/sec (maximum 470 cm/sec) for an average fall of 225 cm, resulting in power dissipation of approximately 0.33kW at a 15,000-g/sec-transfer rate. The energy associated with this transfer will be converted into melt/air interfacial (surface) area during the initial stages of ladle fill when the stream impacts on the ladle bottom.

During later stages of metal transfer, energy dissipation by the metal bath will result in reduced interfacial area generation as potential energy is converted into melt turbulence. This process recurs when molten aluminum is dispensed from the ladle into holding furnaces at the casting units.

Transfer induced surface area generation results in substantial surface renewal. A 14,000 kg quantity of molten aluminum in a 215 cm diameter ladle has 36.3×10^3 cm^2 of surface area that will support hydrogen adsorption and oxidation reactions. If the same quantity of metal produces a 25 cm diameter x 215 cm long stream during a 15,000- g/sec transfer, an *additional* effective surface renewal of 97.4×10^3 cm^2 will be created from the stream alone. The development of transfer induced metal droplets will dramatically increase surface above these values. Transfer induced turbulence also provides excellent bulk transport of surface hydrogen and intimate mixing of surface generated oxides that results in significant melt contamination and melt loss.

Conventional ladle heating methods exacerbate this problem. Traditional transfer ladles are pre-heated directly by flame from a high capacity burner. This causes oxidation of residual metal on the wall and migration of water vapor (product of combustion) into the refractory. When the ladle is full, lid mounted burners heat the surface of the contained metal. A typical 14,000 kg ladle will use a 2 – 3.3 MM BTU/hr input burner operating at 8-15% thermal efficiency.

Top heating and the total absence of beneficial natural convection results in considerable temperature stratification as demonstrated by a recent study. The ladle bath temperature was monitored at 13 cm, 89 cm, and 127 cm levels for a period of 11 hours after the metal was introduced at 645 °C in a cold ladle. At 4 hours the temperatures were 927 °C, 722 °C, and 613 °C at the lowest depth. At 11 hours, the temperatures were 961 °C, 791 °C, and 654 °C, clearly demonstrating the large temperature stratification. The oxidation rate of aluminum at 950 °C is 17.5 times higher than at 732°C.

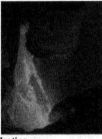

It has been shown that conventional melt preparation methods contaminate molten aluminum through excessive temperature, turbulence induced oxidation, and exposure to products of combustion. The avoidance of such contamination requires the development and implementation of advanced melting, metal transfer, and metal dispensation technologies.

Figure 4: Burner Method of Ladle Heating

Heat Transfer Mechanisms/ Isothermal Melting

Radiation is the dominant primary (source to sink) heat transfer mode in electric or combustion reverberatory furnaces. Heat flux is proportional to the difference between the forth power of source and sink temperatures, and directly proportional to melt emissivity and exposed surface area. A net heat flux of approximately 85 kW-m^2 (27,000 BTU/hr-ft^2) is used in most commercial furnace designs to avoid excess melt surface temperature and optimize thermal efficiency. High surface temperature decreases radiative heat flux and creates metal quality related problems. Secondary heat transfer occurs within the charge and is principally conduction in static furnaces, and a combination of conduction and forced convection in recirculating loop submersion melters. Natural convection cannot occur in a top down heating geometry. Secondary heat transfer is serial to primary heat transfer. Accordingly, the maximum heat flux attainable by primary heat transfer is determined by the secondary heat flux.

Isothermal Melting embodies an array of direct immersion resistance heaters operating at a surface heat flux as high as 217 kW-m^2 (130 w/in^2), transferring heat by predominantly forced convection to a flowing metal stream, and ultimately sinking this heat by a continuous melter charge feed. The effective melt surface heat flux in the array is at least 1210 kW-m^2 (385,000 BTU/hr-ft^2), and heater surface to bulk metal temperature difference is less than 35°C. Sink side heat transfer to the charge is also by predominantly forced convection, and the maximum bulk to charge temperature differential is less than 20°C. Detailed descriptions of this process have been provided in the literature[3-6]

Low temperature differential melting, characteristic of Isothermal Melting, has had a substantial impact on thermal efficiency, melt loss, and metal quality.

6

The specific melting energy (SME) for this process has been measured at 0.396 kW-kg^{-1} (614 BTU/lb) - with holding losses, and 0.5 kW-kg^{-1} (487 BTU/lb) - without holding losses, for Al-0.5% Mg alloys. Typical reverberatory furnaces operate at an industry average of 1.35 – 1.48 kW-kg^{-1} (2100-2300 BTU/lb). Actual Isothermal Melter melt loss is under 0.5% and holding melt loss is 0.08%/month for a furnace vessel containing 680 kg of aluminum. Significantly lower peak process temperature obtained in Isothermal Melting and subsurface anaerobic heat transfers results in a dramatic reduction in oxide generation and hydrogen adsorption.

Metal Transportation/Dispensation Components

The metal transport and dispensation system will essentially eliminate holding furnaces at the casting facility. It consists of two sub-systems: a means to maintain metal temperature and a method to dispense molten metal in a highly controlled manner. Each of these sub-systems will be discussed.

Metal transfer of metals to and from comparatively small in-plant delivery ladles is almost

exclusively accomplished by highly turbulent top pouring. Larger over the road ladles are also typically filled by top pouring, but contain tap holes that avoids the need to tilt the ladle for metal dispensation. A 2 m metallostatic head, however, exacerbates the regulation of metal flow rate. The traditional means for controlling metal flow rate in a gravity flow situation is to vary the area available for metal flow through an orifice positioned at the bottom of the ladle. Area is manipulated by a translating tapered rod or, alternatively, a movable slide gate that exposes a given area for flow. In some cases,

Figure 5: Portable Heated 14,000 kg Ladle

pumps, siphons, or a metal displacement pressure head is used to control flow.

An energy balance can be used to quantitatively describe fluid flow from ladles. In such reservoirs, the mechanical energy balance (or Bernoulli's equation) can be applied wherein the energy terms are the sum of pressure, kinetic, potential, and friction (thermal) energies. The equation written in traditional form is:

$$\int dP/\rho + 1/2(v_2^2/\beta_2 - v_1^2/\beta_1) + g\Delta h + E_f \quad = \quad M^* \tag{2}$$

Pressure   Kinetic   Potential   Friction   Mechanical

Typical ladles in commercial use that contain incompressible molten metal are relatively large vessels with constant or nearly constant cross-section and open to atmospheric pressure. In this situation, it can be shown that Equation (2) reduces to:

$$W = C_D A_O \rho (2gh)^{1/2} \tag{3}$$

where: W is the mass flow rate of metal from the ladle, ρ is molten metal density, A_O is orifice area, h is metal height above the orifice, and C_D is the orifice discharge coefficient.

Metal Ht, cm	Flow Rate, kg/sec
200	14.0
100	9.9
15	3.8

Figure 6: Instantaneous Metal Flow Rate at Selected Metal Height Values

Metal discharge from a conventional gravity-driven flow ladle is described by Equation (3). This equation is solved for several values of metal depth (height) in a situation involving molten aluminum (ρ = 2.3 g/cm^3), a fixed orifice area of 10 cm^2, and a discharge coefficient of 0.97.

Equation (3) can be modified to include the pressure energy term from Equation (2):

$$(P_2 - P_1)\,g/\rho + 1/2V^2 - g\Delta h = 0 \qquad (4)$$

Simplification, rearrangement, and insertion of orifice area and discharge terms yields:

$$W = C_D A_O \rho \,\{2g\,[\Delta h - (P_2 - P_1)/\rho]\}^{\,\frac{1}{2}} \qquad (5)$$

This relationship is similar to Equation (3), except that it includes a pressure energy term that can be manipulated to modify the mass flow rate of fluid issuing from the ladle at any given value for metal height, h. Using the same values as in the previous example, Equation (5) will be solved for the pressure required to maintain a constant mass flow rate of 4.5 kg/sec:

Metal Ht, cm	Flow Rate, kg/s	Δ P, kPa
200	4.5	- 33
100	4.5	- 12.5
15	4.5	+ 9

Figure 7: Pressure Requirements to Maintain Constant Flow with Variable Metal Level

Thus, it is possible to regulate flow from a molten aluminum ladle throughout its operating metal height range while manipulating a relatively modest value for pressure. This can be accomplished by reasonable ladle sealing coupled with high capacity kinetic blowers to create the appropriate positive/negative pressure conditions[7].

Ladle heating is the other challenge. The heating method that has been successfully integrated into 1,200 – 14,000 kg ladles is a conductive panel heating system known as "BSPP". In BSPP heating, multiple heating units are incorporated directly in the metal contacting sidewall of the ladle. The system operates at a variable heat flux up to a maximum of 40 kW-m², and is carefully regulated by sophisticated process control.

A power input under 60 kW is capable of maintaining isothermal conditions in a 14,000 kg ladle due to favorable external surface area to volume ratio of a cylindrical ladle, coupled with effective insulation. The total installed power can be in excess of 120 kW, which allows for rapid temperature recovery if needed. Figure 8 depicts the location of the heating units in a ladle sidewall and a 1,200 kg ladle.

Additionally, conductive heated BSPP troughs will be used as in-plant metal distribution infrastructure to provide heated metal to the casting unit with a tolerance of ±1°C. Due to the inherent nature of BSPP heating, heat transfer is sub-surface and does not result in surface heating as contrasted with conventional radiant heated "hot top" troughs[8]. Heat transfer is optimized with this arrangement to the extent that solid aluminum has actually been melted in a BSPP heated trough.

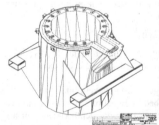

Figure 8: BSPP Heater Positioning and a Complete 1,200 kg Integrally Heated Electric Ladle

8

System Integration and In-Plant Metal Delivery

Aleris International performs most of GM's aluminum melting in their off-site facility. Aleris uses a variety of feedstock; such as P1020, twitch, scrap castings, runners, etc. These are melted and alloyed to specification in a large, somewhat conventional, reverberatory melter. The metal is then pumped to a transfer trough, which flows into a cascade transfer top drop delivery system for over the road 14,000 kg ladles. These ladles are held on standby with top gas heaters until GM requires delivery.

Once at GM, these ladles are then bottom tapped and flow into large 68,000 kg reverberatory holding furnaces. The metal is then laundered to a dual degas station, and then to a separate ladling furnaces of 10,000 kg capacity. At each ladle furnace the metal is recirculated and degassed. Metal is then either pumped or ladled into a mold according to the process.

An extensive system-wide energy tracking system has been developed and implemented. Initial energy audits of this process reveal that a substantial amount of energy is required to maintain metal in a holding furnace metal, which can exceed 0.025 kW-kg^{-1}. There is also redundancy in holding capacity further exacerbating the holding losses. The excess holding also results in an increased metal loss.

Further, in excess of 78,000 kg of molten metal inventory is maintained in-line with this scheme at GM. Such capacity is required to provide a ballast of metal to ensure constant flow conditions, and to allow for sufficient residence time for heating and metal treatment. Real time tailored alloy chemistry adjustment or alloy change is not possible with the current scheme.

A comparison between the current-conventional and fully integrated metal delivery and dispensation schemes is made in Figure 9. Importantly, holding furnaces are eliminated in the advanced scheme, as the delivery ladle essentially becomes the holding furnace. Metal temperature control is provided by close loop feedback between the casting unit, conductive troughs (see Figure 10), and the BSPP ladle. The duplex docking station is equipped with a flow through a rotary phase contactor for continuous metal treatment. Less than 2,000 kg of metal is contained within the distribution system that results in low inertia response for alloy changes. Details of this arrangement have been provided elsewhere[9,10].

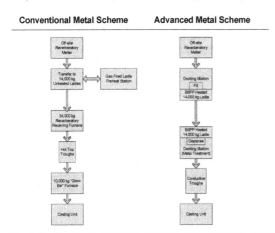

Conventional Metal Scheme   Advanced Metal Scheme

Figure 9: Flowcharts Depicting Current and Advanced Metal Distribution

Summary

Traditional melting practices have been shown to result in high formation of oxides, hydrogen levels, and energy loss.

9

Figure 10: SMCO Conductive Trough Section Installed and Prior to Installation (w/o covers)

These practices, including gas fired holding, gas-fired burner heating of ladles, metal transfer in open atmosphere, and turbulent transfer deteriorates metal quality and thus degrades the quality and related performance of the cast product.

The Isothermal Melting process has demonstrated a revolutionary change in melting technology. This advanced technology changes the paradigm of the foundry-melting department. Metal can now be melted at near theoretical energy values with minimal metal loss. The metal can be transported in high efficiency electric ladles.

Future developments of advanced Turbo Electric Ladle trucks will further demonstrate the ability to efficiently transport metal on grade, on temperature, and at the required specific gravity. Traditional holding furnaces can be replaced. These advanced ladles will show the capability to transfer metal quiescently with minimal exposure to the atmosphere.

These strategies will help the foundry to eliminate virtually all metal holding. Metal can be melted at the highest efficiencies and delivered directly to the casting unit with minimal holding losses.

Acknowledgements

A portion of this work is being supported under the US Department of Energy projects DE-FC07-01ID14021 and DE-FG36-05GO15047. The authors gratefully acknowledge Jacques Beaudry-Losique, Ehr-Ping Huang Fu and Bradley Ring, US Department of Energy – Office of Industrial Technology, and Ray Peterson, Aleris International, for their support, guidance, and advice provided for this work.

References

1. Eckert, C.E.; "History of Metal Treatment in Aluminum"; keynote talk – Cast Shop Technology, TMS Annual Meeting, Nashville, TN; 2000.
2. Eckert, C.E.; Osborne, M.; et. al.; "Preventative Metal Treatment Through Advanced Melting Technology";TMS Campbell Symposium; 2005.
3. "The Isothermal Melting Process – A Success Story", (Washington, DC: U.S. Department of Energy Industrial Technology Program, Friday Apr 16, 2004), http://www.oit.doe.gov/cfm/fullarticle.cfm/id=822.
4. "The Isothermal Melting Process"; Aluminum Now; Vol. 6; No. 4; July/August 2004; Washington, DC; The Aluminum Association; 26-29.
5. US Patents: 5,850,072; 5,850,073; 5,894,541; 6,872,294.
6. US Patents: 5,963,580; 5,968,223; 6,069,910.
7. US Patents 6,049,067; 6,066,289.
8. US Patent 6,444,165.
9. Osborne, M; "Advanced Aluminum Melting and Dispensation – Foundry Applications"; Ohio Technology Showcase – Casting Section"; Cleveland, OH; September 28, 2005.
10. New Technology Spotlight – Isothermal Melting; Light Metals Age; August 2005; 70.

A COMPARISON OF METHODS USED TO ASSESS ALUMINIUM MELT QUALITY

Derya Dispinar[1], John Campbell[2]

[1] Department of Metallurgical and Materials Engineering, University of Istanbul, Turkey
[2] Department of Metallurgy and Materials, University of Birmingham, UK

Keywords: aluminium alloys, melt quality, quality assessment, oxides, bifilms, reduced pressure test

Abstract

This study reviews and compares the techniques such as LIMCA, PoDFA, Prefil, LIAS and RPT that are used to quantify defects in liquid aluminium alloys. In general, it is found that most of the current techniques are not fully satisfactory. Only the Reduced Pressure Test (RPT) appears to give reliable results of the number and size of bifilms (and sometimes, but not infallibly, an indication of hydrogen content). The density (or Density Index) is valuable, but requires the supplement of additional quantification from the number and size of bifilms, which are recommended as additional major quality indicators. The use of a new Melt Quality chart is described.

Introduction

One of the challenges faced by the foundry technologist in the Al alloy foundry is to obtain an adequate and consistent melt quality. The chemical composition of the alloy is relatively easily achieved, but the three remaining features that are important and that define the aluminium melt quality are trace elements, hydrogen and oxides. This paper concentrates on the assessment of oxides, the so-called 'cleanness' of the melt. Since all oxides are expected to be in the form of doubled-over films, the terms oxide and bifilm will be used interchangeably in this account.

Even the precipitation of hydrogen from solution to create gas porosity requires the presence of bifilms (otherwise the hydrogen is forced to remain in solution in a supersaturated state). Thus the assessment and control of bifilms is seen to be the key to the control of casting quality.

The Oxide Film

For pure aluminium, because it is so very reactive with oxygen, oxidation starts by the rapid formation of an amorphous alumina layer within milliseconds, where the film grows by outward diffusion of aluminium ions. This amorphous oxide film has a high impermeability to the diffusion of aluminium and oxygen ions; thus the film cannot thicken rapidly, and so forms a protective layer over the molten aluminium. Because they are entrained quickly and have little time to grow, amorphous films are referred to as '*young oxides*'. They are characterized by extreme thinness, usually measured in tens of nanometers. These films are entrained into the bulk liquid by an enfolding action, necessarily causing them to be double and therefore to act as cracks. These can represent a serious threat to the quality of the final cast product.

Over the years, a number of test methods have been developed for inclusion detection in liquid aluminium [1], but the general experience in the casting industry has been that these were usually slow, inappropriately complicated and/or expensive for use on the foundry floor. The main approaches are listed below.

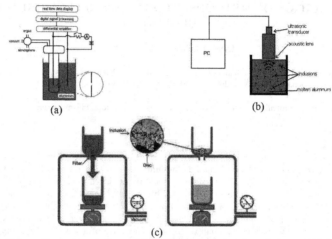

(a)

(b)

(c)

Figure 1: Schematic representation of (a) LiMCA, (b) ultrasonic technique, (c) PoDFA test

LIMCA (Liquid Metal Cleanliness Analyser). A measurement is made of the electrical potential across a small hole (diameter 0.05 to 0.10 mm) through which a sample of liquid is forced to flow (Figure 1a). The approach is one of the few inclusion detection techniques that can be rapidly repeated in situ, and so can monitor quality of metal flowing in a launder for instance. Naturally this technique is limited to inclusions that can enter the hole.

The thinking behind the LIMCA is based on the detection of particles. However, the present work indicates that most inclusions of importance to the quality of shaped castings are in the form of films (actually double films) often around 1 mm diameter (as will be seen below) but sometimes up to 10 mm diameter or more. Thus the LIMCA test is expected to behave curiously for most film-type inclusions. Asbjornsonn [2] has noticed heaps of spirals of oxides in the base of the LIMCA tube, indicating that in practice the films that enter the sampling tube become caught up at the mouth of the tube, and rotate into spirals like a flag tied to the mast by only one corner [3]. Fragments of the films remaining after most of the film has torn away would be expected to build up, eventually blocking the tube, as is often experienced. Thus in contrast with the unfortunately erroneous expectations based on the inclusions taking the form of particles, the signals from LIMCA are almost certainly not straightforward to interpret. Where LIMCA is used to monitor genuine particles in melts, such as borides or carbides present for grain refinement, the interpretation is likely to be unambiguous and valuable.

PoDFA and PREFIL. Both the PoDFA (Porous Disc Filtration Analysis) and PREFIL (Pressurized Filtration) tests involve the pressurizing of an approximately 2 kg sample of melt to force it through an extremely fine filter. The pore size of early filters was controlled as accurately as possible at close to 60 μm, although later developments have used larger pore sizes. The filter and its deposit (the filtrate) can be subsequently sectioned and the inclusions identified by metallographic techniques. The interpretation of the metallographic sections is of course a skilled and somewhat lengthy operation, and not easily quantified.

Unfortunately, techniques using filtration cannot be expected to detect nanometer thick films. Such films are not structurally rigid and so wrap around the filter structure so closely that they will remain undetectable, being only of the order of one thousandth of the thickness of sections of the filter. Thus permeability changes will be expected to be completely insensitive to their presence, and they are likely to remain invisible in metallographic sections.

However, of course, the identification of precise forms and chemistries of those inclusions that can be retained in the filtrate are valuable for guidance on their source.

For PREFIL the rate of blockage of the filter has been proposed as an additional measure of cleanness (Figure 1c). However, Cao [4-7] has recently drawn attention to the difficulty of the control of pore size in the filter, leading to variations in permeability that are greater than the differences generated by the growth of the filtrate. Effectively, he claims the technique is insensitive to inclusions in general because the signal-to-noise ratio is poor. Naturally, these concerns apply even more seriously to the important young bifilms. However, the move to larger pore size is commended, since it is a rational move to improve to some extent the sensitivity of the technique.

Ultrasonic Probes. A number of interesting attempts have been made to monitor reflections from inclusions in the melt using ultrasonics [8-10]. The method can be clearly used on single samples or continuously (Figure 1b). Signals can be obtained and counted, but seem to correspond to inclusions of size only up to 0.1 mm diameter [1]. This is possibly because echoes from crumpled areas of films, would give the appearance of clusters of small reflectors instead of one large, but not very flat, film. Thus, once again, despite the addition of sophisticated data processing, the interpretation of signals from the technique is not straightforward and in all probability does not correspond meaningfully to the bifilms present in the melt.

Summary of Quality Assessment Techniques

Thus all of our current techniques for inclusion detection and monitoring give cause for concern. In addition many of the techniques have become highly sophisticated and therefore involve expensive hardware and software. In addition, the accuracy of some of the techniques has to be questioned when alumina films can exceed 10 mm across, but are only nm thick.

All major techniques for the assessment of the quality of melts tend to measure either hydrogen or inclusion content. However, as mentioned above, hydrogen alone is not the major factor controlling quality; it is actually the combination of hydrogen and bifilms together that plays the effective role.

The Reduced Pressure Test (RPT)

Having carried out this critique of the techniques for quality assessment of liquid aluminium alloys, in contrast to other quality assessment techniques, the Reduced Pressure Test (RPT) was regarded as having the best potential for development. It was simple and rugged, and sensitive to both gas and inclusion content, particularly double oxide films.

The Straube-Pfeiffer vacuum solidification test, giving the RPT the full name of its German originators, has been generally denigrated as a result of its failure to be a reliable hydrogen test. In fact, however, it has always been a good test of the *porosity potential* of the melt. Interestingly, the humble RPT is precisely what is required to meet the required conditions for the effective control of porosity, since, as we have noted previously, we need to monitor both inclusions and hydrogen content. Neither alone is sufficient.

13

In addition to being fundamentally appropriate, the test is easy, robust, simple, low cost, and relatively quick, assessing the effect of both detrimental defects: (i) hydrogen and (ii) bifilms [3].

The test simply consists of solidification under a reduced pressure. The reduction of pressure serves to magnify the effect of dissolved gas on the opening of bifilms. For instance, if the pressure is reduced to $1/10^{th}$ of an atmosphere (100 mbar), it is expected that the residual air layer between the films would be expanded by approximately 10 times, thus converting an essentially invisible defect into one that is visible on an image produced by X-ray radiography, and to the unaided eye on a polished section.

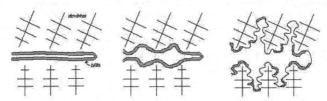

Figure 2: Schematic representation of the formation of porosity from a bifilm as might occur in an RPT sample or in a real casting [3]

The RPT is a simulation of the common condition inside a casting (Figure 2), where the effects of shrinkage and gas act in combination to grow porosity; the application of the vacuum additionally cooperates to expand the gap between the bifilms. This accurate simulation makes it a particularly appropriate test of the quality of the melt.

Quantification of the RPT

The quantification of the RPT has been widely carried out simply by density measurement, often comparing the density of a substantially solid control sample to give a 'density index'. This technique is, however, of no use for the detection of bifilms, particularly if the hydrogen content is low (it is assumed, hydrogen precipitates *only* in entrainment defects such as bifilms and bubbles). This is because bifilms can be so thin that they have practically no influence on the measured density, which remains high. However, the melt can be extremely poor for the production of castings requiring high and reliable properties, because of the potential presence of large populations of bifilms acting as cracks.

The quantification technique proposed here is the sectioning and polishing of the RPT sample. The polished section is then subjected to quantitative metallographic analysis to obtain two parameters:

1. The total number of defects on the surface of the sample; and

2. The total length of defects. This is the sum of the maximum length dimension of each defect. It is proposed that this length gives a measure of bifilm length [11, 12]. In fact, of course, if the hydrogen is high the expansion of the bifilm can continue to create a bubble of diameter larger than the length of the originating bifilm. However, the order of magnitude of the measurement is not likely to be greatly in error. In practice, it seems that useful assessments can be made using this approximation.

14

Examples of the use of the RPT

An extensive study on the reduced pressure test, using these quantification concepts, was carried out by Dispinar [13]. The effect of temperature, hydrogen content, binder type, binder content and vacuum pressure was investigated. Several different casting conditions and different alloys were examined. Finally, thousands of samples were collected [11-16]. A thorough examination of the results ultimately, proposed a 'bifilm index' which quantified the most deleterious defects in aluminium castings.

One of the studies was carried out in the electric furnaces with alloy a secondary Al-7Si-2Cu alloy. Two different gas flow rates were investigated during fluxing and degassing in a 2.5 ton electric furnace [14]. In the lance degassing stage, a high gas flow rate was used in the normal production conditions. This rate caused severe disturbance to the surface of the melt. A second experiment was carried out at a lower gas flow rate, so that the surface of the melt was hardly disturbed. With the high flow rate, the density of the RPT samples was increased but hydrogen remains high. On the contrary, with low flow rate, when surface of the liquid metal was not disturbed, the hydrogen levels were lowered but the density values were scattered (Figure 3a).

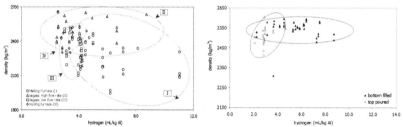

Figure 3: (a) Degassing studies with high gas flow rate and low gas flow rate (Al-7Si-2Cu) (b) density-hydrogen relationship in samples poured from different heights.

A study was carried out in a commercial alloy producer [14]. Two different casting devices that were designed to transfer the melt from the launder into the ingot moulds were investigated. The first device involved a free fall from the launder of approximately 250 mm giving turbulent filling of the ingot mould. Ingots resulting from this technique were referred to as 'Top Poured'. The second device filled the mould from approximately 10 mm height without disturbing the surface, yielding ingots referred to as 'Bottom Filled'. Although the average hydrogen content of the top pouring system had lower hydrogen content compare to bottom filled castings, they gave lower density results (Figure 3b). It is important to note that this showed that it was bifilms (turbulent filling) that decreased the quality, not the gas content.

Another study involving degassing was performed in the holding furnace with alloy Al-8Si-3Cu-Fe [15]. This time, two diffusers were sited at the bottom of the holding furnace. Three conditions were tested; both off, one on and one off, and both on. When no diffuser was used, the densities of the reduced pressure test samples fell towards the end the casting process. Once the diffusers were active, the density stayed almost unchanged at high values and the results were less scattered.

In another study, the rate of the removal of the stopper at the tap hole (at the exit of holding furnace) was controlled slowly such that the turbulence at the entrance to the launder was minimized. This helped the liquid to flow slowly in the launder. In addition the filling of the liquid in the mold was controlled quiescently by lowering the launder close to the casting mold, reducing the height of the final fall of the melt into the mold. The quality of the casting was significantly increased with the achievement of quiescent conditions [15].

An examination of mechanical properties of test specimens obtained from the above castings trials was also investigated [16]. It was extremely inconvenient and costly to machine bars directly from the ingots. Therefore, initially, different test bar patterns were investigated by real-time video X-Ray radiography with a view to optimizing the design to achieve quiescent filling. Then, the Weibull distributions of the properties were compared with bifilm index. Interestingly, as the bifilm index (the total oxide length) increased ultimate tensile strength increased somewhat but ductility fell. This observation points to a surprising and perverse action of bifilms to improve strength slightly, although ductility pays a high price. In contrast, with the achievement of quiescent conditions that gave a lower bifilm index, the elongation values were increased but UTS values fell slightly. The results confirm that the dominant benefit from quiescent filling is the ductility. The change in Weibull modulus of UTS and elongation showed that the scatter was increased with increased bifilm index (Figure 4).

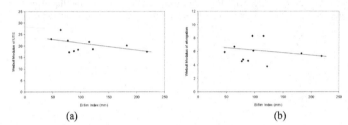

(a)   (b)

Figure 4: Bifilm index change with Weibull modulus of;
(a) elongation values, (b) UTS values

Number of Pores versus Length of Pores

For the quantification of RPT, a series of different quality indices were investigated [12]. Many different parameters and their combined effects were studied. It was observed that the number of pores was a dominant factor in some cases. Since the approach proposed in this study takes as a fundamental assumption that bifilms are the initiators of pores the number of the pores has to directly correspond with the number of bifilms.

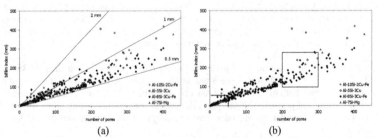

(a)   (b)

Figure 5: The relationship between the total length and number of pores showing (a)
average size range, and (b) use as a control chart for different qualities of melt.

16

For the alloys used in this study, the relation between bifilm index and number of pores appears to be reasonably linear (Figure 5a); it suggests that in this particular batch of material there were no single large pores and there were no cases of hundreds of fine pores. All data seemed to lie in between extremes of 0.5 and 2 mm diameter, (thus most of the bifilms in this sample were approximately 1 mm in diameter to within a factor of 2).

Anson and Gruzleski [17] draw attention to the fact that a single shrinkage pore in a casting can be observed as many small pores in two dimensions. In the RPT the action of shrinkage is negligible, so that the number of pores corresponds reasonably accurately to the number of bifilms. However, in a casting a single shrinkage pore could be initiated by a single bifilm, but appear to result in a multiplicity of pores. An attempt at assessment of quality from the casting is therefore somewhat problematic, whereas the strength of the RPT assessment is that it is unambiguous. In terms of a quality diagram for future practical use, Figure 5b shows how the quality field can be divided into target zones. A high quality melt will target the bottom left hand corner of the figure, with separate limits on numbers and lengths of bifilms. A melt required to be 'gassy' with well-distributed fine pores might chose a zone as illustrated by the box. In addition, of course, such a melt would also require specified limits for hydrogen content, that would in principle, be assessable from the RPT result.

Measurement of Hydrogen Content of the Melt

Turning now to the problem of the measurement of hydrogen and relating it with the density of RPT samples. This has been the central issue pursued with mixed success by previous researchers. In addition, of course, such efforts have often suffered from the lack of a reliable hydrogen analysis technique.

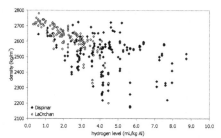

Figure 6: Comparison of LaOrchan's RPT result [18] with the current work

In the studies described in this paper, because of practical difficulties experienced with Alscan, the Hyscan technique was used. Hyscan measures the pressure of hydrogen evolved into a known evacuated volume from a known weight of solidifying alloy. From one melt sampled repeatedly over a period of time, the hydrogen analyses were found to be neither constant nor consistent, possibly reflecting to some extent real changes, but also probably influenced by the changes of nucleation difficulty known to affect this technique. The scattered results have been plotted against the density of the RPT sample (Figure 6). Interestingly, however, these results mirror those by LaOrchan *et al* [18] who had used Alscan, a fundamental technique for hydrogen analysis. Figure 6 underlines the conclusion that density alone is of practically no use as an indicator of hydrogen content.

It is important to keep in mind that bifilm length and number are, of course, independent of the hydrogen measurement, and its errors. Ultimately, the RPT test and the bifilm index are the direct measure of defects that have the potential to degrade the properties of casting, even though, of course, a high hydrogen level will enhance this degradation by tending to open bifilms.

Conclusions

1. The common criticism of the RPT as a poor technique for the assessment of hydrogen content is seen to be accurate for fundamental reasons.
2. The RPT is an excellent *porosity assessment technique*, indicating the combined effects of hydrogen and bifilms to cause porosity in the casting.
3. At the present time no other inclusion assessment method appears satisfactory to detect bifilms (possibly the most important defects in Al alloys).
4. The new interpretation of the test as an assessment technique for bifilms appears to be valuable and robust.
5. Recommended practice for determining the quality of the aluminium melts includes the use of the RPT to determine the (i) Density (or better, Density Index); (ii) the Bifilm Index (the total length of bifilms on the sectioned surface of the RPT sample); and (iii) the Number of bifilms. The Length and Number data are conveniently presented on a Quality Chart that can display separate limits for total lengths and numbers.
6. An example of the use of the RPT demonstrates that the more quiescently the melt is controlled, the higher the quality of the products. Good control includes reduced fall heights of the liquid at each transfer stage, and the reduction of turbulence at all stages, particularly in the final mold filling stage of the casting process.

References
1. Makarov, S., D. Apelian, and R. Ludwig, AFS Trans 1999 **107** 727-735
2. Asbjornsonn, E.J. 2001 *Dispersion of grain refiner particles in molten aluminium*, PhD Thesis, Department of Materials, University of Nottingham, UK.
3. Campbell, J., *Castings*. 2nd ed. 2003 Butterworth-Heinemann.
4. Cao, X., Scripta Materialia, 2005 **52** (9) 839-842.
5. Cao, X., Materials Science and Engineering: A, 2005 **403** (1-2) 94-100.
6. Cao, X., Materials Science and Engineering: A, 2005 **403** (1-2) 101-111.
7. Cao, X. and M. Jahazi, Materials Science and Engineering A 2005 408 (1-2) 234-242.
8. Doutre, D.A. and I.L. Guthrie, US Patent 4,555,662. 1985, LIMCA Research Inc. p 1-16.
9. Eckert, C.E., US Patent 4,563,895 1986: ALCOA, Pittsburgh, PA, USA. p 8.
10. Stiffler, R.C., R.C. Wojnar, M.F.A. Warchol, L.W. Cisko, J.M. Urabnic, US Patent 5,708,209, 1998: ALCOA, Pittsburgh, PA, USA. p 12.
11. Dispinar, D. and J. Campbell, Internat. J. Cast Metals Research 2004 **14** (5) 280-286.
12. Dispinar, D. and J. Campbell, Internat. J. Cast Metals Research 2004 **14** (5) 287-294.
13. Dispinar, D. 2005 *Determination of Metal Quality of Aluminium and its Alloys* PhD Thesis, School of Metallurgy and Materials, University of Birmingham, UK p 192.
14. Dispinar, D. and J. Campbell, J. of Inst. Cast Metals Engineers 2004 **178** (3612) 78-86.
15. Dispinar, D. and J. Campbell. *Effect of melting and casting conditions on aluminium alloy quality. 67th World Foundry Congress.* 2006. Harrogate, UK.
16. Dispinar, D. and J. Campbell, Internat. J. of Cast Metals Research 2006 **19** (1) 5-17.
17. Anson, J.P. and J.E. Gruzleski, Materials Characterization, 1999. **43** (5) 319-335.
18. LaOrchan W., M H Mulazimoglu, X-G Chen and J E Gruzleski; AFS Trans 1995 **103** 565-574.

Shape Casting: The 2nd International Symposium *Edited by Paul N. Crepeau, Murat Tiryakioğlu and John Campbell*
TMS (The Minerals, Metals & Materials Society), 2007

INITIAL FILTRATION BEHAVIOR OF LIQUID ALUMINUM ALLOYS

X. Cao[1,2]

[1]Dept. of Metallurgy and Materials, University of Birmingham, Birmingham, B15 2TT, UK
[2]Aerospace Manufacturing Technology Center, Institute for Aerospace Research, National Research Council Canada, 5145 Decelles Avenue, Montreal, Quebec, Canada, H3T 2B2

Keywords: Pressure filtration, Initial transient stage, Filtration mechanism, Aluminum alloy.

Abstract

There has been an increasing tendency to assess liquid metal quality using pressure filtration methods. Recently, with the introduction of derivative methods and the classification of three flow stages (i.e. initial transient, steady and terminal transient), some new insights into pressure filtration behavior of liquid aluminum alloys have been obtained. However, very little is known about the filtration behavior of this initial transient stage before the formation of a filter cake at the top surface of a filter medium. Clearly, this stage is important because it appears prior to the establishment of cake mode filtration and thereby it has significant influences on the structure and properties of the cake formed, the filtration behavior and the detection of liquid metal quality. This work discusses some fundamental occurring in this stage.

Introduction

Quality control of liquid metal is crucial to obtain sound castings. There has been an increasing tendency to detect liquid metal quality using Prefil Footprinter tests since it can provide both real-time filtrate weight vs. filtration time curves and highly concentrated samples for metallographic investigations of the solid inclusions. It has been generally believed that cake mode filtration dominates the filtration behaviors of the Prefil Footprinter tests [1-3]. To better understand the filtration mechanism, classic equations of cake mode filtration are briefly introduced here.

During the filtration test, the fluid passes through the filter medium with area A (m^2), which offers resistance to its passage, under the influence of a force which is the pressure differential across the medium. The cake thickness increases from 0 to L_c (m), corresponding to an increase in filtrate volume from 0 to V (m^3) or weight from 0 to W (kg) during filtration time t (s). According to Darcy's and Poiseuille's laws the flow velocity of the filtrate u (m/s) through the cake and filter medium is proportional to the pressure difference ΔP (Pa) imposed over the cake and the filter medium; the filtrate velocity is inversely proportional to the viscosity of the flowing fluid μ (Pa × s, or N-s/m^2) and the resistance of the cake R_c (m^{-1}) and the filter medium R_m (m^{-1}) [4]:

$$u = \frac{1}{A}\frac{dV}{dt} = \frac{\Delta P}{\mu(R_c + R_m)} \tag{1}$$

A material displaying constant cake concentration is incompressible and this type of filtration is known as incompressible cake filtration [4-6]. In constant pressure filtration we have [7, 8]:

$$\frac{dt}{dW} = \frac{\mu\sigma\alpha}{\rho^2 A^2(\Delta P)}W + \frac{\mu R_m}{\rho A(\Delta P)} \tag{2}$$

$$\frac{t}{W} = \frac{\mu\sigma\alpha}{2\rho^2 A^2(\Delta P)}W + \frac{\mu R_m}{\rho A(\Delta P)} \tag{3}$$

Where ρ is density of filtrate (kg/m^3), parameter α is specific cake resistance (m/kg), and σ is solid inclusion weight captured per unit filtrate volume (kg/m^3).

In 1956, Grace [9] proposed a general mathematical expression to correlate the filtrate volume (V) –filtration time (t) data obtained in constant pressure filtration:

$$\frac{d^2t}{dV^2} = K(\frac{dt}{dV})^n \tag{4}$$

Where the value of constant n defines the mode of filtration and the value of constant K for a particular mode of filtration depends on the filtration system. In simple terms, Equation (4) states that the rate of change of filter resistance is proportional to the instantaneous filtration resistance raised to a power which is dependent on the mode of filtration. When n = 0, the equation at constant pressure drop is given [9]:

$$\frac{t}{V} = \frac{K_1}{2}V + \frac{1}{Q_0} \tag{5}$$

Where, K_1 is the rate constant, and Q_0 the initial volumetric flow rate (m^3/s). Dividing both sides of Equation (5) by filtrate density ρ, the cake mode filtration Equation (3) can be obtained.

For the deep-bed filtration, n = 3/2 [9]. At constant pressure we have [9]:

$$\frac{t}{V} = \frac{K_2}{2}t + \frac{1}{Q_0} \tag{6}$$

Constant K_2 is rate constant. The mode of deep-bed mechanism, although not so well established experimentally as cake mode filtration, has been found to fit the data for a significant portion of the filtration cycle with many dilute suspensions [9]. Based on Equation (6), the linear relationship between t/V and t has been widely used to correlate the filtration behavior of viscose and cellulose acetate through various types of filter media which function by filtration in depth [9].

In this work, the above equations will be used to examine the filtration behavior occurring at the initial transient stage.

Experimental Procedures

The material used in this research was nominal Al-7/11.5Si-0.4Mg commercial alloys with various Fe and Mn contents. The experimental details were described in earlier publications [10, 11]. A weight of 3 kg of the experimental alloy was melted in an induction furnace and some were subjected to precipitation and sedimentation processing to sediment primary α-Al(MnFe)Si phase and oxide inclusions [12-14]. Approximately 2 kg metal from the experimental alloys was poured into a crucible to conduct the Prefil Footprinter test. The crucible was made of a low heat capacity, highly insulating fibrous material equipped with a filter of an average pore size 90 μm. When the metal in the crucible cooled to 700 or 720°C, the pressure chamber was closed and pressure applied to the liquid metal, pushing it through the filter. As filtered metal fell into the weigh ladle it was weighed every 3 seconds. After the pre-set test time (150 seconds) the pressure was automatically released from the pressure chamber and the filtration test was terminated. During the filtration test, two stage pressures were employed. A high-pressure of approximately 0.21 MPa (30 psi) was used to start metal flow through the filter at the beginning. When a filtrate weight of more than 20 gram was recorded at the loadcell the pressure was switched to a low value of approximately 0.08 MPa (12 psi) to maintain metal flow through the filter for the remainder of the filtration time [15].

Results and Discussions

Initial Transient Stage

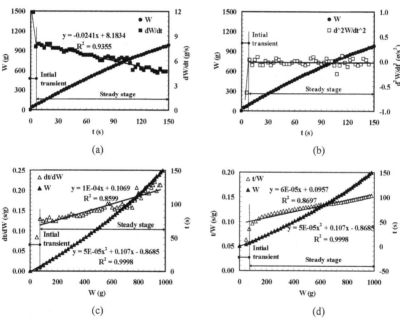

(a)

(b)

(c)

(d)

Figure 1. Analyses of a filtration curve for Al-11.52Si-0.38Mg-0.57Fe-0.59Mn alloy.

The filtration curve was analyzed using the 1st and 2nd derivative methods [10, 11]. A typical example is shown in Figure 1 for Al-11.52Si-0.38Mg-0.57Fe-0.59Mn alloy filtered at 700°C. The first derivative value (dW/dt) of the filtration curve in Figure 1a is the slope indicating the weight flow rate of liquid metal during the filtration test. The initial flow rate starts at a maximum, followed by a progressive fall. At 6 seconds the flow rate has reached a steady state. Thus, the moment at 6 seconds is identified to mark the end of the initial transient and the beginning of the steady stage. The transitional point derived from the 2nd derivative is seen to occur at 9 seconds. The 3 seconds delay is due to the 3-second intervals in the recording of the

21

filtrate weight-time data. Shorter intervals would assist with a convergence of the two points. During the initial transient stage the 2nd derivative values increase indicating that the maximum decrease in weight flow rate occurred at the start of the filtration tests and then the decrease in weight flow rate gradually becomes slowly. As indicated in Equation (2), a linear relationship between dt/dW and W is expected for an incompressible cake filtration mode. As shown in Figure 1c, the linearity between dt/dW and W was between 6 and 150 s indicating that incompressible cake mode filtration is the dominating mechanism over the steady stage [7]. However, the linearity deviates before 6 s. Therefore, the linear relationship between dt/dW and W can also be used to distinguish the initial transient stage from the steady stage if the filtration operates in incompressible cake mode. The deviation at the initial transient stage is further confirmed in the t/W vs. W plot as shown in Figure 1d. For an incompressible cake filtration at constant pressure, the linear relationship between t/W and W as shown in Equation (3) has been widely used to examine and verify the cake filtration mode [4, 6] but the transitional points are difficult to discern using this method. Therefore, it is suggested that the linear relationship between dt/dW and W, or the 1st derivative method is the better approaches to determine the transient points.

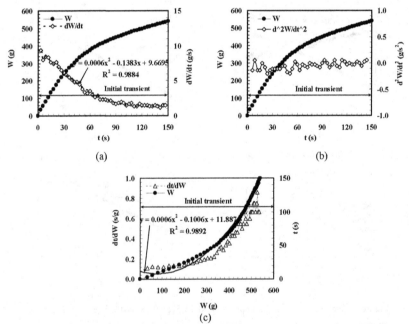

(a)

(b)

(c)

Figure 2. Analyses of a filtration curve for Al-11.52Si-0.38Mg-0.57Fe-0.59Mn-0.17Ti alloy.

It is found that the initial transient stage may last various lengths of time during the filtration test. In most cases, the initial transient stages are less than 10 s or even disappear [11]. Occasionally, this stage may last the whole period of the 150-filtration operation as shown in Figure 2. The 2nd derivative values are found to be around zero but the weight flow rate decreases gradually over the whole filtration test and a linear steady stage is not reached indicating that the test operation has been in an initial transient state. Further examination of the linearity between dt/dW and W (Figure 2c) indicates that an incompressible cake filtration has not been reached. In

22

addition, the best-fitted t vs. W equation has a high constant term (11.887 in Figure 2c) instead of zero as indicated in Equation (3). All these examinations indicate that the transient stage seems to extend over the whole period of this filtration test.

The occurrence of the initial transient state clearly demonstrates that the flow of liquid metal is unstable at the initial stage of a filtration test. The time interval and the flow mode occurring at the initial transient state might be used to evaluate the liquid metal quality. Clearly, this stage is important because it appears prior to the establishment of cake mode filtration and thereby it has significant influences on the structures and the properties of the cakes formed, the filtration behavior and the detection of liquid metal quality. It was reported that the high initial flow rate could influence the structure of the filter cake, resulting in the variation in specific resistance [6]. If the initial flow rate is too high, some of the particles may be forced into the filter medium and partly blind it [16]. The passage of fine particles into and through the medium can increase medium resistance, leading to gradual drop in flow rate [6]. In addition, prior to the occurrence of cake mode filtration, solid particles may also be sieved out onto the surface of the medium, blocking certain pores partially or completely, i.e. in effect reducing the open flow paths in the filter medium [6, 16]. These two mechanisms are termed as bed-deep filtration and surfaces straining, respectively, as discussed in detail below.

Deep-Bed Filtration (Standard Blocking)

The term "standard blocking" was used in filtration practice because it was observed to be by far the most common in the absence of cake filtration [17]. Deep-bed filtration describes a removal phenomenon where the effective pore size of the filter medium is much larger than the diameter of the particles that it will retain. The separation process occurs within the whole filter medium. Deep-bed filtration begins when the first particle adheres to or attaches to a pore wall, capillary, or to other particles previously retained. This can progressively reduce the internal diameter of the pore, cause a sudden or gradual increase in medium resistance, and eventually cease when either surface straining or cake filtration commences [17-19]. The solid particles are deposited suddenly, discontinuously and non-uniformly over or inside the filter medium, associated with the early stage of cake establishment [6, 17, 18]. The particles penetrate into the porous medium and deposit at various depths. The solid inclusions deposit onto the filter medium due to diffusion, direct interception, gravity and/or surface forces [5]. If an inclusion is to be captured by depth-bed mechanism, it is necessary that adhesive forces be higher than the dynamic effect of metal, which attempts to entrain the inclusions in the flow. It is evident that the chance to capture inclusions will increase with higher adhesive forces and lower flow rates. The locations with very low rates of flow are easier to capture solid inclusions. Figure 3 shows an example of deep-bed mechanism where solid inclusions deposit inside the filter medium. The primary α-Al(FeMn)Si attached to the pore wall inside the filter medium is due to its *in situ* sedimentation during or after the filtration test.

According to Equation (6), a linear relation between t/V and t (here between t/W and t) can be found if deep-bed filtration is the dominating mechanism. For the test as indicated in Figure 2, the result is presented in Figure 4. It is found that an excellent linearity between t/W and t is obtained over the whole filtration operation indicating that the deep bed filtration is the dominating mechanism for this test. In this case, a cake filtration mode has not been reached during the 150-s test. It was also claimed that Equation (6) stemmed from common differential Equation (4) can even be adapted to any portion of the filtration cycle for constant pressure filtration [9, 20]. Therefore, Equation (6) is further used to examine relatively short initial transient stages for Al-11.52Si-0.38Mg-0.57Fe-0.59Mn-0.17Ti alloy (Figure 5a) and Al-7.5Si-0.4Mg-0.5Fe-0.3Mn-0.13Ti (Figure 5b). The good linear relationship between t/W and t is found at the initial transient stage indicating that the deep-bed filtration is the dominating mechanism at this stage. Clearly, cake mode is an extreme case of deep-bed filtration where no blocking occurs. Another extreme for deep-bed filtration is surfaces straining as discussed below [21].

Figure 3. Internal deposition of solid inclusions onto pore wall of filter medium.

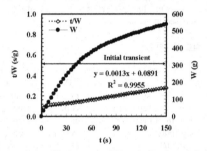

Figure 4. Good linear relationship between t/W and t for the test indicated in Figure 2.

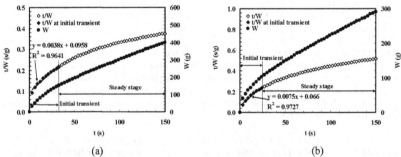

(a)   (b)

Figure 5. Analyses for (a) another filtration curve of Al-11.52Si-0.38Mg-0.57Fe-0.59Mn-0.17Ti and (b) a filtration curve of Al-7.5Si-0.4Mg-0.5Fe-0.3Mn-0.13Ti alloy.

Surfaces Straining

This mechanism is also termed as absolute blocking, filtration by straining, complete blocking, direct sieving, or medium filtration [4, 6, 16, 21]. Surfaces straining occurs when single particles larger than the holes in the inlet of the filter media capture and plug the individual pores or capillaries [4, 16, 19]. This method can capture only big particles, in particular those inclusions that form films [16]. Surfaces straining is a consequence of pore plugging, with a number of channels still available for fluid flow but the flow rate of the filtrate passing is proportionally reduced [4]. Figure 6 shows a typical surfaces straining appearing in Al-11.12Si-0.36Mg-1.14Fe-1.09Mn alloy filtered at 720°C. Clearly, the surface straining of all the medium pores is unlikely to occur in the pressure filtration test for aluminum alloys if the test is successfully carried out.

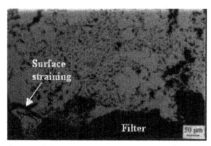

Figure 6. Surface straining for Al-11.12Si-0.36Mg-1.14Fe-1.09Mn alloy.

During pressure filtration tests, more than one filtration mode may occur. The filter medium incorporates both small and large holes and the filtered suspension may be composed of a wide range of particle sizes. Thus both surfaces straining and deep-bed mechanisms probably take place simultaneously, especially in the early stage of cake buildup or in the filtration of dilute suspensions [6, 19]. The particular combination of filtration mechanisms that occur will depend on the relative size of particles and pores, and concentration and particle approach velocity [6]. It is likely that the mechanism of surfaces straining dominates the clogging of small passages, while the deep-bed filtration dominates the clogging of large passages. Concurrent with the clogging of the passages in the filter medium, bridging of particles on the surface of the filter medium takes place. Most of the filtered particles are probably used for bridging and only a small fraction of the particles is used for clogging. When bridging is completed no more clogging occurs and cakes filtration begins [19]. In addition, surface filtration usually proceeds after this initial depth-bed filtration [16]. However, deep-bed filtration is more likely to dominate the filtration behavior at the initial transient stage than surfaces straining since good linear relationship between t/W and t is well found as discussed above. In some cases, however, the deep-bed mechanisms may play an insignificant role if the initial transient stage is short, say less than 6 s. This has been well confirmed in an earlier work where the effective resistance of the filter media is almost the same as the resistance of the new filter media [8]. In conventional filtration practice, many workers have demonstrated that the effective resistance of the filter medium during cake filtration is considerably greater than the resistance of the medium alone due to the internal deposition of solid particles (deep-bed mechanism) [9].

Conclusions

1. The initial transient stage can be best determined using the 1st derivative method of the filtrate weight (W) vs. filtration time (t) curves. The linearity between inverse weight flow rate (dt/dW) and filtrate weight (W) can also be used to determine the initial transient stage for incompressible cake mode filtration.
2. The initial transient stage may last for a variety of time periods ranging from its absence to several seconds. In some extreme cases it may last the whole time of the filtration process.
3. During the initial transient stage the flow rate usually decreases rapidly with filtration time.
4. Both deep-bed filtration and surfaces straining may appear but the deep-bed mechanism may dominate the filtration behavior at the initial transient stage during Prefil Footprinter tests.

References

1. P.G. Enright and I.R. Hughes, "A Shop Floor Technique for Quantitative Measurement of Molten Metal Cleanliness of Aluminum Alloys," *Foundryman*, 89 (11) (1996), 390-395.
2. P.G. Enright, "Molten Metal Quality Measurements on the Shop Floor" (Paper presented at Ensuring the Highest Quality and Reliability in Light Alloy Castings, The Institute of British Foundrymen, Suttow Coldfield, UK, 28-29 Oct. 1996), 6.1-6.18.
3. A.A. Simard, "Cleanliness Measurement Benchmarks of Aluminum Alloys Obtained Directly at-line Using the Prefil-Footprinter Instrument," *Light Metals*, ed. R.D. Peterson (The Minerals, Metals & Materials Society, 2000), 739-744.
4. A. Rushton, A.S. Ward and R.G. Holdich, *Solid Liquid Separation Technology* (Weinheim, Germany, 1996), 7, 33-83.
5. D. Apelian, "Tutorial: Clean Metal Processing of Aluminum," Proc. from Materials Solutions Conf.'98 on Aluminum Casting Technology, ed. M. Tiryakioglu and J. Campbell, 1998, 153-162.
6. M.J. Matteson and C. Orr, *Filtration Principles and Practices*, 2nd ed. (Marcel Dekker Inc., New York, 1987), 133-161.
7. X. Cao and M. Jahazi, "Examination and Verification of Filtration Mechanism of Cake Mode during the Pressure Filtration Tests of Liquid Al-Si Cast Alloys," *Mater. Sci. Eng. A*, 408 (2005), 234-242.
8. X. Cao and M. Jahazi, "Estimation of Resistance of Filter Media Used for Prefil Footprinter Tests of Liquid Aluminum Alloys," *Mater. Sci. Tech.*, 21 (10) (2005), 1192-1198.
9. H.P. Grace, "Structure and Performance of Filter Media," *A.I.Ch.E. J.*, 2 (1956), 307-336.
10. X. Cao, "A New Analysis of Pressure Filtration Curves for Liquid Aluminum Alloys," *Scripta Mater.*, 52 (2005), 839-842.
11. X. Cao, "Pressure Filtration Tests of Liquid Al-Si Cast Alloys. I. Flow Behavior," *Mater. Sci. Eng. A*, 403 (2005), 101-111.
12. X. Cao and J. Campbell, "Effect of Precipitation of Primary Intermetallic Compounds on Tensile Properties of Cast Al 11.5Si 0.4Mg Alloy," *AFS Trans.*, 109 (2001), 501-515.
13. X. Cao and J. Campbell, "Nucleation of Fe-rich Phases on Oxide Films in Liquid Al-11.5Si-0.4Mg Cast Alloy," *Metall. Mater. Trans.*, 34A (2003), 1409-1420.
14. X. Cao, N. Sanders and J. Campbell, "Effect of Iron and Manganese Contents on Convection-free Precipitation and Sedimentation of Primary α-Fe Phase in Liquid Al-11.5Si-0.4Mg Cast Alloy," *J. Mater. Sci.*, 39 (2004), 2303-2314.
15. P.G. Enright and M. Lovis, private Communications with author, N-Tec., October 2004.
16. D.B. Purchas, *Industrial Filtration of Liquids* (London: Leonard Hill Books, 1971), 40-43, 427.
17. P.M. Heertjes, "Studies in Filtration: Blocking Filtration," *Chem. Engineering Science*, 6 (1957), 190-203.
18. P.M. Heertjes, "Filtration," *Trans. Instn. Chem. Engrs*, 42 (1964), 266-274.
19. E. Kehat, A. Lin and A. Kaplan, "Clogging of Filtration Media," *I & EC Process Design and Development*, 6 (1967), 48-55.
20. Peter R. Johnston, *A Survey of Test Methods in Fluid Filtration* (Houston, Texas: Gulf Publishing Company, 1995), 116-123.
21. R.J. Wakeman, "The Formation and Properties of Apparently Incompressible Filter Cakes Under Vacuum on Downward Facing Surface," *Trans. I. Chem. E.*, 59 (1981), 260-270.

Shape Casting: The 2nd International Symposium *Edited by Paul N. Crepeau, Murat Tiryakioğlu and John Campbell*
TMS (The Minerals, Metals & Materials Society), 2007

RADIOACTIVELY LABELLED PARTICLE TRACKING IN STEEL CASTINGS

Y. Beshay[1], W. D. Griffiths[1], D. Parker[2] and X. Fan[2]

[1]The Department of Metallurgy and Materials,
[2]School of Physics and Astronomy,
The University of Birmingham,
Edgbaston, Birmingham, B15 2TT, United Kingdom.

Keywords: Steel, Inclusion, Radioactivity, Particle Tracking, Casting.

Abstract

It is well established that non-metallic inclusions are detrimental to the mechanical properties of metals. However, the movement of such inclusions during filling and solidification of a casting is difficult to determine experimentally. This paper describes the development of a technique by which the movement of inclusions can be tracked. Alumina particles of size 355-425 μm were radioactively labelled using a cyclotron and placed on a steel mesh at the entry of a ceramic shell mould. A low carbon steel was cast into the moulds and, after the casting had solidified, the position of the radioactive particle was determined using a γ-ray positron camera. The co-ordinates of the particle within the casting were obtained to an accuracy of ± 2-3 mm. This technique has been shown to be a valuable tool for any application where the presence of inclusions is critical.

Introduction

Non metallic inclusions are detrimental to mechanical properties. They act as stress raisers and hence sites for crack initiation and, when a metallic component is in service, may lead to catastrophic failures. There are two types of inclusions in steel castings, exogenous and indigenous inclusions. Exogenous inclusions are larger (≥80 μm) [1] and result from the incorporation of slags and refractories into the molten metal. Indigenous inclusions are finer inclusions that result from chemical reactions taking place in the melt, such as deoxidation of steel using Al, which results in alumina. Both inclusion types adversely affect fatigue, creep and corrosion properties [1]. Although there are several metallurgical treatments to reduce such adverse effects, such as the Ca-treatment to improve the morphology of MnS inclusions in steels, the presence of inclusions in castings is far from ideal.

In shape castings, inclusions may not be eliminated completely from the casting due to a lack of knowledge of how such particles move with the molten metal. There are several ways of diverting inclusions away from the casting, such as using dross traps and ceramic foam filters in the running system, but these methods are applied according to rules of thumb and experience. The exact mechanism by which such inclusions move within the melt, and the governing thermo-physical properties that affect this movement, have not been experimentally identified. Therefore this paper introduces an experimental procedure that can be used to track the movement of inclusions in steel castings by radioactively labeling them.

27

This procedure was based on Positron Emission Particle Tracking (PEPT) and Positron Emission Tomography (PET). PEPT is a process where a fluid or particle is radioactively labeled by changing one of its oxygen atoms into the radioactive isotope F^{18} [2]. The movement of a radioactive particle can then be tracked using a positron camera, which produces a real time path of the positron emitting particle. This procedure has been used extensively to obtain dynamic information about granular flow in the chemical, food and pharmaceutical industries [3-5].

Experimental Procedure

The experimental procedure used during this project was continuously developed to establish a working PET technique for steel castings that could serve as the foundation for future research in this area. The first attempt to apply this technique used a sand mould, which was not sufficiently robust, subsequently investment casting moulds were used and developed until repeatable results were obtained. These attempts are described in detail as follows.

The geometry chosen for the resin-bonded sand mould is shown in Figure 1. Alumina particles of size 355-425 μm were radioactively labeled using the cyclotron. The alumina particles were directly irradiated by a 35 MeV, ^{3}He beam. A few of the oxygen atoms in the particles are converted into radioactive F^{18} via the reactions $^{16}O+(^3He, p) \rightarrow {}^{18}F$ or $^{16}O+(^3He,n) \rightarrow {}^{18}Ne \rightarrow {}^{18}F$ [3-5]. The radioactivity of the irradiated particles ranged from 64 μci to 135 μci with a half life of 118 minutes. Typically an irradiated particle would still be detectable by the positron camera up to 3 hours after the particle was irradiated, although this varied with varying particle radioactivity. A stainless steel mesh (of mesh size 213 μm) was fixed at the top of the downsprue opening. The irradiated alumina particles were glued to a pin head which was placed through the mesh, hence fixing the particle at the centre of the downsprue opening at a known position.

A low carbon steel (EN3B; shown in Table I) was melted in a 45 kW induction furnace, and tapped into a graphite crucible at approximately 1700°C for pouring into the mould. When cast the molten steel instantly melted the glue on the pin head releasing the particle into the metal stream.

Table I The Chemical Composition of EN3B Steel

Element	C	Si	Mn	P	S
Wt %	0.15-0.25	0-0.35	0.3-0.9	0-0.05	0-0.05

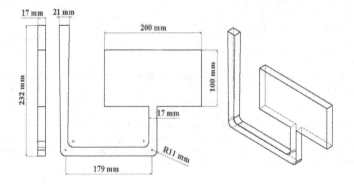

Figure 1. The sand cast steel plate dimensions.

28

The tracking experiment was repeated with investment casting moulds as these were more thermally stable. The geometry of the casting is shown in Figure 2. The ceramic shell moulds were built in layers, with a zircon primary coating, followed by layers of Molochite 30/80 and Molochite 16/30 respectively. The secondary coating was an Al_2O_3-SiO_2 based slurry and the average shell thickness was 15-17 mm.

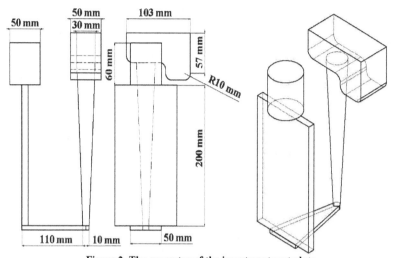

Figure 2. The geometry of the investment cast plate.

The solidified castings were then taken to the positron emission scanning camera. Figure 3 shows the setup of the casting between the camera faces. The emissions from the embedded radioactive alumina particle in the casting were recorded by computer and translated to coordinates with reference to the camera datum. In order to determine the exact particle position within the casting, another radioactive particle (the reference particle) was placed at several known positions and the coordinates of both particles were recorded with reference to the camera datum. This reference particle was attached to the corners of the plate, (as shown in Figure 3) and allowed the co-ordinates of the embedded particle to be determined relative to the co-ordinates of the reference particle with an accuracy of ± 2-3 mm.

This experimental method was used to study the effect of various parameters on the final particle location, namely, the effect of the initial particle position, the particle size, density and finally the use of cast iron instead of steel. These experiments have been summarized in Table II.

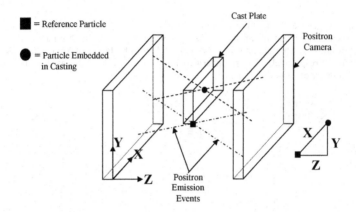

Figure 3. The casting setup between the camera faces. The positron emission events from both the reference and embedded particles form straight lines and the camera locates the particles at their intersection.

Table II The experiments carried out to investigate the particle tracking results.

Number of Castings	Number of Particles Per Casting	Cast Metal	Particle Type	Initial Particle Position in Downsprue	Particle Size (μm)	Filter Size
2	1	Steel	Al_2O_3	Centre	355-425.	
1	1	Steel	Al_2O_3	Edge	355-425	
3	1	Steel	Al_2O_3	Centre	355-425	20 mm; 10 ppi 15 mm;10 ppi
2	1	Steel	SiO_2	Centre	355-425	
1	1	Steel	Al_2O_3	Centre	≈ 600	
1	5	Steel	Al_2O_3	5 mm from centre	355-425	
1	1	Cast Iron	SiO_2	Centre	355-425	

Results

The particle tracking results shown in Figures 4 and 5, show the location of the particle in a plate cast in the sand moulds. Each point in the Figures shows one result obtained from the placement of the reference particle in a certain position. As the reference particle was moved to a new position on the casting, a new location for the radioactive particle inside the casting was obtained, with each location having an associated error. Figures 4 and 5 show five possible positions for the particle inside the casting were determined, and that the location of the particle was approximately halfway along the runner bar in both castings. However, some locations were determined to be actually outside the casting. The results showed good reproducibility, and these positions were verified using a Geiger counter shortly after the castings cooled down. The average error in all X, Y and Z directions was ± 2.9 mm, with the error arising mostly from the method used to place the reference particles.

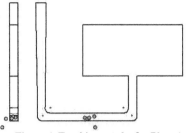 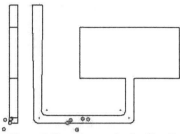

Figure 4. Tracking results for Plate A.   **Figure 5. Tracking results for Plate B.**

Figures 6 to 16 show the particle tracking results obtained in the ceramic shell moulds. Figures 6 and 7 are the results obtained when an alumina particle was placed at the centre of the downsprue opening. Two possible locations for the particle were found, as indicated in Figure 6, the bottom left corner of the plate and the centre, but in Figure 7 the particle was detected at the top right corner of the plate. In other words, the final location of the particle was not reproducible in this case. The results shown in Figure 6 were obtained from an early experiment, and the more clearly identified location shown in Figure 7 may reflect the fact that better results were obtained as the technique became more developed. Figure 8 shows the result when the particle's initial position was changed to the edge of the downsprue opening (from being placed at the centre). The particle came to rest at approximately the midpoint of the right side of the plate. In this case the average error was ± 2.8 mm.

Figures 9-11 show the results obtained when a ceramic foam filter was used. Two sizes were used, a 20 mm thick 10 ppi filter (Figure 9) and, in two experiments, 15 mm thick filters, also 10 ppi (Figures 10 and 11). The particles were found to be in the pouring basin in all three experiments, and it seems that the presence of the filter prevented the particles from reaching the casting without physically trapping them. The large scatter obtained when the particle is in the pouring basin is a result of the large thickness obstructing the positron emissions.

Figure 12 shows the results obtained when cast iron was used instead of steel. The particle's final location was somewhere around the centre of the plate. Although there was a large scatter in the results, the average error in the particle co-ordinates was only ± 3 mm. Figure 13 shows the result produced from an experiment where 5 radioactive alumina particles were cast in the same mould. They were initially placed within a radius of 5 mm at the downsprue opening. The particles seemed to have agglomerated and came to rest at the lower part of the downsprue. This was confirmed with a Geiger counter and no radiation was detected from elsewhere in the casting. As a result all 5 particles were detected as 1 particle by the positron camera. The average error in location was ± 3 mm. (The point recorded from, higher up in the downsprue was a result of misplacing the reference particles). Figures 14 and 15 were the tracking results for the experiments where a silica particle was used instead of alumina, to investigate the effect of particle density on the particle's movement. At this stage of the development of the technique, most errors associated in placing the reference particles where minimized and hence the scatter observed on Figures 14 and 15 was minimal. Although the results did not show good reproducibility, the average error was also a minimum of ± 2.4 mm.

Finally Figure 16 shows the tracking results obtained when a 600 μm alumina particle was used in order to investigate the effect of changing the particle size. The particle was found to be in the pouring basin and therefore the accuracy and scatter of the measurements was poor.

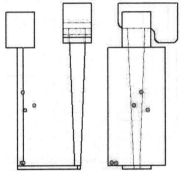

Figure 6. Al$_2$O$_3$ particle placed at the centre of the sprue opening.

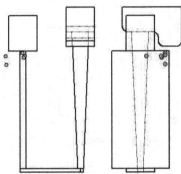

Figure 7. Al$_2$O$_3$ particle placed at the centre of the sprue opening..

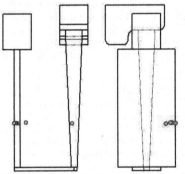

Figure 8. Al$_2$O$_3$ particle, at edge of sprue opening.

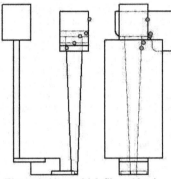

Figure 9. 20mm thick filter, 10ppi.

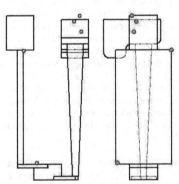

Figure 10. 15 mm thick filter, 10ppi.

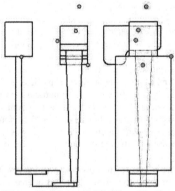

Figure 11. 15 mm thick filter, 10ppi.

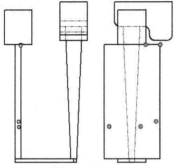

Figure 12. SiO₂ particle and cast iron.

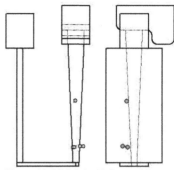

Figure 13. Multiple Al₂O₃ particles.

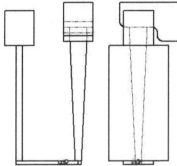

Figure 14. SiO₂ particle at centre of sprue opening.

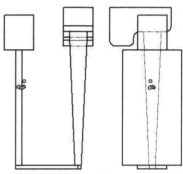

Figure 15. SiO₂ particle at centre of sprue opening.

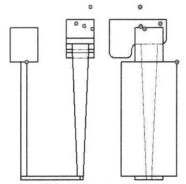

Figure 16. Al₂O₃ particle, 600 μm, placed at the centre of the sprue opening.

Discussion

The particle tracking results showed how complex and random the casting process is. The movement of the particle and its final location in the casting was mainly affected by the fluid flow of the liquid metal but, because the casting process is not very reproducible, reproduction of the final particle location was not often observed, as shown by Figures 6 and 7, and 14 and 15. The errors in the particle co-ordinates and the scatter observed were mainly due to the difficulty in placing the reference particles. Some attempts in placing the reference particles caused them to shatter, leaving fragments on the rough casting surface as they were moved from one point to the other. This led to some points being recorded outside the castings, or widely separated from the majority of the points indicating the actual particle location. Also, any movement of the casting between the camera faces resulted in misleading recordings of the particles position. Therefore, towards the end of the development of this technique, the castings were aligned more carefully with the cameras co-ordinate system and the reference particles were handled very carefully to produce the accurate results (error ± 2.4 mm) shown in Figures 14 and 15.

The effects of changing the particle size (Figure 16) should be governed by Stoke's Law. The larger the particle size, the easier it should be for it to float upwards towards the pouring basin. The effects of changing the particle's initial position, density or the metal cast (steel or cast iron) were difficult to deduce. This was because it was not known whether the changes in the final particle locations were the result of the variables or because of the irreproducibility of the casting process. This might be improved, in future, by exercising more control during mould filling, for example, by using a stopper in the pouring basin. In the case of the castings that contained a filter, (Figures 9 to 11), the particle was consistently found in the pouring basin. It is possible that, as the metal slowed down to overcome the obstructing filter, the less dense particle was allowed to float upwards and remain in the pouring basin.

Summary

The PET technique has demonstrated its capability in producing accurate particle tracking information, providing the experimental procedure was carried out well. It has been shown to be a useful tool that could be used, for example, to test casting modeling software and their ability to predict the movement of inclusions in castings. This technique could also be used to provide the track of an inclusions movement in real-time, if the casting occurs between the camera faces.

Acknowledgments

Dr. J-C Gebelin, Dr. S. Jones and Mr.A. Caden are thanked for their technical assistance.

References

1. R.Kiessling, *Non-Metallic Inclusions in Steel, Part III*, (The Institute of Materials, London, 1989).
2. X. Fan, D. Parker and M. D. Smith, "Enhancing 18F uptake in a single particle for positron emission particle tracking through modification of solid surface chemistry", *Nuclear Physics and Methods in Physics Research*, A558 (2006), 542-546.
3. P. W. Cox, S. Bakalis and D. J. Parker, "Visualisation of three-dimensional flows in rotating cans using positron emission particle tracking (PEPT)", *J. of Food Eng.*, 60 (2003), 229-240.
4. R. W. Barley, J. Conway-Baker and R. D. Pascoe, "Measurement of the motion of grinding media in a vertically stirred mill using positron emission particle tracking (PEPT)", *Part II, Min. Eng.*, 17 (2004), 1179-1187.
5. R. D. Wildman, S. Blackburn and D. M. Benton, "Investigation of paste flow using positron emission particle tracking", *Powder Technology*, 103 (1999), 220-229.

Shape Casting: The 2nd International Symposium *Edited by Paul N. Crepeau, Murat Tiryakioğlu and John Campbell*
TMS (The Minerals, Metals & Materials Society), 2007

THE EFFECT OF HOLDING TIME ON DOUBLE OXIDE FILM DEFECTS IN ALUMINIUM ALLOY CASTINGS

[1]W. D. Griffiths, [2]R. Raiszadeh[1] and A. O. Omotunde

[1] Dept. Metallurgy and Materials Science, School of Engineering, University of Birmingham, Birmingham, United Kingdom. B15 2TT
[2] Dept. of Materials Science and Engineering, Shahid Bahonar, University of Kerman, Jomhoori Eslami Blvd., Kerman, Iran.

Abstract

Double oxide film defects have been shown to have a deleterious effect on the properties of light alloy castings, and have been proposed as one reason why casting properties may not be as reproducible as required. Once double oxide film defects have formed in liquid Al, it is anticipated that they should undergo further development due to interaction with the surrounding bulk liquid metal. This may modify their effect on the mechanical properties of a casting, and therefore requires further study. The experiments reported in this paper involved casting into ceramic investment moulds, which were then held in the liquid state for different periods of time before solidification. Examination of the oxide films found on the fracture surfaces of tensile test bars from these castings showed changes in structure with time, such as bonding, which may be interpreted to gain a better understanding of the effect of double oxide film defects on casting properties.

Introduction

Double oxide films have been identified by Campbell as a serious defect in light alloy castings, contributing to reduced and variable mechanical properties [1]. They form as the oxidised surface of the liquid alloy is disturbed and folded over onto itself, trapping a layer of gas, which is subsequently submerged into the bulk metal. This can occur due to poor melt handling procedures, and a poor running system design, both leading to splashing of the liquid metal and entrainment of the oxidised surface. The characteristic nature of the defect is therefore that of two oxidised surfaces with a layer of gas between them [1], acting as a crack in the final casting.

It has been speculated that after a double oxide film defect has formed, the interior atmosphere of the defect can continue to react with the surrounding liquid alloy [2]. This has been demonstrated experimentally to be true [3,4], with first oxygen being consumed to form additional aluminium oxide at the interface of the defect and the surrounding melt, and then nitrogen being consumed to form AlN. Once all of the oxygen and nitrogen have been consumed in this way, the remaining gas would be argon, which forms 1 vol.% of air, but this is unreactive and insoluble in Al. In addition, hydrogen has been shown to diffuse into the atmosphere of the double oxide film defect until it reaches equilibrium with the surrounding melt [3,4].

It was further speculated that, once the interior gases of the double oxide film defect were consumed, the two sides of the defect might bond together in some way [2], reducing their deleterious effects on mechanical properties. The results reported here are from experiments that attempted to explore this effect by holding double oxide film defects in liquid aluminium alloys for extended periods of time to find any changes in their morphology.

Experimental Procedure

To study the effect of holding time of oxide film defects it was first necessary to ensure that the oxides observed were of known ages. This was accomplished by initially melting and holding liquid alloy under a vacuum before casting. Since the double oxide film defects in the melt consisted of a doubled-over oxide film that was expected to contain a small volume of trapped atmosphere, by holding the melt under a vacuum the volume of the atmosphere would expand, causing the defect to become more buoyant. Therefore, before casting, 9 kg of the liquid alloy was held in a chamber containing a furnace under a reduced pressure of 80 mbar; this should have resulted in an expansion of the atmosphere within an oxide film defect by about 12.5 times. Estimates of the velocity of buoyant particles in a melt based upon Stokes Law suggested that 1 hour was more than sufficient to allow any expanded oxide film defects to float to the surface of the melt [3]. This experimental arrangement has been shown in Figure 1.

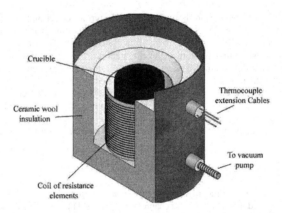

Figure 1. Diagram of the vacuum furnace in which the liquid Al was held before casting.

After this treatment the surface oxide was skimmed from the melt, and it was then cast into preheated ceramic shell moulds. These moulds had been designed in order to produce entrainment of the oxidised surface of the melt during filling, and their design has been shown in Figure 2. The liquid metal was poured into the conical pouring basin, from where it flowed into the horizontal runner bar, and then passed down in to the rectangular test bars below. This method of filling has been shown by real-time X-ray examination to be accompanied by considerable surface turbulence and entrainment of the oxidised surface skin of the liquid metal. Some moulds were allowed to solidify, (a process that took about 90 seconds), while others were placed in a resistance-heated furnace at a temperature of about 800°C, in order to prevent solidification. These moulds were subsequently removed after holding for different periods of time and then allowed to solidify, thus preserving the character of the oxide film defects present at that time. Therefore, the holding of the melt under vacuum prior to casting should mean that any oxide film defects found in the solidified castings would have been created during mould filling, and would be of known age, depending upon how long they were held in the furnace before solidification. These experiments were carried out with both commercial purity Al, (>99.7wt.%Al), and an Al-7wt.%Si-0.3wt.%Mg alloy, (2L99). For a summary of the experiments see Table I.

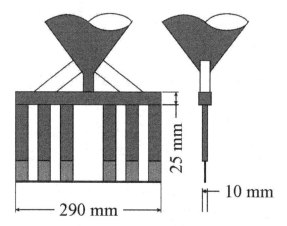

Figure 2. Diagram of the mould used to create oxide film defects during mould filling.

Table I. Summary of the casting experiments.

Alloy	0 mins.	10 mins	20 mins	30 mins	40 mins.
Commercial purity Al	√		√		
Al-7wt.%Si-0.3wt.%Mg	√	√	√	√	√

The cast test bars were then machined and tested using a Zwick 1484 tensile testing machine with a strain rate of 1 mm min^{-1}. The fracture surfaces of the test bars were then searched using a Philips XL-30 SEM for the presence of oxide film defects.

Results

Figure 3 shows an example of an oxide film defect found in a commercially pure Al casting that was allowed to solidify after the mould was filled, (meaning that solidification occurred in about 90 s). This defect appeared to be about 2 mm in length, and slightly less than 1 mm in width. Figure 4 shows an oxide film defect found on the fracture surface of a tensile test bar from a casting of commercially pure Al held in the furnace, in a liquid condition, for 20 minutes. The oxide film defect shown in Figure 3 lay across the fracture surface, but in Figure 4 the oxide film appears to enter the structure beneath the fracture surface, and the sides of the oxide film appear to have connections or bonds between them. Between the bonds the two sides of the defect seem to have expanded, giving the impression of a series of elliptical holes leading into the test bar, normal to the fracture surface. Figure 5 shows a high magnification view of one of the bonds between the two oxide film surfaces, and Figure 6 shows an EDX analysis taken from this point that suggested it was Al_2O_3.

Figure 3. An oxide film defect (arrowed) on the fracture surface of a tensile test bar from a casting of commercially pure Al solidified immediately after pouring.

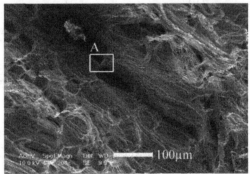

Figure 4. An oxide film defect found on the fracture surface of a tensile test bar from a casting of commercially pure Al solidified 20 minutes after pouring.

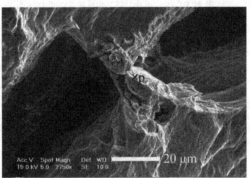

Figure 5. Higher magnification view of the region marked A in Figure 5 showing the bonded surfaces of the two sides of the oxide film defect.

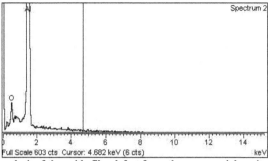

Figure 6. EDX analysis of the oxide film defect from the commercial purity Al alloy casting shown in Figure 3.

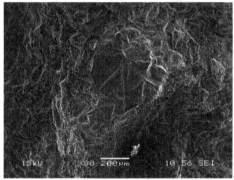

Figure 7. An oxide film defect on the fracture surface of a tensile test bar from a casting of Al-7wt.%Si-0.3wt.%Mg alloy that solidified immediately after pouring.

Figure 7 shows an SEM image of an oxide film defect found on the fracture surface of a tensile test bar from a casting of the Al-7wt.%Si-0.3wt.%Mg alloy that was allowed to solidify after pouring. Here the oxide film defect was about 600 μm in size. EDX analysis of this oxide film, shown in Figure 8, revealed peaks associated with oxygen and magnesium, suggesting a composition of $MgAl_2O_4$ spinel. Figure 9 shows an oxide film defect from a fracture surface of test bar taken from a casting held for 10 minutes before solidification, which suggests it was associated with a pore. Figure 10 shows an example of a symmetrical defect, which EDX analysis suggested was Fe-rich phase, but which indicated the presence of oxygen and magnesium also, suggesting the presence of an associated oxide film. This was observed on the fracture surface of a test bar from a casting that had been held for 20 minutes after pouring before solidification.

In other words, no bonded oxide film defects were seen in these Al-7wt.%Si-0.3wt.%Mg alloy castings, similar to those shown in Figures 4 and 5. Instead, oxides seemed to be associated with other types of defects such as pores and Fe-rich phase.

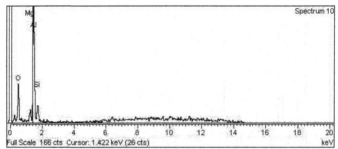

Figure 8. EDX analysis of the oxide film defect from an Al-7wt.%Si-0.3wt.%Mg alloy casting shown in Figure 7.

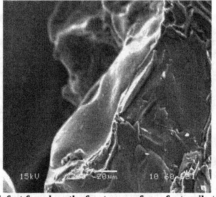

Figure 9. Oxide film defect found on the fracture surface of a tensile test bar taken from an Al-7wt.%Si-0.3wt.%Mg alloy casting held for 10 minutes before solidification.

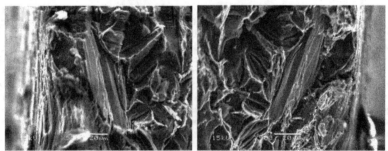

Figure 10. Two images of Fe-rich phase defects on opposing sides of the fracture surfaces of a test bar taken from an Al-7wt.%Si-0.3wt.%Mg alloy casting solidified after being held for 20 minutes after pouring.

Discussion

The SEM images of the oxide film defects showed that the defects found on the fracture surfaces had different characteristics in the two different alloys. In the commercially pure Al the two sides of the oxide film defect appeared to have bonded together to some extent. (See Figures 4 and 5; this was the only example of an oxide film defect found in these samples). In contrast, the oxide film defects in the Al-7wt.%Si-0.3wt.%Mg alloy held in the liquid state did not shown any evidence of bonding, and were associated with other defects.

In an earlier study of the effect of these defects on mechanical properties it was proposed that some bonding of the opposing sides of a double oxide film defect could take place, once the interior atmosphere, (which would probably be primarily oxygen and nitrogen) had been consumed by reaction with the surrounding melt [2]. In previous work it was demonstrated that this reaction of the interior atmosphere with the surrounding melt could take place within a few minutes at most, or at least in a time much less than the 20 minutes for which the casting of the commercial purity Al alloy was held before solidification [3,4]. Figures 4 and 5 may therefore be the first experimental evidence of the bonding of the opposing sides of an oxide film defect suggested by Nyahumwa et al. [2]. Previous research has demonstrated experimentally that hydrogen can diffuse into the interior of a double oxide film defect [4] and it is tempting to interpret the elliptical holes of the defect seen in Figure 4 as an indication that hydrogen had diffused into the defect from the melt and caused it to expand where it was not constrained by the bonds.

The behaviour of the oxide film defects in the Al-7wt.%Si-0.3wt.%Mg alloy was different, with large oxide film defects only found in the casting solidified soon after pouring. One explanation may be that, in this experiment, the double oxide films had fragmented into smaller sizes with time. Alternative explanations are possible, such as that with increased holding time the oxide film defects have greater time in which to migrate out of the region of the test bar due to density differences with the liquid alloy. However, this may be considered less likely, since the oxide film defect found in the commercial purity Al alloy after a 20 minute holding period was a relatively large defect and might therefore have been expected to be removed if this process was operating.

The different behaviour of the defects in the two alloys, bonding in the commercial purity Al alloy but not in the Al-Si-Mg alloy, may be due to the different oxides formed in each case. EDX analysis of the defect in the commercially pure Al alloy, shown in Figure 6, suggested that the oxide was alumina, while the EDX analysis of the film in the Al-7wt.%Si-0.3wt.%Mg alloy, shown in Figure 8, suggested that in this case the oxide was a spinel. Thermodynamic assessments of the oxide forming on Al alloys with varying Mg content also showed that alumina was to be expected in the case of the pure Al alloy, and spinel was expected in the case of the Mg-containing alloy [5]. It is known that alumina undergoes transitions to different crystal structures with time, unlike spinel, and this has been suggested as a way in which bonding of the two surfaces can occur [2].

It was not possible to relate the changes in the oxide film defects to the tensile properties of the test bars in which they were found. The oxide film defect in the commercial purity Al alloy , shown in Figure 4, appeared to lie normal to, rather than on the fracture surface, perhaps indicating it may not be the fracture initiator in this case. As the oxide film defects change their morphology with time their effect on mechanical properties would also change, but this relationship is still to be understood.

41

Conclusions

1. Experiments in which Al alloys were held in the liquid for periods of time before solidification showed that double oxide film defects found on fracture surfaces of tensile test bars can alter their nature with time.

2. Evidence has been found that, in commercial purity Al, the two surfaces of the oxide film defect can begin to bond together.

3. In contrast, in an Al-7wt.%Si-0.3wt.%Mg alloy, it appears that oxide film defects are reduced to smaller sizes, and are associated with other defects.

Acknowledgements

The authors would like to gratefully acknowledge the technical assistance of Mr. A. Caden at the University of Birmingham.

References

1. J. Campbell, *Castings*, 2nd ed., (Butterworth-Heinemann, Oxford, 2003).
2. C. Nyahumwa, N. R. Green and J. Campbell, "Effect of Mold-Filling Turbulence on Fatigue Properties of Cast Aluminium Alloys" *AFS Trans.*, 118, (1998), 215-223.
3. R. Raiszadeh, "A Method to Study the Behaviour of Double Oxide Film Defects in Aluminium Alloys", (Ph.D. thesis, University of Birmingham, United Kingdom, 2006).
4. R. Raiszadeh and W. D. Griffiths, "A Method to Study the History of a Double Oxide Film Defect in Liquid Aluminium Alloys", (Paper presented at the Shape Casting: John Campbell Symposium, 2005 TMS Annual Meeting, San Francisco, CA, Feb. 13-17, 2005), 13-22.
5. M. P. Silva and D. E. J. Talbot, "Oxidation of Liquid Aluminum-Magnesium Alloys", (Paper presented at the TMS Light Metals Committee 118th Ann. Mtg., Las Vegas, Nevada, Mar. 1989), 1035-1040.

HOW TO MEASURE VISCOSITY OF LIQUID ALUMINUM ALLOYS?

Mohammad Minhajuddin Malik[1], Guillaume Lambotte[2], Mohammed S Hamed[1], Patrice Chartrand[2], Sumanth Shankar[1],

[1]Centre for Solidification & Thermal Processing (CSTP), McMaster University, Hamilton, ON, Canada L9H 2E8.

[2] Centre for Research in Computational Thermochemistry, Department of Chemical Engineering, École Polytechnique, Montréal, QC, Canada H3C 3A7.

Keywords: Viscosity, Rheometer, Aluminum, and Al-Si alloy

Abstract

Viscosity of liquid Al and Al-Si alloys are critical to better understand the solidification and porosity formation in the mushy zone towards the end of solidification of the alloy. In this paper, viscosity data from the literature for Al and Al-Si alloys are presented along with the drawbacks of the most prevalent technique used to evaluate viscosity – the oscillating vessel viscometer. Experimental techniques using an alternative rotational rheometer equipped with a cone and plate measuring geometry are presented along with results of flow characterization of molten Zn and Al-Si alloys. The results show that molten Zn exhibits Newtonian flow characteristics and molten Al-Si eutectic alloy exhibits a non-Newtonian behavior at various melt superheats and low shear rate regimes

Introduction

Al-Si alloys are the most widely used alloys in the automotive and aerospace casting industry. It is critical to fully understand the flow characteristics of these alloys since the rheological properties of the liquid influences the filling of the mold cavity and feeding the mushy zone during solidification. Understanding the rheological properties of Al-Si alloys will help develop robust predictive models simulate the evolution of microstructure, porosity formation and hot tearing tendencies in the mushy zone during the final stages of alloy solidification. One of the key rheological properties of liquid metals is apparent viscosity.

Figure 1 presents a comprehensive graph tabulating most of the available date for the viscosity of pure Al. It is evident in Figure 1 that there is a 400% spread in the data available in the literature today [1]. Figure 2 presents similar tabulation of viscosity data for various compositions of Al-Si alloys. Figure 2 shows a 100% spread in the data from literature [1]. Over ninety percent of the data shown in Figures 1 and 2 were obtained using the oscillating vessel viscometer. Although the data was acquired by critically controlled experiments, there is a large spread in the reported values of melt viscosity. This may not be fully attributed to differences in the mathematical formulation and experimental errors. The shear rate experienced by liquid metal in the oscillating vessel viscometer is not uniform both in spatial and temporal regimes. However, the viscosity is measured over a range of shear rates and reported as one value for the liquid metal over the entire spatial and temporal regimes for the experiments. This reported value of melt viscosity is meaningful if and only if the liquid exhibits Newtonian flow behavior. In other words, the reported values of viscosity in Figures 1 and 2 [2] are valid only if the viscosity is independent of the shear rate experienced by the molten metal. However, if the liquid metal exhibits non-Newtonian behavior, a singular value of viscosity for an entire shear rate regime is not accurate. Therefore, we believe that the spread in reported viscosity values shown in Figures 1 and 2 may be due to an innate flaw in the assumption that liquid Al and Al-Si alloys exhibit Newtonian flow behavior. Further, there is no experimental evidence in the literature to justify an assumption of Newtonian or non-Newtonian flow behavior of Al and Al-Si alloys. Hence, the use of the oscillating vessel viscometer may not lend itself to accurate evaluation of melt viscosities of liquid metals exhibiting non-Newtonian flow characteristics.

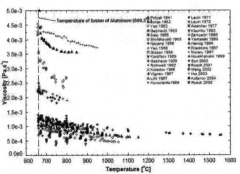

Figure 1: Reported viscosity data of pure liquid aluminum at various temperatures [2]

In this paper, we will present experimental evidence to show that, contrary to popular belief, pure Al and Al-Si alloys exhibit non-Newtonian flow behavior. Further, an alternate experimental scheme using a rheometer fitted with a cone and plate measurement geometry will be presented to evaluate, more accurately, viscosities of molten metals exhibiting non-Newtonian flow characteristics.

Viscosity of Molten Metal

Prevalent viscosity measurement techniques for molten metals and alloys include the capillary method, oscillating vessel method, rotational method, oscillating plate method, and

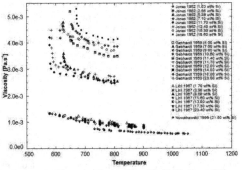

Figure 2: Viscosity data of liquid Al-Si alloys [2]

acoustic method. The oscillating vessel viscometer is the most popular among these techniques accounting for over 90% of the reported data for Al alloys shown in Figures 1 and 2. [1]

In the oscillating vessel viscometer the fluid sample in the crucible or vessel is set in oscillatory motion about a vertical axis by applying a set torque pulse. The torque will create an oscillatory motion of the vessel, which will eventually damp out, primarily by frictional energy absorption and viscous dissipation within the liquid. The viscosity of the liquid is then calculated by solving the governing second order differential equation of motion shown in Equation (1) as presented by Wang et al [3].

$$I_0\omega_0\left(\frac{d^2\beta(\tau)}{d\tau^2} + 2\Delta_0\frac{d\beta(\tau)}{d\tau} + (1+\Delta_0)^2\beta(\tau)\right) = 0 \qquad (1)$$

Where I_0 is the moment of inertia of the empty vessel or crucible, ω_0 is the angular frequency, $\beta(\tau)$ is the angular displacement, Δ_0 is the damping parameter and τ is dimensionless time.

44

It is the most preferred method for viscosity measurement of liquid metals due to the ease and accuracy in measuring the time period and decrement of oscillations. In addition, the geometry of the cylindrical crucible used is in range of 50 mm to 125 mm in height and 14 mm to 50 mm in internal diameter, which also leads to a relatively stable temperature profile throughout the sample [1][4].

Despite simplicity of construction, the mathematical equations used to describe the oscillatory movement of the crucible are complex and presently there is no reliable mathematical procedure available for solving them [5]. One of the major reasons for the spread in viscosity could be differences in mathematical models used to relate the experimental parameters to viscosity. The major mathematical models are Knappwost's Equation [6], Shvidovskii's Equation [7], Roscoe's Equation [8] and the Kestin & Newell model [9]. Among these mathematical models for the oscillating vessel viscometer, according to Iida et al [4], the Roscoe Equation provides the most accurate values of viscosity. However, the absolute Roscoe equation needs further modification by means of correction factor to account for the end effects caused by the curvature of the liquid metal at the free surface [4].

One of the main assumptions in using the oscillation vessel viscometer and developing the mathematical models is that the liquid metal exhibits Newtonian flow behavior. Figure 3 shows a typical schematic graphical representation of variation of shear rate of the liquid metal as a function of time and position within the oscillating vessel. Figure 3 (a) shows that the shear rate periodically decreases over the entire time of experiment due to the viscous damping of the liquid. Figure 3 (b) shows that shear rate continuously increases from the centre of the oscillating vessel to the wall of the vessel. Hence, the shear rate of the liquid in the oscillating vessel is not a constant at any instant in time or position in space. To present a singular value of viscosity for such an experimental set up will entail the assumption that viscosity of the liquid metal is independent of the shear rate (Newtonian flow behavior). Hence, to accurately evaluate viscosities of liquid metal showing non-Newtonian dependence on shear rate an alternate experiment technique will have to be adopted. A suitable technique is the rotational rheometer apparatus equipped with cone and plate measurement geometry.

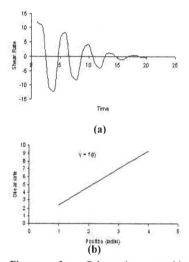

Figure 3: Schematic graphical representation of the variation of shear rate in an oscillation vessel viscometer. (a) Temporal variation and (b) Spatial variation.

Rotational Rheometer

The basic principle is to apply a known rotational torque to a fixed volume of liquid metal and measure the angular velocity attained by the liquid volume. The value of angular velocity is then converted to shear rate experienced by the liquid volume by a simple mathematical evaluation. The shear stress distribution in the known volume of liquid metal is evaluated from the rotational torque applied in the experiment. A graphical plot between the shear stress and the shear rate termed as the 'flow curve' is obtained as output from the rotational rheometer experiments [10]. The slope of the flow curve represents the 'apparent viscosity' of the liquid metal sample. The term 'apparent viscosity' was used instead of 'viscosity' because of the dependence of viscosity on the shear rate. The flow curve for liquids exhibiting Newtonian flow behavior is a straight line with slope (apparent viscosity) independent of shear rate. However, the flow curve for liquids exhibiting non-Newtonian flow behavior is not a straight line and the slope (apparent viscosity) varies as a function of shear rate.

Figure 4 compares schematics of an oscillating viscometer and a rotational rheometer equipped with cone and plate measurement geometry [11]. Table 1 lists governing mathematical equations for this measurement geometry and for two others, the cone and plate, and the cup and bob.

A predetermined volume of metal is subjected to shear between the measurement components of the rotational viscometer. Shear can be held constant or varied with time depending on rheometer and desired experimental conditions. This shear causes a laminar flow in the liquid sample [12]. The measured torque, T and the measured angular velocity, ω are then converted to shear stress, τ and

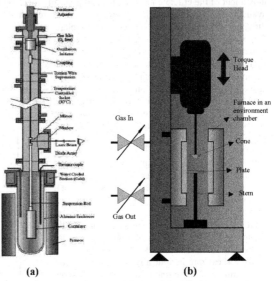

(a)   **(b)**

Figure 4: Schematics of (a) Oscillating vessel viscometer [11] and (b) Rotational rheometer equipped with a cone and plate measurement geometry.

shear rate, γ, respectively. (Equations (2) to (5) show the mathematical formulations for a cone and plate measuring geometry.

$$T = \frac{2}{3}\pi R^3 \tau \qquad (2)$$

$$\gamma = \frac{\omega \times r}{r \times \tan\theta} \cong \frac{\omega}{\theta} \qquad \text{(For small } \theta\text{)} \qquad (3)$$

Newtonian fluids $\rightarrow \tau = A\gamma \qquad (4)$

Non-Newtonian fluids $\rightarrow \tau = A\gamma^n \qquad (5)$

The slope of the plot between τ and γ (flow curve) represents the apparent viscosity of the liquid metal. For Newtonian fluids, A in Equation (4) is the apparent viscosity and for non-Newtonian fluids $(A\gamma^{(n-1)})$ in Equation (5) represents the apparent viscosity. Table I and Equation (3) show that the cone and plate measuring geometry is unique when compared to the other two geometries in that the value of shear rate, γ is constant across the volume of the liquid metal. It is only dependant on the angular velocity and the angle of the cone. In the cone and plate geometry, shear rate is constant in both the spatial and temporal regimes of the experiment.

The cone and plate measuring geometry in a rotational rheometer is considered as the ideal measuring geometry for evaluating flow behavior of liquid metals and alloys [12][13]. Apart from the advantages of evaluating the entire flow characteristics of the liquid metal, the mathematics in a rotational rheometer with cone and plate geometry is simpler than that in an oscillating vessel viscometer and presents an exact analytical solution for of the governing equations.

Table I: Measuring geometries and respective governing equations for rotational rheometer. γ is the shear rate experienced by the fluid sample, θ is the cone angle, ω is the velocity with which the cone is rotated, R is the radius of the cone, T is the torque and, τ is the shear stress.

Measuring Geometry	Governing Equation
Cone and Plate Setup	$\gamma = \dfrac{\omega \times r}{r \times \tan\theta} \cong \dfrac{\omega}{\theta}$ For small θ $T = \dfrac{2}{3}\pi R^3 \tau$
Parallel Plate Setup	$\gamma = \dfrac{\omega \times r}{h}$ $T = \dfrac{\pi R^4 \mu\omega}{2h}$
Cup and Bob Setup	**For annular gap** $\gamma = \dfrac{\omega \times R}{a}$ $T = \dfrac{2\pi R^3 \mu\omega H}{a}$ **For bottom gap** $\gamma = \dfrac{\omega \times r}{b}$ $T = \dfrac{\pi R^4 \mu\omega}{2b}$

Experiments and Results

Shankar et al [14] carried out experiments to evaluate the flow curves of pure Zn and Al–12.5wt%Si alloy using a rotational rheometer equipped with cone and plate. The alloy was made from 99.999% purity raw materials by melting and casting. The aim of this work was to evaluate the flow characteristics of the interdendritic eutectic liquid at the final stages of solidification of the mushy zone.

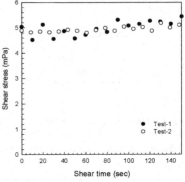

The time dependency of the flow behavior of near eutectic composition Al-Si alloys was characterized by measuring the shear stress produced at a given shear rate as a function of shear time and calculating the corresponding torque. Figure 7 shows that shear stress at constant shear rate was constant during the experiment which verifies that the measurement sample remained 100 % liquid.

Figure 8 shows the flow cures and apparent viscosity data for 99.99% purity zinc. Figure 8(a) shows that the relationship between shear stress and shear rate at constant superheat is linear, thus demonstrating Newtonian behavior. In Figure 8 (b), the plot for apparent viscosity versus temperature (evaluated by Equation (4)) agreed with published values, Figure 8(b).

Figure 7: Relationship between shear stress and shear time for the Al-12.5wt%Si alloy tested at $\gamma = 0.5$ s^{-1} and 598°C.

Figure 9 shows measured data and power-law fits of flow curves and apparent viscosity data (evaluated using Equation (5)) for Al-12.5 wt% Si alloys measured with a rotational rheometer equipped with cone and plate, thus demonstrating, contrary to popular belief, that molten Al-12.5wt% Si exhibits non-Newtonian flow behavior. Figure 9 also shows that Al-Si eutectic alloy is a shear thinning liquid.

Figures 8 and 9 illustrate the difference in experimental results for Newtonian and non-Newtonian fluids. It is to be noted that the measured data shown in Figures 8 and 9 are in the low shear rate regimes characteristic of the interdendritic regions of the mushy zone during the final stages of solidification.

Fundamental physics suggests that if a binary alloy exhibits non-Newtonian flow behavior then one of the pure components should also behave similarly. In Figure 10, it is evident that 99.999% pure Al liquid metal exhibits a non-Newtonian behavior at a given melt superheat in the low shear rate regimes [15]. Moreover, in Figure 10, pure Al shows a shear thinning behavior in apparent viscosity as a function of shear rate and also shows a hysterisis in the ramp up and ramp down stages of shear rate cycle.

Summary

The following can be summarized from this publication.

- Viscosity data for molten metals and alloys using the oscillation vessel viscometer is valid if and only if the liquid has been proven to be a Newtonian fluid, unlike in the case for Al and Al-Si alloys.

- Pure Al and Al-Si alloys are non-Newtonian fluids with shear thinning behavior. Hence, only experiments which measure viscosities as a function of shear rates will present accurate data for viscosities of these alloys.
- Rotational rheometer with a cone and plate geometry has been shown to be reliable equipment to measure viscosity of molten metals (both Newtonian and non-Newtonian) because the measurement technique does not assume a fluid behavior.

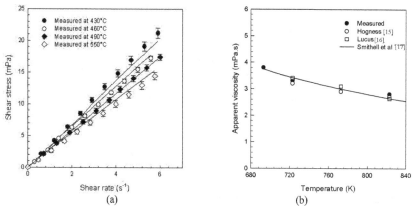

(a)   (b)

Figure 8: Results from rotational rheometer experiments with cone and plate for 99.99 % purity molten Zn. (a) Flow curves and (b) Apparent viscosity as a function of temperature [16,17,18].

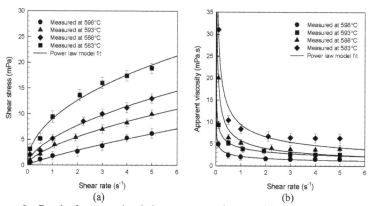

(a)   (b)

Figure 9: Results from rotational rheometer experiments with cone and plate for molten Al-12.5 wt% Si alloys. (a) Flow curves at four melt superheats. (b) Apparent viscosity data as a function of shear rate and melt superheats.

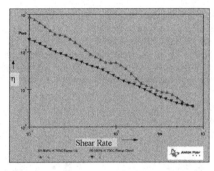

Figure 10: Apparent Viscosity curve for 99.999 % pure Al melt showing non-Newtonian (shear thinning) behavior [18].

References

1   Assael et al., J. Phys. Chem. Ref. Data, Vol. 35, No. 1, 2006.
2   Lambotte. G. and Chartrand P., Private Communications with authors, École Polytechnique, Montreal, QC, Canada.
3   Wang, D., and Overfelt R.A, Int. J. Thermophysics, v23, p1063–76.
4   Iida, T. and Guthrie R I L., *The Physical Properties of Liquid Metals*, (Clarendon Press, 1988), p283.
5   Morita Z., Iida T. and Ueda M., Inst. Phys Conf. Ser, Bristol, 1977, n30, , p 600
6   Knappwost, A., Z. Phys. Chem., 1952, v200, p81.
7   Shvidkovskiy, Ye G., *Certain Problems Related to Viscosity of Fused Metals* (State Publishing House for Technical and Theoretical Literature, Moscow, 1955).
8   Roscoe R., *Viscosity determination by the oscillating vessel method: 1. Theoretical considerations*, Proc. Phys. Soc. 1958, v72, p576–84.
9   Kestin J. and Newell G F, Z. Angew. Math. Phys. 1957, v8, p433.
10 Schramm G., *A practical approach to rheology and rheometry - second edition*, (Thermo Haake, 2000).
11 Brooks R F., Dinsdale A T. and Quested P N., Meas. Sci. Technol. 2005, v16, p354–362.
12 Barnes H.A., Hutton J.F. and Walters K., *An introduction to rheology*, (Elsevier, 1989).
13 Whorlow R. W., *Rheological Techniques - second edition*, (Ellis Horwood Limited, 1992).
14 Shankar S. and Makhlouf M.M., Private Communciations with authors, MPI-WPI, Worcester, MA, USA.
15 Paroline G., Brown E. F., and Langenbucher G., *A New Oven for Characterizing the Rheological Behavior of Molten Metals*, poster presented by Anton Paar USA, Ashland, VA 23005 USA.
16 Spells K.E.: Proc. Phys. Soc, 1936, v48B, p299.
17 Menz W. and Sauerwald F.: Acta. Met., 1966, v14, p1617.
18 De Waele A., J. Oil Colloid Chem. Assoc., 1925, v6, p33.

NEW APPROACHES TO UNDERSTAND MODIFICATION AND NUCLEATION MECHANISMS OF HYPOEUTECTIC Al-Si ALLOYS

Kazuhiro Nogita[1, 2], Stuart D. McDonald[2] and Arne K. Dahle[1, 2]

[1]The ARC CoE for Design in Light Metals, Australia
[2]Materials Engineering, The University of Queensland Brisbane, Qld 4072, Australia

Keywords: modification, casting, electron microscopy, synchrotron radiation

Abstract

Recently we have developed methods to enable the application of state-of-the-art analysis techniques for the determination of modification mechanisms, nucleation and growth during solidification in non-facet/facet hypoeutectic alloys. This work aims to use this unique method to develop grain refiners while keeping eutectic modification for use in the casting industry. The application focuses on the development of eutectic grain refiners for modified hypoeutectic aluminium-silicon alloys, which will result in major material property and processing improvements. New experimental approaches are μ-XRF(X-ray fluorescence) in Synchrotron radiation, EBSD (electron back-scattering diffraction) in SEM and FIB (focusing ion beam) in TEM.

Introduction

The demands for the performance of cast aluminium alloys continue to increase, while at least with respect to current knowledge, aluminium-silicon alloys are already on their performance limit. Solidification of the eutectic is one of the last reactions to occur and this is where most casting defects, such as porosity and hot-tears are formed. These defects reduce the mechanical properties, compromise the ability of the casting to contain liquid or gas under pressure, interfere with coating and machining operations and can detract from the appearance of the product. Although it is frequently acknowledged that the solidification of the eutectic is critical for porosity formation, currently there is a lack of understanding regarding how the eutectic evolves and how it can be controlled. Our research is primarily focused on the eutectic nucleation stage by investigating alloys containing different levels of impurities and with additions of different potential modifying elements, but also the resulting growth and silicon morphology. As a result, three different eutectic solidification modes have been identified in Al-Si foundry alloys, with the operation of each mode being controlled by the chemical composition and casting conditions[1]. Recent developments in the understanding of eutectic solidification in hypoeutectic Al-Si foundry alloys will be discussed in this paper. New experimental approaches, μ-XRF/Synchrotron radiation, EBSD/SEM and FIB/FE-TEM to engineering eutectic solidification are also explored. These new approaches and results are likely to form the basis for future developments in the metallurgy of Al-Si foundry alloys.

Experimental

Samples

The samples used in the experiments were hypoeutectic Al-Si alloys with Si in the range of 7-10wt%, with and without modifier elements additions. Castings were made in stainless steel cups with a thermocouple in the centre of the sample, 15 mm from the bottom, in order to monitor the temperature during solidification. A more detailed description of the experiments can be found elsewhere[2]. A melt temperature of 720°C was used throughout the study. The sample was placed in an insulating sleeve and allowed to cool, resulting in a cooling rate in the liquid at 610°C, just prior to nucleation of the solid, of approximately 1°C/s, and a total solidification time of approximately 300 seconds. The castings were sectioned perpendicular to the axis of the cylinder, 15 mm from the bottom, and polished for metallographic, μ-XRF/synchrotron, EBSD/SEM and FIB/TEM analyses.

Synchrotron μ-XRF

μ-XRF with synchrotron radiation allows for elemental mapping of trace level distributed elements for a depth of several micrometers. Experiments were performed at undulator beamline 47XU of the SPring-8 synchrotron in Japan[3]. A schematic diagram of the experimental set-up is shown elsewhere[4]. The undulator radiation was monochromatized at 17.9keV by passing through a liquid-nitrogen-cooled Si 111 double-crystal monochromator. The diameter is 155 μm and the focal length at the X-ray energy of 17.9 keV is 223.75 mm. The outermost zone width is 100 nm. In this setup, the beam size was 180 nm (vertical) x 150 nm (horizontal) and the total flux of the focused probe was ~2 x 10^9 photons/sec. The focused X-ray beam was used as the probe. The samples were mounted on a translation scanning stage with a motion accuracy of better than 10 nm. The XRF spectra were measured with a Si drift diode detector (Rontec Xflash D301). For elemental analysis, aluminium Kα1 (1.49keV), silicon Kα1 (1.74keV), strontium Kα1 (14.17keV), potassium Kα1 (3.31keV), titanium Kα1 (4.51keV), vanadium Kα1 (4.95keV), and iron Kα1 (6.40keV) were used. The use of the strontium Kα avoids any confusion that may occur from the overlap of the strontium Lα1 (1.81keV) and silicon Kα1 (1.74keV) peaks.

EBSD/SEM

Crystallographic relationships between primary aluminium and eutectic aluminium and silicon, which is key to understanding the eutectic solidification mode, were measured by EBSD mapping[1]. To obtain a flatter surface to enable area mapping, rather than point analysis, an ion-milling method was developed. The sample was thinned (thickness of less than 0.5mm) and polished to a 1 micron finish. Ion-milling was carried out with Ar+ ions with the conditions of: Voltage: 6 kV, Gun tilt angle: 15°, Time: 1.5 hours, followed by: Voltage: 4 kV, Gun tilt angle: 12°, Time: 0.5 hours, and with cooling by liquid nitrogen during milling. The samples were inspected in a Philips XL30 and JEOL 6460LA SEM at a tilt angle of 70° and acceleration voltage of 20 kV. The spatial resolution with these parameter settings is in the sub-micron range. A stationary electron beam was focused on the sample and the diffracted electrons that formed the backscattered image were collected and analyzed with the OXFORD Instruments EBSD system (Link OPAL camera) and HKL EBSD system. Mapping data were obtained with an automatic EBSD collection system.

FIB/TEM

FIB milling can be used to physically locate the nuclei for eutectic silicon and in combination with TEM allows for determination of the crystallographic relationships between the nuclei and silicon, and elemental analysis to be performed [5]. Eutectic grains were examined using scanning electron microscopy and those with internal particles were selected for further preparation using FIB. The advanced techniques of FIB milling combined with in-situ sample micro-manipulation has become available to selectively extract a TEM sample from any desired region within a larger sample. The FIB sample preparation was performed with a Hitachi FB-

2000A with 30kV Ga liquid metal ion source. TEM observations, micro-selected area diffractions (SAD) taken from the area of about 60 nm in diameter, small probe convergent beam electron diffraction pattern (CBED) taken from the area of about 10 nm in diameter and EDX analysis were conducted using a 200kV Technai20 TEM.

Results and Discussion

Eutectic modification mechanism

Theoretically, the Impurity Induced Twinning (IIT) model relies on the formation of a growth twin at the interface, which is theoretically favourable when the atomic radius of the modifying element relative to silicon (r/r_{si}) is 1.65[6]. Elements including Sr, Na, Ca, Ba and Eu have been reported to cause fibrous eutectic modification, and all these elements have an atomic radius ratio close to the theoretical ideal[1, 7]. However, the IIT mechanism is somewhat controversial as there are published examples showing well-modified silicon fibres without a high density of twins[8, 9], demonstrating that other mechanisms are also important in the modification of silicon, for example altered nucleation frequency of eutectic grains or altered surface tension.

Hampering efforts to confirm and expand on the IIT theory for silicon modification are the fundamental difficulties involved in analyzing the inter- and intraphase concentrations and distributions of typical modifying elements. For instance, strontium, the most common commercial modifier is often present in concentrations ranging from only 20ppm to 600ppm. The low concentrations used are below the detection limits of more conventional analysis techniques such as EDS and WDS (energy and wavelength dispersive spectroscopy). In fact, the only successful study examining the interphase distribution of strontium was performed by selectively dissolving the aluminium and silicon eutectic phases and performing atomic absorption spectroscopy on the solutions[10]. It was shown that strontium is strongly partitioned to the eutectic silicon phase, however it was not possible to determing the distribution of strontium within the silicon.

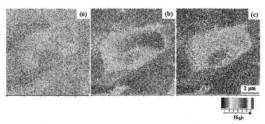

Figure 1. Synchrotron μ-XRF elemental map taken by fine-scan of (a) Al, (b) Si, and (c) Sr. (Scan pitch: 100 nm, integration time: 0.4 sec.)

A μ-XRF elemental map of a eutectic region (containing eutectic aluminium and silicon) taken with fine-scan (scan pitch: 100 nm, integration time: 0.4 sec) is shown in Figure 1. It is clear from the mapping results that strontium is present in the eutectic silicon, and is of negligible concentration in the eutectic aluminium phase. This map confirms that strontium is relatively homogeneously distributed throughout the eutectic silicon fibers. The concept of the IIT mechanism was developed to support TEM observations revealing high-density twinning within modified silicon fibres[6]. For the impurity strontium atoms to cause such a high-density of twins would require them to be relatively uniformly distributed (at least finely dispersed on a sub-micron scale) throughout the silicon fibre as observed in the current synchrotron μ-XRF

53

mapping results. These results therefore lend support the operation of the IIT mechanism in modification of Al-Si alloys with Sr. The IIT mechanism cannot however, explain well-modified silicon fibres occurring in the absence of high-density twinning[8, 9]. It is therefore acknowledged that other mechanisms must operate that are supported by a homogenous distribution of strontium inside the fibres. From the results in this paper it can be concluded that any proposed mechanisms of modification that involve segregation of strontium ahead of the advancing silicon interface are not acceptable.

Eutectic growth modes and nucleation site of silicon

There are three eutectic solidification modes (or macroscopic growth patterns) that operate in aluminium-silicon alloys[1]. Figure 2 shows an illustration of the three modes, with supporting evidence of their existence from the SEM and corresponding EBSD crystallographic orientation relationships between primary Al dendrites, eutectic Al and eutectic Si of fully solidified (uninterrupted) samples, and optical microscopy of samples quenched during eutectic solidification. These are: (a) nucleation on or adjacent to primary aluminium dendrites, (b) independent heterogeneous nucleation of eutectic grains in interdendritic spaces and (c) growth of the eutectic solidification front opposite to the thermal gradient. Growth mode (b) occurs in the presence of the most common modifying element, strontium and is typically characterised by a limited number of nuclei and a coarse eutectic grain size. The coarse eutectic grain size occurs because eutectic nuclei that are potent in unmodified alloys are poisoned or removed by additions of strontium[2].

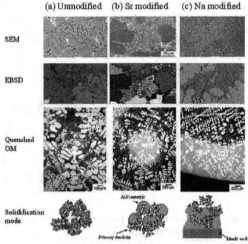

Figure 2. The three eutectic Al-Si growth modes; note in particular that the eutectic grain size in mode (b) is much larger than that in mode (a).

During microstructural analysis, it is not uncommon to observe centrally located particles within eutectic silicon (usually polyhedral near the centre of the eutectic grains), and these particles are often suspected of having a nucleating role. It has often been suggested that such particles in the silicon phase of aluminium-silicon alloys are aluminium-phosphide (AlP) and theoretically this is supported by AlP and silicon have near identical lattice parameters. Conventional EDX techniques do show the presence of aluminium and phosphorus, however oxygen is almost always present in large quantities[5] raising the possibility that the particle is actually aluminium

54

phosphate (AlPO$_4$), a common contaminant of aluminium based melts. Unfortunately, conclusive proof of nucleation cannot be obtained by observing an intimate physical relationship between phases, but is reliant on crystallographic evidence provided by transmission electron microscopy. This makes it difficult to identify a nucleus with confidence for two reasons, firstly, because a great number of TEM samples would need to be prepared before a suitable particle could be found, and, secondly, because the sample can be damaged during conventional preparation procedures. Because eutectic grain boundaries in Al-Si alloys are indistinct, it is necessary to interrupt solidification by quenching to observe the grains. Once eutectic grains are identified in this way, it is possible to locate the approximate center of the grain (unlike dendrites many eutectic grains are roughly spherical) in three dimensions and it can be assumed that this geometric center is where the initial nucleation event occurred. It is only recently that the advanced techniques of focused ion-beam (FIB) milling combined with in-situ sample manipulation has become available to selectively extract a TEM sample from any desired region within a larger sample. This technique allows the researcher to locate a particle and selectively mill a TEM sample from the desired location.

FIB milling followed by FEG-TEM analysis has been used to conclusively prove that the active nucleus for eutectic silicon in *unmodified* commercial alloys is AlP[5], a conclusion that has long been suspected in the literature. The procedure that was used is outlined in Figure 3 (a)-(c), which shows the sample; prior to milling (a), being manipulated with micro-tweezers after FIB milling (b) and ready for TEM analysis (c). At the exposed surface of the TEM sample the nucleus contains oxygen, however deeper in the sample the nucleus was confirmed as being AlP. It was concluded that the oxygen is an artifact from the sample preparation procedure. A high resolution FEG-TEM micrograph of the AlP/Si interface is shown in Figure 3 (d), along with selected area diffraction patterns from regions (i) – Si, and (ii) – AlP, in Figure 3 (e) and (f), respectively. There is no crystallographic mismatch between the AlP and Si, conclusively proving that the AlP particle acted as a nucleation site for the eutectic grain.

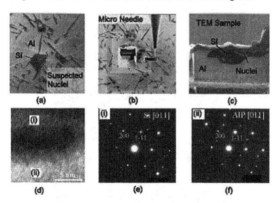

Figure 3 (a) SEM micrograph of a silicon particle containing a suspected nucleus, (b) in-situ micro-manipulation of a TEM sample that has been milled from the area of interest, (c) the sample ready for TEM analysis, (d) high resolution TEM image showing crystal lattice, SAD patterns for the Si (e)and AlP (f) phases show an identical crystallographic orientation.

Locating potential eutectic nuclei in *modified* alloys is very difficult due to the extremely low area density of eutectic grains (around 100 times fewer per unit area than are present in an unmodified alloy). This can be appreciated by observing a typical macrograph of a quenched, strontium modified Al-Si sample as shown in Figure 4 (a) which shows fewer than twenty independent eutectic grains. The probability of finding a nucleus located within any of these grains is quite low as many of the grains are not sectioned close to their center. Despite this, the preparation of many samples occasionally allows centrally located particles to be found, as shown in Figure 4 (b) and (c) or alternatively serial sectioning can be performed to find the center of the grain. Figure 4 (d) shows the results of TEM observation of a suspected nucleus in a strontium modified alloy (taken from the particle shown in Figure 4 (b) and (c) using FIB milling). This nucleus has some crystallographic orientation relationships with surrounding silicon and contains the elements Ti and V as found by EDX analysis. However, the crystallography measured by selected area diffraction pattern did not match any reported Ti-V crystal systems. This suggests the possibility of ternary alloy composition, such as Ti-V-B or Al(Ti)-V-B, where boron cannot be measured by EDX. Figure 5 shows synchrotron μ-XRF images taken from the nucleus region of strontium modified eutectic silicon. Potassium and titanium are clearly observed in the silicon, while small amounts of vanadium and iron are also detected. From the results from FIB and μ-XRF we cannot conclude about the composition and crystallography of the nuclei. However, it is expected that titanium must play an important role in the formation of these nuclei, either alone or in solution with aluminium. It is also seems that neither the modifier element itself (strontium) or AlP, the common nucleus for unmodified eutectic silicon, act as nuclei in strontium modified alloys.

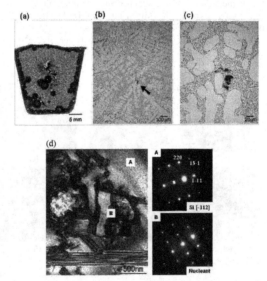

Figure 4. (a) Eutectic grains identified by quenching in an Al-10wt%Si alloy that has been modified with 300 ppm strontium. Occasionally it is possible to identify particles in the center of these grains (arrowed in (b) and higher magnification in (c)). (d) FIB/TEM micrograph and electron diffraction patterns of (A) eutectic silicon. (B) nucleant particle eutectic silicon. EDX shows Ti and V in the nucleant particle.

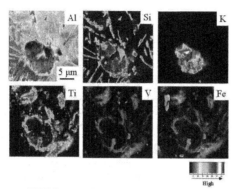

Figure 5. Synchrotron μ-XRF images of Al, Si, K, Ti, V and Fe from Sr modified eutectic. (scan pitch: 250 nm, integration time: 0.3 sec).

<u>Manipulation nucleation while simultaneously maintaining a modified structure</u>

From the above evidence it is reasonable to assume that the mechanisms of eutectic nucleation and modification are related, but that the operating window for eutectic grain size is not specified to obtain full modification. This raises the possibility of manipulating these events individually to tailor the microstructure. Based on the results from the FIB and μ-XRF analyses showing nuclei containing Ti, V and B, individual additions of Ti, V and B were made to an Al-Si melt modified by Sr with the aim of increasing the amount of eutectic grains while maintaining eutectic modification. The results are shown in Figure 6 and indicate the success of this approach, where a significant reduction in the eutectic grain size occurs as the proportion of Ti, V and B increase.

(a)   **(b)**   **(c)**

Figure 6. Macrographs of aluminium-10mass% silicon samples that were quenched approximately halfway through the eutectic reaction. Eutectic that solidified prior to quenching is dark in appearance. (a) 300ppm Sr, (b) 200ppm Sr + low level of addition of the elements Ti-V-B (total < 900ppm), (c) 200ppm Sr + high addition of elements Ti-V-B (total <1,700ppm).

Conclusions

In summary, aluminium-silicon casting alloys need further development to compete against emerging materials. As in other alloy systems, development is being hindered by the limits of traditional research techniques. Currently, the composition of nuclei that are active in strontium modified aluminium-silicon alloys are unknown, and, as a result, more than fifty percent of the microstructure of most aluminium-silicon foundry alloys solidifies with limited control. It is possible that some problems associated with the commercial use of strontium in foundries, the most important being increased tendency for porosity formation, could be eliminated with proper control of eutectic solidification. The research presented in this paper used advanced analysis techniques to understand the modification mechanism and locate and identify the composition of nuclei that are effective in the presence of strontium. The effectiveness of these nuclei and the validity of this research approach were verified using simple quenching experiments. The results also demonstrate for the first time that strontium is indeed uniformly distributed in eutectic silicon fibres, supporting the operation of the impurity induced twinning model.

Acknowledgements

The synchrotron radiation experiments were performed at SPring-8 with the approval of Japan Synchrotron Radiation Research Institute (JASRI) as Nanotechnology Support Project of the Ministry of Education, Culture, Sports, Science and Technology (MEXT), Japan (Proposal No. 2006A1645 / BL No. 47XU). The FIB experiments were supported by the Kyushu University High Voltage Electron Microscopy Program under MEXT (Proposal No. T-Kyudai-H15-013). The authors also acknowledge technical support for electron microscopy of SEM, EBSD and TEM from the Center for Microscopy and Microanalysis (CMM) and The University of Queensland.

References

[1] K. Nogita, S.D. McDonald, C. Dinnis, L. Lu, A.K. Dahle. Proc. of a symposium sponsored by the Solidification Committee of the Materials Processing and Manufacturing Division (MPMD) of TMS. 2004. p.93.
[2] S.D. McDonald, K. Nogita, A.K. Dahle, Acta Materialia 52 (2004) 4273.
[3] K. Nogita, H. Yasuda, K. Yoshida, K. Uesugi, A. Takeuchi, Y. Suzuki, A.K. Dahle, Scripta Materialia 55 (2006) 787.
[4] A. Takeuchi, Y. Suzuki, H. Takano, Journal of Synchrotron Radiation 9 (2002) 115.
[5] K. Nogita, S.D. McDonald, K. Tsujimoto, K. Yasuda, A.K. Dahle, Journal of Electron Microscopy 53 (2004) 361.
[6] S.Z. Lu, A. Hellawell, Metall. Trans. A 18A (1987) 1721.
[7] K. Nogita, S.D. McDonald, A.K. Dahle, Mater. Trans. 45 (2004) 323.
[8] J.Y. Chang, H.S. Ko, J. Mater. Sci. Lett. 19 (2000) 197.
[9] K. Nogita, J. Drennan, A.K. Dahle, Mater. Trans. 44 (2003) 625.
[10] L. Clapham, R.W. Smith, Journal of Crystal Growth 92 (1988) 263.

UNINTENTIONAL EFFECTS OF Sr ADDITIONS IN Al-Si FOUNDRY ALLOYS

Stuart McDonald, Matthew Dargusch, Guangling Song, David StJohn

CAST Cooperative Research Centre, School of Engineering, The University of Queensland, Brisbane, Queensland, Australia

Keywords: modification, porosity, eutectic grain size, corrosion, surface finish

Abstract
Strontium additions are commonly made to aluminium-silicon based foundry alloys to promote fibrous eutectic silicon morphology. Although this structural modification is the primary goal, it has become apparent that there are several concomitant changes. This paper examines this issue, reviewing the well-known alterations in porosity distribution and eutectic grain size that accompany strontium additions and introducing new research on the effects of strontium on corrosion resistance and as-cast surface finish. It is shown that strontium additions significantly affect both surface finish and corrosion resistance, through a combination of physical and chemical mechanisms. It is concluded that despite being present at only trace levels, strontium has a powerful effect on a variety of casting properties that extends beyond a transition in the morphology of the eutectic silicon phase.

Introduction

It is common practice to modify the morphology of eutectic silicon in hypoeutectic aluminum-silicon foundry alloys using trace element additions, most commonly strontium. The change in eutectic silicon from a coarse plate-like to a fine fibrous microstructure results in improved mechanical properties, particularly tensile elongation. A review of the literature reveals that in addition to a structural modification of the silicon phase, a range of other changes are induced by strontium additions, some of which are reviewed below.

Many of the non-modification related changes associated with strontium additions have been discovered as a result of research into porosity formation. This huge volume of research has arisen in an attempt to understand the increase in the volume percent of porosity, pore size and pore number that accompanies strontium modification [1]. Theories proposed to explain these observations, including increased hydrogen absorption [2], decreased liquid surface tension [3], increased number of inclusion [4], increased freezing range for primary solidification [5], altered eutectic solid-liquid interface morphology [6], altered nucleation and growth patterns of the eutectic [1] and many others.

Strontium additions are also known to affect the morphology and occurrence of other phases within the microstructure besides the targeted eutectic silicon phase, including both iron and copper containing [7-9] phases. When strontium concentrations exceed approximately 250ppm an additional phase is introduced into the microstructure. This phase is a compact polyhedral crystal of which the reported stochiometry varies, but is most commonly Al_2Si_2Sr [10-12]. The practical significance of this phase is uncertain however it has been implicated in both a deterioration of mechanical properties [13] and a poisoning of aluminium-silicon eutectic nuclei [14]. Strontium is also reported to decrease the oxidative losses of magnesium containing foundry alloys, due to the rapid formation of a protective oxide on the melt surface [15] (in contrast, without strontium the oxidation is a continuous process).

In addition to modification, strontium additions also result in a massive decrease in the nucleation of eutectic grains[1]. In a typical commercial unmodified alloy, eutectic nucleation is prolific, however strontium additions result in a decrease in nucleation and a resulting increase in eutectic grain size [14]. This increase in eutectic grain size is dependent on strontium concentration [16] and alloy purity [17] and can be as large as one order of magnitude. The decrease in solid-liquid surface area results in an increased eutectic growth velocity and contributes to the commonly observed depression in eutectic growth temperature [14]. The increase in eutectic grain size and redistribution of lower-melting point phases may be a fundamental reason for the changes observed in iron and copper bearing intermetallics.

The above review, while not comprehensive is indicative of the unintentional effect strontium additions can have on the properties and microstructure of aluminium-silicon based alloys. This is particularly remarkable given that in the majority of cases, less than 300 parts per million of strontium is added to the melt. In this paper we further investigate the effect of strontium additions on the as-cast surface finish and corrosion resistance.

Experimental Methods

Experiments were performed with unmodified and strontium modified (50, 100 and 150 ppm) nominal composition aluminium-10wt% silicon alloys. In addition, a series of copper containing aluminium-10wt% silicon alloys (nominal copper contents of 0, 1, 2 and 3 wt%) were produced at two different strontium levels (0 and 150ppm nominal). Melts were produced in clay-graphite crucibles in a 20kV induction furnace by alloying commercial purity aluminium (major impurities: 0.08wt% iron, 0.03 wt% silicon), silicon (major impurities: 0.18wt% iron, 0.012 wt% titanium and 0.04wt% calcium) and copper (all individual impurities < 0.01 wt%). When required, strontium additions were made by adding an aluminium – 10 wt% strontium master alloy, in rod form, to the fully molten alloy. The melt was held at approximately 700°C for alloying followed by a 20 minute degassing period using high purity argon, before the temperature was raised to 725°C for pouring. Each composition was created using a separate melt poured into a sand-mould, the geometry of which is given elsewhere [14].

Samples were prepared for corrosion testing by machining a small specimen of approximately 30x10x3mm from each casting and grinding the surface flat using silicon carbide abrasive. X-radiographic analysis was performed on some samples prior to testing. Samples were thoroughly cleaned and dried using ethanol and the dimensions and mass of each specimen were then accurately measured before immersion in a solution of 100g ferric chloride in 900ml of distilled water ($FeCl_3 \cdot 6H_2O$). Specimens were completely submersed in glass beakers containing 16 ml of the above solution in such a way that there was minimal contact of the sample with the beaker. Samples were removed after a total of 72 hours and then were thoroughly washed with distilled water and brushed using a soft-bristled nylon brush to remove any corrosion products. The corrosion products were further removed by submerging samples in concentrated nitric acid for 2 minutes followed by thorough washing in distilled water. A final weight measurement was then used to calculate an average corrosion rate.

---

[1] For the purposes of this article eutectic grains are considered to be connected regions of eutectic that originated from a common location.

60

Results and Discussion

The average analyzed composition of the alloys used in this research is shown in Table 1. The average copper analysis was within 3% of the total target level. As seen in Table 2 there is some evidence that strontium recovery decreased with increasing copper levels although the reasons for this interaction are unclear. Strontium was not present in detectable amounts (> 10 ppm) when not added intentionally. For simplicity, nominal compositions are referred to in the remainder of this paper.

Table 1 Average composition of the aluminium-10wt% silicon-X wt% Cu sand cast alloys.

Element	Si	Fe	Cu	Mg	Mn	Ti
wt %	9.63	0.12	1-3	<0.01	< 0.01	0.005

Table 2 Average strontium recovery at different nominal copper levels

Nominal copper (wt%)	0	1	2	3
Average Sr recovery (% of addition)	69	53	57	35

Immediately apparent on knocking out the solidified castings from the sand mould was the difference in appearance between those that were strontium modified and those that were unmodified. As seen in Figure 1 (a), typical unmodified aluminium-10 wt% silicon castings had a relatively uniform surface finish with no discernable macroscopic features. In contrast, strontium modified aluminium-10wt% silicon castings had clearly visible 'spots' scattered across the surface of the casting. As shown in Figure 2, these spots were particularly noticeable in castings once the amount of strontium exceeded approximately 100 ppm.

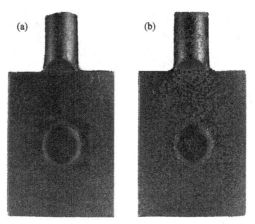

Figure 1 The as-cast surface of an (a) aluminium-10 silicon alloy and (b) identical alloy with a nominal addition of 150ppm strontium. The width of the casting is 110mm.

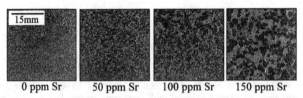

0 ppm Sr   50 ppm Sr   100 ppm Sr   150 ppm Sr

Figure 2 Close-up of the surface finish of aluminium-silicon castings with nominal additions of 0 to 150ppm strontium. The spots become larger with increasing strontium, becoming clearly visible after 100 ppm strontium is exceeded.

The internal structure of the alloys has been discussed elsewhere in detail [16] and only a brief summary of the macrostructures is presented here. Macrographs of samples are shown in Figure 3 and it can be seen that while there are no discernable features in the unmodified sample, those samples containing strontium had a large number of roughly circular features, which increase in size with strontium concentration. A rational interpretation of these circular features has been presented previously and for this paper it is essential only to recognize they are eutectic grains [14]. The eutectic grains are not macroscopically visible in the unmodified sample (Figure 3 (a)) due to their small size, which is estimated at a few hundred micrometers.

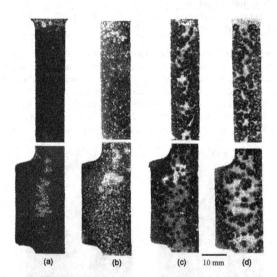

(a)   (b)   (c)   10 mm   (d)

Figure 3 The etched internal structure of the Al-10wt% silicon alloys shows an increasing eutectic grain size with increasing strontium concentration. Nominal strontium concentration (a) 0 ppm, (b) 50 ppm, (c) 100 ppm, (d) 150 ppm. These compositions correspond to the external surface shown in Figure 2.

When copper was present in the alloys, the distribution of the Al_2Cu phases is dramatically altered by the difference in eutectic grain size between unmodified and strontium modified alloys. In unmodified alloys (Figure 4 (a)) the Al_2Cu is distributed in small isolated pockets throughout the microstructure, and is never present in concentrations large enough to be visible using X-radiographic analysis. In contrast, due to the increased size of the eutectic grains in the modified alloys (Figure 4 (b)) the Al_2Cu is segregated into large connected networks, clearly visible on X-radiographs. Changes were also observed in the type and morphology of intermetallics that were visible. In the unmodified alloys the iron phases are present as either α or β as indicated in (Figure 4 (b)) and copper phases were present as either fine Al₂Cu (short black arrows) or massive or 'blocky' Al₂Cu (short white arrows). In the modified alloys all Al₂Cu was of the massive morphology and β-plates are the dominant iron bearing phase.

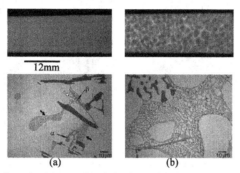

Figure 4 X-Radiographs (upper) and optical micrographs (lower) of aluminium-10 silicon alloys containing nominal additions of 2wt% copper and (a) 0 ppm strontium and (b) 150 ppm Sr. In the unmodified alloys the iron phases are present as either α or β as indicated and copper phases as either fine Al₂Cu (short black arrows) or massive or 'blocky' Al₂Cu (short white arrows). In the modified alloys all Al₂Cu was of the massive morphology and β-plates are the dominant iron bearing phase.

The corrosion rate as measured by weight loss is shown in Figure 5. From these results it is apparent that the severity of corrosion increased with the copper content of the alloy, such that in the 3wt% copper alloys the rate of corrosion was approximately double that of the copper-free alloys. In alloys containing copper, additions of strontium marginally decreased the overall corrosion rate. The morphology of the corrosive damage to the samples is shown in Figure 6. The damage to the unmodified alloys was fairly uniform, with small pits distributed across the surface of the sample. In the modified alloys the frequency of pitting was less, however the pits were generally deeper and the damage could be considered more localized.

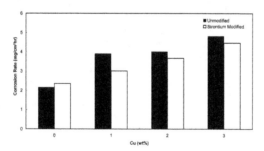

Figure 5 The corrosion rate measured as weight loss per unit area (mg/cm².hr).

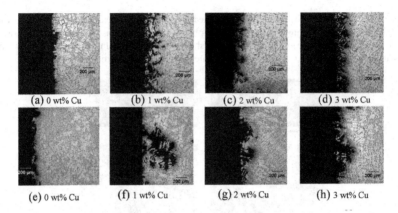

(a) 0 wt% Cu   (b) 1 wt% Cu   (c) 2 wt% Cu   (d) 3 wt% Cu

(e) 0 wt% Cu   (f) 1 wt% Cu   (g) 2 wt% Cu   (h) 3 wt% Cu

Figure 6 Typical morphology of the corrosion damage that resulted during the test period. Micrographs (a) to (d) showing unmodified alloys of increasing copper concentration and (e) to (h) showing corresponding modified alloys (150ppm nominal addition) of increasing copper concentration.

Discussion and Conclusions

From the results it can be concluded that strontium modification results in a deterioration of as-cast surface finish and alters the corrosion properties of the alloy. These very different observations are both fundamentally linked by the increase in the eutectic grain size that occurs with modification. Theories concerning why the eutectic grain size increases with modification are discussed elsewhere [14, 16, 17] and only the implications with respect to surface finish and corrosion resistance are discussed here.

The appearance of 'spots' on the surface of the modified castings can be explained with reference to Figure 7, showing a two dimensional schematic of a solidifying casting. During eutectic solidification the microstructure contains both dendrites and eutectic grains, some of which, depending on the alloy and thermal conditions, may be located along the wall of the casting (Figure 7(a)). If there is a demand for feed liquid from elsewhere in the casting, hydrostatic pressure will be generated and weak areas at the casting surface may undergo localized slumping ((Figure 7 (b)). As pre-existing eutectic grains will provide reinforcement, slumping is more likely to happen around their circumference, leaving the casting with a spotted appearance similar to that seen in Figure 1 (b). The prolific eutectic nucleation that occurs in unmodified aluminium-silicon alloys will result in more solid being present locally along the mould wall for the same degree of overall solidification and therefore lower likelihood of these surface defects forming. Furthermore, if localized slumping does occur in an unmodified alloy, due to the small size of the eutectic grains, it is less likely to be noticeable.

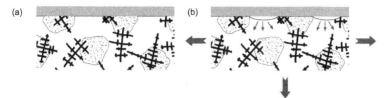

(a)   (b)

Figure 7 (a) A schematic of a partially solidified casting, showing eutectic grains, dendrites and the mould wall (grey). (b) A demand for feed liquid may result in hydrostatic pressure (large arrows) and localized slumping at weak areas at the mould wall (small arrows).

The increase in corrosion that occurs with an increased copper content, is already a well known phenomena [18]. The highly cathodic Al_2Cu phase leads to corrosion along anodic paths that will be comprised primarily of aluminium solid solution. The well connected large Al_2Cu networks that form in the modified alloys did not lead to increased corrosion and, in fact, the overall corrosion rate was decreased despite more localized damage occurring. This can be explained with the aid of <u>Figure 8</u> where it is proposed that each Al_2Cu precipitate has a finite influence on the local microstructure with respect to its galvanic effect. As such, a fine distribution of Al_2Cu (<u>Figure 8</u> (a)) will result in a larger overall galvanic affected zone than a more coarse distribution (<u>Figure 8</u> (b)). Based on this figure, it is understandable that the overall corrosion damage of an aluminium alloy containing finely dispersed Al_2Cu particles will be greater than that caused by large connected Al_2Cu networks.

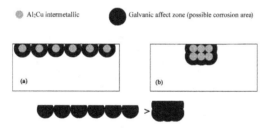

Figure 8 Schematic illustration of the corrosion damage caused by (a) dispersed isolated Al_2Cu particles and (b) connected Al_2Cu networks. For the same volume fraction of Al_2Cu the combined size of the galvanic affected zones is smaller but more localised the coarser the distribution.

The observations above for the deteriorated surface finish and altered corrosion properties that occur with strontium modification, are, to the best of our knowledge, previously unreported. They add to the extensive list of unintentional strontium-induced changes to the casting properties of aluminium-silicon alloys. They illustrate well, the disproportionate effect that the presence of trace elements can have on the processing and properties of materials and underscore the importance of research into the concomitant effects of compositional changes.

Acknowledgements

The assistance of Mr Michael Birmingham in corrosion testing and interpretation is greatly appreciated. The authors acknowledge financial support from the CAST Cooperative Research Centre, Australia.

References

1. S.D. McDonald, A.K Dahle, J.A. Taylor and D.H. StJohn: Modification Related Porosity Formation in Hypoeutectic Aluminium-Silicon Alloys, Metallurgical and Materials Transactions B, **35B** (2004), pp. 1097-1106.

2. J.R. Denton and J.A. Spittle: Solidification and Susceptibility to Hydrogen Absorption of Al-Si Alloys Containing Strontium, Materials Science and Technology, **1** (1985), pp. 305-311.

3. D. Emadi, J.E. Gruzleski and J.M. Toguri: The Effect of Na and Sr Modification on Surface Tension and Volumetric Shrinkage of A356 Alloy and Their Influence on Porosity Formation, Metallurgical Transactions B, **24B** (1993), pp. 1055-1063.

4. G.K. Sigworth, C. Wang, H. Huang and J.T. Berry: Porosity Formation in Modified and Unmodified Al-Si Castings, AFS Transactions, **102** (1994), pp. 245-261.

5. J.E. Gruzleski: The Art and Science of Modification: 25 Years of Progress, AFS Transactions, **100** (1992), pp. 673-682.

6. D. Argo and J.E. Gruzleski: Porosity in Modified Aluminium Alloy Castings, AFS Transactions, **96** (1988), pp. 65-73.

7. M.H. Mulazimoglu, N. Tenekedjiev, B.M. Closset and J.E. Gruzleski: Studies on the Minor Reactions and Phases in Strontium-Treated Aluminium-Silicon Casting Alloys, Cast Metals, **6** (1993), pp. 16-28.

8. M. Djurdjevic, T. Stockwell and J. Sokolowski: The Effect of Strontium on the Microstructure of the Aluminium-Silicon and Aluminium-Copper Eutectics in the 319 Aluminium Alloy, International Journal of Cast Metals Research, **12** (1999), pp. 67-73.

9. L. Wang and S. Shivkumar: Strontium Modification of Aluminium Alloy Castings in the Expendable Pattern Casting Process, Journal of Materials Science, **30** (1995), pp. 1584-1594.

10. G. Chai and L. Bäckerud: Factors Affecting Modification of Al-Si Alloys by Adding Sr-Containing Master Alloys, AFS Transactions, **100** (1992), pp. 847-854.

11. G. Laslaz, Dual Macrostructure in Hypoeutectic Al-Si Alloys, Dendrites and Eutectic Grains. Its Effects on Shrinkage Behaviour, Proceedings of the 4th International Conference on Molten Aluminum Processing, (Orlando, Florida, U.S.A., 1995), pp. 459-480.

12. L. Wang and S. Shivkumar: Influence of Sr Content on the Modification of Si Particles in Al-Si Alloys, Z . Metallkde, **86** (1995), pp. 441-445.

13. B. Closset and J.E. Gruzleski: Structure and Properties of Hypoeutectic Al-Si-Mg Alloys Modified With Pure Strontium, Metallurgical Transactions A, **13A** (1982), pp. 945-951.

14. S.D. McDonald, A.K Dahle, J.A. Taylor and D.H. St. John: Eutectic Grains in Unmodified and Strontium Modified Hypoeutectic Aluminium-Silicon Alloys, Metallurgical and Materials Transactions A, **35A** (2004), pp. 1829-1837.

15. K. Dennis, R.A.L. Drew and J.E. Gruzleski: Effects of Strontium on the Oxidation Behavior of an A356 Aluminum Alloy, Aluminum Transactions, **3** (2000), pp. 31-39.

16. S.D. McDonald, K. Nogita and A.K Dahle: Eutectic Grain Size and Strontium Concentration in Hypoeutectic Aluminium-Silicon Alloys, Journal of Alloys and Compounds, **422** (2006), pp. 184-191.

17. S.D. McDonald, K. Nogita and A.K Dahle: Eutectic Nucleation in Al-Si Alloys, Acta Materialia, **52** (2004), pp. 4273-4280.

18. H.P. Godard, W.B. Jepson, M.R. Bothwell and R.L. Kane: The Corrosion of light Metals, (New York: Wiley, 1967).

SHAPE CASTING:
2nd International Symposium

Process Design/Analysis

Session Chair:
Makhlouf Makhlouf

ADVANCED INTERMETALLIC MATERIALS AND PROCESSES: OVERVIEW OF THE IMPRESS INTEGRATED PROJECT

D. J. Jarvis, D. Voss, N. P. Lavery

European Space Agency
ESA/ESTEC, Physical Sciences Unit, Directorate of Human Spaceflight, Microgravity and Exploration, Keplerlaan 1, 2201AZ, Noordwijk, The Netherlands
URL: www.spaceflight.esa.int/impress
E-mail: david.john.jarvis@esa.int

Keywords: Intermetallics, turbine blades, casting, alloy solidification, microgravity, modelling

Abstract

IMPRESS is part of the European Commission's 6th Framework Programme and stands for Intermetallic Materials Processing in Relation to Earth and Space Solidification.

The main scientific objective of the project is to gain a better understanding of the complex links between materials processing, material structure and resultant properties of intermetallic compounds. In order to achieve this objective, the IMPRESS project combines a wide range of activities including improved casting technology, fundamental studies of solidification both terrestrially and in space, the creation of reliable thermodynamic and kinetic databases, thermophysical property measurements, multiscale computer modelling, structure characterisation, mechanical property testing and industrial product development.

With the knowledge generated, the project aims to develop and test 40cm investment-cast TiAl turbine blades for aero-engines and stationary gas turbines. For blade manufacturing the principal casting processes under study are tilt casting, counter-gravity casting and centrifugal casting. All these processing routes involve a phase transformation to the solid, and it is the control of this solidification that represents one of the most important challenges in the project.

Introduction

Due to the useful properties of intermetallic compounds, in particular the TiAl family of intermetallics, a number of niche industrial applications have emerged over the past decades; including structural applications in static and rotating turbine engine components, as well as space applications in rocket and re-entry systems. In both these cases, weight reduction is the primary driver for their use. The attractiveness of TiAl specifically can be summarised as follows: (a) low density of about 4 g/cm3 due to the high Al content, (b) a high melting point due to an ordered lattice structure, (c) high stiffness at 700°C of 150 GPa, (d) specific strength similar to cast Ni superalloys, and (e) good oxidation resistance [1].

However, despite the usefulness of these structural intermetallic materials, full-scale casting production and real industrial implementation have been rather slow. Some of the typical reasons for this include (i) the difficulties associated with reproducibly casting highly reactive TiAl melts, (ii) an incomplete design database of mechanical, chemical and physical properties and (iii) the lack of a cost-effective material supply chain, from raw material to end-product.

These problems are at the core of the 5-year IMPRESS Integrated Project, which started in November 2004 and involves 150 of the leading academic and industrial experts in the field in Europe [2]. Although the problems mentioned above are not insurmountable, considerable effort needs to be made in the coming years to provide viable industrial solutions, as emphasised in a recent TiAl review paper by Wu [3].

From a technical standpoint, the IMPRESS Project has set for itself the ambitious goal of developing and testing high-quality 40cm-long investment-cast fine-grained TiAl turbine blades for aero-engines and stationary gas turbines, in particular for low-pressure turbine stages as shown in Figure 1. The work currently being undertaken involves the optimisation of the casting and subsequent heat treatment processes, as well as thorough property testing over a wide range of conditions.

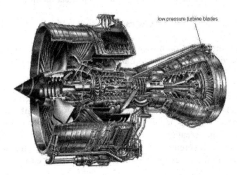
low pressure turbine blades

Figure 1. Cross-section through a Trent 900 aero-engine, showing potential TiAl component applications (copyright of Rolls-Royce plc, UK).

The focus of this paper is primarily on the challenges faced in the area of solidification processing of TiAl. Consequently, the paper first addresses alloy selection and fundamental issues of TiAl solidification. Secondly, aspects of microgravity research, that support the optimisation of solidification processing, are highlighted. The next section introduces the industrial solidification processes required to mass-produce large high-quality turbine blades, and the final section covers a brief introduction to the life-cycle analysis which is also being performed within the project.

Alloy Selection

One of the key developments in TiAl intermetallics in recent years was the addition of niobium as an alloying element. Liu and co-workers [4] found that, in TiAl alloys with the same microstructures, increasing the Nb content up to 10 at% and decreasing the Al content down to 45 at% resulted in substantial increases in yield strength at 900°C. Further work by GKSS Research Centre, Germany in this area has led to a patented high strength and creep-resistant alloy with high niobium content [5]. Based on this work and more recent work on post-solidification heat treatment [6], which allows grain refinement via massive transformation without the need for boron additions, the IMPRESS team have chosen a Ti45.5Al8Nb (at%) intermetallic for further solidification research and industrial casting work.

70

Fundamentals of Solidification

In terms of the fundamentals of solidification, there are a number of important links that need to be made between solidification processing and microstructural evolution. To this end, classical Bridgman-type experiments with unidirectional heat flow are being carried out to analyse the morphological changes of the solid-liquid interface as a function of growth conditions, namely temperature gradient G and growth rate R. A natural extension of this work includes a thorough investigation of the columnar-to-equiaxed transition (CET), commonly observed in industrial castings when the G/R ratio changes during freezing.

Segregation of alloying elements in the TiAlNb system is another important area of fundamental research under study in IMPRESS. Given that the three constituent elements all have different densities in liquid form, with aluminium being the least dense and niobium the densest, one can expect gravity-induced solutal convection and macrosegregation in large castings of the order 40cm. At a microscopic level, it is also critical to understand the partitioning behaviour and the way in which solute segregates in, and around, individual dendrite grains. Figure 2 shows a typical WDS concentration map of a dendrite in a Ti45.5Al8Nb intermetallic, with Al segregating into the interdendritic regions due to kTi/Al <1 and Nb segregating towards the centre of dendrites due to kTi/Nb >1 [7]. The implications of this microsegregation are not fully understood, but it is generally known to have some effects on post-solidification heat treatment processes.

In support of the work above, significant attention has been given to the creation of multicomponent thermodynamic databases for the TiAlNb system plus minor elemental additions. Commercial software packages like ThermoCalc [8] and MTDATA [9] are being regularly used by process developers and computer modellers to predict important sections of the equilibrium phase diagram, solidification paths and fraction solid evolution. An example of a ternary phase diagram section at 1750K is shown in Figure 3.

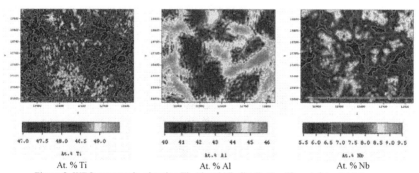

| At. % Ti | At. % Al | At. % Nb |

Figure 2. WDS cartography showing Ti, Al and Nb distribution (from left to right) in a Ti45.5Al8Nb intermetallic dendrite (courtesy of INPL, F).

Microgravity Experimentation

A unique part of the IMPRESS project is the capability to perform benchmark experiments in space, to elucidate the role of gravity on solidification [10]. A number of solidification phenomena are affected by gravity, for example, the settling or flotation of grains/inclusions in the melt, the columnar-to-equiaxed transition, gravity-induced melt convection and segregation

71

of alloying elements. By carrying out identical experiments on ground (1g conditions) and in space (very close to 0g conditions) and comparing results, it is possible to get deeper insights into gravity's role during an industrial casting processes. To this end, an IMPRESS space experiment involving the directional solidification of TiAlNb is under preparation, first onboard an ESA sounding rocket and later onboard the International Space Station (ISS). The team will study the influence of gravity/microgravity on CET and macrosegregation, while also investigating magnetic field effects on microstructural evolution. These benchmark results will be run through numerical models to predict the TiAlNb solidification process under diffusive and convective conditions. An example of such an exercise is shown in Figure 4, which adopts the 3D polycrystalline phase-field approach [11].

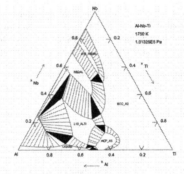

Figure 3. A section of the TiAlNb ternary phase diagram (courtesy of NPL, UK).

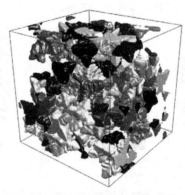

Figure 4. 3D phase field model results of TiAl equiaxed solidification, showing over a hundred dendrite grains with different crystallographic orientations (courtesy of SZFKI, H).

Equally important in IMPRESS to the development of reliable computer models of solidification is the accurate measurement of thermophysical properties of liquid metals. In this area, use is being made of a containerless melt processing technique known as electromagnetic levitation (EML), shown schematically in Figure 5 [12], and regularly used in microgravity on the

ESA/DLR parabolic flight campaigns and TEXUS sounding rockets (later onboard the ISS). Due to the unique microgravity conditions of space, certain thermophysical properties can be measured more accurately than with equivalent ground-based methods [13].

So far, various liquid-state properties have been measured for the Ti45.5Al8Nb intermetallic, (often as a function of temperature) such as: viscosity, surface tension, thermal expansion, density, thermal conductivity, specific heat capacity and melting range. This thermophysical property database is in current use by software developers (e.g. CALCOM ESI, University of Greenwich, ACCESS e.V.) to help predict solidification and defect formation in TiAl investment casting.

Industrial Solidification Processing

From an industrial standpoint, investment casting seems to be the only cost-effective route for processing TiAl turbine blades with complex geometries. Within IMPRESS, three different techniques are under development, namely centrifugal, tilt and counter-gravity casting. Tilt casting and counter-gravity casting both offer quiescent mould filling, thus minimising the entrainment of deleterious oxide films, but usually with low superheats. Centrifugal casting, although a more turbulent melt process, offers complete mould filling suitable for thin blade sections. It should be added that, for all of these casting techniques, substantial effort is also being made to tackle the inter-related issues of melting, ceramic shell mould production, post-solidification heat treatment and recycling.

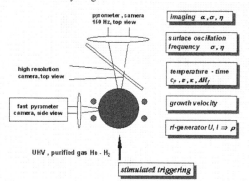

Figure 5. Schematic diagram of the EML technique and its capabilities (courtesy of University of Ulm, D).

Typical defects, familiar to the foundry industry, also apply to TiAl intermetallics. These include misrun, hot-tearing and cracking, contamination, porosity and surface dimples. Misrun can occur when the superheat prior to casting is too low; hot-tearing and cracking can come from mould restraint during the solidification and cooling process; contamination can easily occur in TiAl since the melt is highly reactive and can pick-up elements like oxygen, nitrogen, hydrogen and carbon; porosity can be caused by either solidification shrinkage or the entrapment of gas during turbulent mould filling; while surface dimples can appear after the HIPing process when the aforementioned pores collapse [14].

Much of the industrial work in the project is being strongly supported by computer modelling teams across Europe, each focusing on the different process steps: melting, casting, heat treatment and HIPing. As an example, Figure 6 shows the good agreement between X-ray imaged tilt casting experiments performed by the University of Birmingham and the respective computer simulations done by Calcom ESI using ProCAST.

Due to the strong industrial focus of IMPRESS and the fact that the project is being driven by the end-users, special attention is now being given to process up-scaling. In addition to many support companies, the project includes the biggest titanium casting company in Europe, with expertise in TiAl and a vacuum arc melting capacity of 1 tonne. This set-up offers unique opportunities for IMPRESS to produce and test prototype turbine blades in the near future.

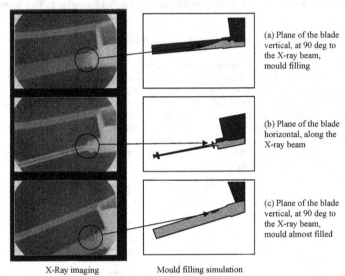

| X-Ray imaging | Mould filling simulation |

Figure 6. Agreement between tilt casting results derived from X-ray imaging (University of Birmingham, UK) and computer simulation by ProCAST (CALCOM-ESI, CH).

Developing a Life-Cycle Analysis in IMPRESS

The function of a life cycle analysis (LCA) is to gain an overview of resources used during the manufacturing processes and product's in-service life, with the goal of assessing the environmental impact of both of these phases. International standards exist (ISO 14000), but there are various competing indicators defining the environmental impact. The LCA methodology also serves a commercial purpose when coupled to financial cost, as clearly defined rules for gathering and assessing data-quality lend assistance to decision-making on the commercial viability of a product by considering for example the effects of fluctuating raw-material prices or new legislations on recycling.

With regards to the IMPRESS project, the LCA data collection has commenced with raw material prices [15, 16]. Historic prices (1989 to 2006) for titanium sponge, aluminium, niobium, nickel, chromium and molybdenum are shown in Figure 7. Increases in the price of titanium can generally be attributed to increased demand from aerospace and automotive sectors, coupled with a relatively constant supply from titanium sponge producing countries such as Russia. The long-term forecast is for titanium prices to come down due to improving processes (e.g. FFC-

74

Cambridge/Armstrong/Electrolytic Reduction Processes), as well as large reserve supplies in countries like China.

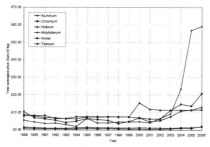

Figure 7. Elemental prices of Ti, Al, Nb, Ni, Cr and Mo since 1989.

A basic cost comparison is made between IMPRESS-selected intermetallic Ti45.5Al8Nb and a commonly-used INCONEL nickel superalloy, IN713LC in Table I, providing a basic comparison in the raw-material costs between the two turbine blade materials.

Table I. Nominal composition (in wt%) of alloying elements and the weighted constituent cost (Euros/kg of alloy) for Ti45.5Al8Nb and IN713LC superalloy[*]

Alloy	C	Al	B	Co	Cr	Fe	Mo	Nb	W	Ti	Zr	Ni	Total
IN713LC	0.062	6.11	0.008	0.02	11.8	0.04	4.3	2	0.03	0.59	0.057	75	100
(Euros/kg)	(-)	(0.17)	(-)	(0.01)	(0.36)	(-)	(3.84)	(0.35)	(-)	(0.19)	(-)	(13.19)	(18.11)
TiAlNb	-	29.3	-	-	-	-	-	17.7	-	53.0	-	-	100
(Euros/kg)		(0.82)						(3.06)		(16.71)			(20.59)

* Table footnote: (-) cost is considered negligible; - element not used in alloy

It can be seen that the two materials are rather cost-comparable, both being of the order of 20 €/kg, but the analysis needs to be extended to obtain cost and environmental indicators for all downstream processing steps related to turbine blade manufacturing. A comprehensive LCA inventory is currently being generated using the IMPRESS-selected TiAlNb alloy, and will be compared to an equivalent nickel superalloy processes, providing a comparative benchmark.

Conclusion

Solidification processing of TiAl intermetallics is particularly challenging, due to the high reactivity of the melt with shell mould systems and the difficulty of achieving high superheats prior to casting. Typical casting defects in TiAl include misrun, gas and shrinkage porosity, ceramic inclusions, entrained oxide films, hot-tears and chemical inhomogeneities. Within the first year of IMPRESS, many of these problems have been well studied and technical casting solutions are now becoming evident. By combining theory, experiments, space research, numerical modelling, and industrial product development into an integrated project, it is believed that substantial progress can be made, leading to high-value cost-effective cast products like TiAl turbine blades.

Acknowledgments

The authors would like to thank the IMPRESS partners for their contributions to this paper. The IMPRESS Integrated Project (Contract NMP3-CT-2004-500635) is co-funded by the European Commission in the Sixth Framework Programme, the European Space Agency, the Swiss

Government and the individual partner organisations. The project is coordinated by the European Space Agency.

References

[1]   W. E. Voice, et al., "Gamma Titanium Aluminide, TNB", *Intermetallics*, 13 (9) (2005), 959-964.

[2]   D. J. Jarvis and D. Voss, "IMPRESS Integrated Project - An Overview Paper", *Materials Science & Engineering A*, 413 (2005), 583-591.

[3]   X. Wu, "Review of Alloy and Process Development of TiAl Alloys", *Intermetallics*, 14 (10) (2006), 1114-1122.

[4]   Z. C. Liu, et al., "Effects of Nb and Al on the Microstructures and Mechanical Properties of High Nb Containing TiAl Base Alloys", *Intermetallics*, 10 (7) (2002), 653-659.

[5]   M. Oehring, et al., "Titanium Aluminide Based Alloy", WIPO, WO/2006/056248.

[6]   D. Hu, et al., "The Effect of Boron and Alpha Grain Size on the Massive Transformation in Ti-46Al-8Nb-xB Alloys", *Intermetallics*, 14 (7) (2006), 818-825.

[7]   D. Daloz, Private communication with author (2006).

[8]   B. Jansson, et al., "The Thermo-Calc Project", *Thermochimica Acta*, 214 (1) (1993), 93-96.

[9]   R. Davies, et al., "MTDATA - Thermodynamic and Phase Equilibrium Software from the National Physical Laboratory", *Calphad*, 26 (2) (2002), 229-271.

[10]   D. J. Jarvis and O. Minster, "Metallurgy in Space", (Paper presented at 4th International Conference on Solidification and Gravity, Miskolc-Lillafüred, Hungary, 2004).

[11]   L. Gránásy, et al., "A General Mechanism of Polycrystalline Growth", *Nat Mater*, 3 (9) (2004), 645-650.

[12]   R. Wunderlich and H. J. Fecht, "Thermophysical Property Measurements by Electromagnetic Levitation Methods under Reduced Gravity Conditions", *Journal of the Japan Society of Microgravity Application (JASMA)*, 20 (3) (2003), 192-205.

[13]   T. Hibiya and I. Egry, "Thermophysical Property Measurements of High Temperature Melts: Results from the Development and Utilization of Space", *Measurement Science and Technology*, 16 (2) (2005), 317-326.

[14]   R. A. Harding, M. Wickins, and Y. G. Li, "Progress Towards the Production of High Quality Gamma-TiAl Castings", (Paper presented at Structural Intermetallics (TMS) 2001, 2001).

[15]   "Annual U.S. Mineral Commodity Summaries (1996-2006)." U.S. Geological Survey, 2006.

[16]   T. Kelly and G. Matos, "Historical Statistics for Mineral and Material Commodities in the United States": U.S. Geological Survey, 2006.

Shape Casting: The 2nd International Symposium *Edited by Paul N. Crepeau, Murat Tiryakioğlu and John Campbell*
TMS (The Minerals, Metals & Materials Society), 2007

QUANTITATIVE XCT EVALUATION OF POROSITY IN AN ALUMINUM ALLOY CASTING

Joseph M. Wells

JMW Associates, 102 Pine Hill Blvd, Mashpee, MA 02649
jmwconsult1@comcast.net

Keywords: Porosity, Diagnostics, Casting, X-ray Computed Tomography, XCT.

Abstract

Traditional destructive sectioning and NDE techniques including radiography and ultrasound provide insufficient 3D qualitative and quantitative porosity diagnostics, characterization, and visualization capabilities. Improved methods are required for efficiently assessing the detailed nature, extent and distribution of porosity in research, product development, prototyping and first-items manufactured inspection stages of engineering materials and structures. One quite powerful and recent NDE modality that has been demonstrated to be most advantageous in the 3D diagnostics of volumetric porosity is industrial x-ray computed tomography, XCT. With this non-invasive approach, very large numbers of irregular shaped pores can be individually characterized according to voxel or object spatial coordinates, surface area, pore volume, aspect ratio, and size distributions. In addition, virtual transparency and pseudo-color image processing techniques permit the 3D visualization of all pores in-situ within the object casting. Examples of such porosity characterization and the cognitive visualization of same using the XCT modality in a representative aluminum casting are presented and discussed.

Introduction

Porosity may be defined as the quality of being porous, or alternately, as the ratio of a cumulative volume of interstices or voids in a material to the volume of its mass. Two basic porosity types are: a. open pore (interconnected individual voids) and b. closed pore (isolated pores without interconnectivity). Sometimes, controlled porosity levels can be considered useful in engineering materials for design applications including: liquid or gas absorption, filtering, energy absorption, heat exchanging, porous electrodes, etc. Undesired porosity limits the mechanical properties and may be quite detrimental to the functionality and/or structural integrity of aluminum castings [1]. Whether such porosity is intended or undesirable, there is, nonetheless, an evolving engineering requirement for the effective detection, characterization, and quantification of that porosity. The details of such a challenging requirement may vary with the subject component, material, porosity size, morphology, amount, distribution and locations, and the attendant risk factors associated with its presence. The fulfillment of such a porosity diagnostic requirement likely depends upon whether one utilizes destructive or non-destructive diagnostic approaches. The acceptable tolerance limit for inadvertent porosity levels will generally depend on the material, applicable manufacturing processes, and the design/service application protocol for the component object.

The best available diagnostic approaches are necessary for the effective control or mitigation of porosity introduced during various manufacturing processes. All types of engineering materials, whether monolithic metals, ceramics, polymers or composites, are susceptible to porosity in various commercial manufacturing consolidation processes. Several representative consolidation processes and engineering/manufacturing applications in which porosity is typically a significant concern are listed in Table 1. This listing is intended primarily to be instructional and is not meant to be all inclusive, comprehensive, and complete.

Table I. Example Technology Areas Where Porosity is a Significant Interest

Material	Consolidation Processes	Typical Engineering/Mfg Application
Metals & MMCs	Solidification Cold Isostatic Pressing, CIP Hot Isostatic Pressing, HIP Metal Spray Processes Electroplating	Castings Metal Joining (Weld, Braze, Solder) Powder Metallurgy Coatings for Environmental Protection Controlled Porosity Materials: 　*Metal Foams, Porous Electrodes* 　*Energy Absorbing Materials* 　*Thermal Control* Biomedical Implant Surfaces
Ceramics & CMCs	Sintering Hot Pressing Crystal Growth	Refractory Ceramics Technical Ceramics & Coatings Armor Ceramics: Transparent/Opaque Optical Ceramics & Glasses Concrete & Construction Brick
Polymers & OMCs	Autoclave Curing Injection Molding Pultrusion/Extrusion	Monolithic Polymers Organic Matrix Composites Coatings & Adhesives

It is quite possible to inadvertently introduce extrinsic porosity manifestations on the sectioned and polished surface during preparation with irreversible and destructive metallographic techniques. These techniques provide one or more 2D physical planar surfaces from which two dimensional (2D) pixilated photographic images are used to record and measure the observable porosity manifestations. The actual sizes of many pores intersected by this technique are not readily discernable since different apparent pore sizes can also be attributed to various statistical cross-sections of a given geometrical pore size. The spatial location and volume measurements of irregular-shaped pores are also complicated with this approach. Furthermore with the destructive approach, the practical volume of the target object actually sampled is somewhat limited and the further utilization of the component sectioned is compromised. The discussion in this paper will focus on the non-invasive detection, characterization, visualization, and statistical analysis of porosity via the diagnostic modality of industrial x-ray computed tomography, XCT.

X-Ray Computed Tomography Diagnostics

The 3D nondestructive XCT diagnostic modality permits the interrogation of both the external and internal design surfaces and/or damage features within the bulk of the original object. A completely digitized 3D "density" map of a solid object is constructed by the triangulation of volumetric x-ray absorption data during which the dimensional and structural features of that object are retained. The nominal resolution level for the meso-scale tomography of a modest size ($\sim$1.5x10^{7}mm^3) medium density object is $\sim$0.250 mm. The specific resolution level depends upon both the object size and density and the x-ray source and detector system. Higher resolution levels of < 20 microns are achievable with micro-tomography techniques, but only on relatively less dense and/or smaller objects of $\sim$1.5 x10^4 mm^3. Considerably higher resolution levels require destructive sectioning and high resolution electron techniques, which, if necessary, may be used to augment the results of XCT.

A characteristic feature of porosity is the very low density of the gas contained within compared to the considerably greater density of the surrounding solid material. This characteristic is ideally suitable for the use of x-rays as the diagnostic nondestructive evaluation, NDE, modality where

78

the contrast in x-ray photon absorption/attenuation for a pore volume compared to a surrounding solid material is substantial. The essential difference in porosity detection and analysis with XCT is the high resolution 3D voxel data used versus the 2D pixel data utilized with 2D destructive sectioning or with 2D film or digital x-rays. Additionally, the XCT scanning data of the entire object volume digitized, as well as the computer rendered virtual images and their associated measurements, are fully recoverable and can be conveniently digitally archived. Because of space limitations, the XCT scanning process will not be discussed here and the reader unfamiliar with this technique is referred to the following references [2-4]. A previous report on the use of 3D Computed Tomography for quality control of aluminum castings was presented by Simon & Sauerwein [5].

Following the acquisition of a series of contiguous digital XCT scans, several advanced image processing tools and techniques are available with commercially available software for various diagnostic purposes. For the present porosity interrogative purpose, those image processing capabilities of primary interest include:
- Initial preview of 2D XCT axial slices
- 3D solid object reconstruction
- Virtual sectioning on both orthogonal and arbitrarily oriented planes
- Local in-situ metrology for linear, area, and angular features of interest
- Defect analysis of specific features (eg. Individual pore volume, surface area, gray level, aspect ratios, and spatial position)
- Statistical histogram of number vs pore volume
- 3D visualization of porosity distribution

XCT digitized voxel scan data files are easily reconstructed into 3D virtual solid objects representing either the entirety of the original object or a localized sub-set region of interest using the commercially available Volume Graphics StudioMax v1.2.1 (VGSM) voxel analysis and visualization software [6]. The author and his collaborators have conducted several detailed XCT damage diagnostic studies, especially of impact-induced damage in terminal ballistic targets [7-12]. Impact-induced porosity has been observed and characterized in some of these studies. However, this current paper will address the more commercially challenging issue of the effective diagnostic characterization, quantification, and visualization of as-manufactured porosity in more familiar engineering materials. Here - because of space limitations - porosity characterization, quantification, and 3D visualization results are shown on a representative aluminum alloy casting. This prototype casting contains considerable porosity of varying sizes and widely distributed throughout the volume of the casting. The initial 3D solid object visualization reconstructed from the axial XCT scans is shown in figure 1.

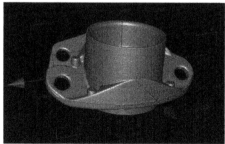

Figure 1. Opaque 3D solid object reconstruction of XCT scan data to provide a high quality virtual visualization of the subject aluminum alloy casting

79

In-situ dimensional metrology of this casting reveals the overall dimensions of W= 63.0 mm, L=125.1mm, and H= 53.9 mm, as shown in figure 2.

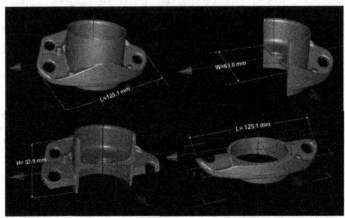

Figure 2. 3D solid object reconstructions of aluminum alloy casting indicating overall dimensions in full and virtual half section visualizations.

Porosity Visualization Techniques

It is important for the reader to recognize that all of the casting images produced from the XCT diagnostic analysis are not photographic images but rather are digital reconstructions of XCT scan data. These images are thus capable of revealing features of interest on the surface or at any position within the interior bulk of the original casting. An obvious limitation of the opaque 3D solid object visualization is the inability to observe internal features of interest such as porosity and inclusions within the bulk casting material. There are essentially two convenient ways of directly observing the internal defects, namely: (1) virtual sectioning on either orthogonal or arbitrary orientated planes, or (2) gray level segmentation and virtual transparency. Examples of these direct observations of internal porosity follow:

Orthogonal Sections
The three traditional orthogonal section views are the axial (xy–plane), the frontal (xz-plane) and the sagittal (yz-plane). These orthogonal planes are perpendicular to the virtual Z, Y, and X axes respectively. Each of these planes can be viewed independently of the others, or in the case of a specific localized defect, they may be selected collectively so that each includes a view of that defect. Any specific defect can be marked so as to be readily identifiable and observable in each orthogonal view. A related capability allows the 3D solid object to be sectioned along one of the selected orthogonal planes to help visualize that defect within the bulk object. An example of a specific defect distinguishable amongst multiple surrounding defects is shown circled on the three orthogonal planes A-198, F-499, and S-360 in figure 3. The same circled defect is also shown in the axial sectioned oblique view of the 3D solid object figure 3. This latter view makes the spatial localization of this defect within the overall casting bulk much more easily visualized.

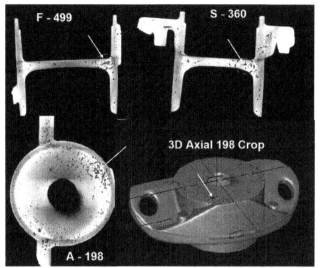

Figure 3. Virtual sections showing a common circled defect in the three orthogonal planar views and in the truncated axial sectioned view of the 3D solid object view.

Sections of Arbitrary Orientation

On occasion, one desires to view features on sectioned planar views other than the standard axial, frontal, or sagittal orthogonal planes. The capabilities exist with this technology to view any arbitrarily selected virtual sectioned plane. An example of such an arbitrary sectioned plane for this casting is observed in figure 4. Several clusters of porosity are circled and observable on the expanded view of this plane.

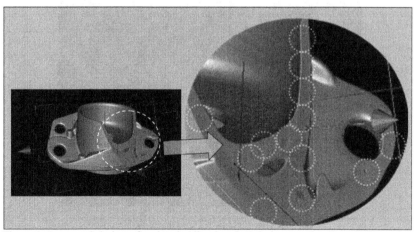

Figure 4. An expanded view of a virtual obliquely oriented planar section through the casting showing several circled areas revealing porosity.

81

Segmented Porosity with Pseudo-colored Transparent Imaging

The detection and characterization of the porosity content of the object casting is accomplished with the aid of the defect analysis tool. In the automatic defect detection mode, the gray level threshold must be established between the material and the surrounding air. Internal porosity that is completely surrounded by the casting material is included in this analysis, but any pores directly exposed to either external or interior design surfaces are excluded. The minimum and maximum values of the desired defect volume range are specified along with the quality limit index for the analysis beforehand. The quality index, QI, is a threshold indicator of the data quality; the higher the Q.I., the greater the probability that the defect detected is real and not an artifact. The resulting analysis of the casting with a specified range of defect volume from 0.029 to 50.0 mm^3 and a Q.I. = 1.0 revealed a total of 4,346 individual defects for an overall porosity level of 1.11%. A statistical histogram of the size distribution of the identified defects is shown in figure 5 with added fiducial circles indicating the larger volume defects. A transparent 3D visualization of this in-situ porosity distribution is presented in figure 6 using a pseudo-color scale to reflect the pore volume.

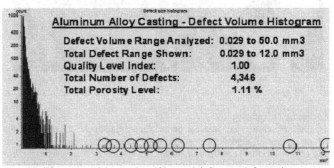

Figure 5. Histogram of the number of internal defects vs defect volume

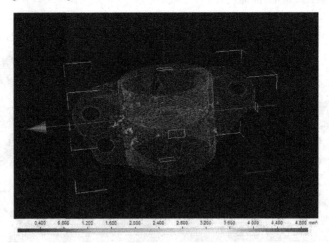

Figure 6. A semi-transparent visualization of the casting revealing the internal porosity calibrated by pore volume to the pseudo-color scale shown along the bottom of the figure.

82

Discussion

Selected additional images are further revealing of defect characteristics not obvious from the information presented above. The gray level of any defect feature can be easily measured and reflects the relative density of that defect. A very low gray level value < 5 is attributed to a pore (or a crack) whereas a high gray level defect feature is attributed to an inclusion. The contrast between such high versus low gray level features is clearly observed in each of the views in figure 7.

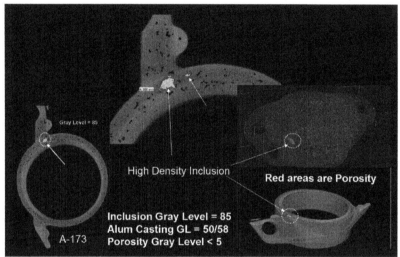

Figure 7. Axial (A-173) sectional views with multiple low gray level defects (pores) surrounding one large high gray level -light appearing – defect identified as an inclusion.

The presence of such defects may be detrimental to the mechanical/structural performance of the casting depending upon the location within the casting of a critical concentration of defects and the anticipated service loading. An engineering assessment could be conducted using finite element stress analysis to establish the risk severity of a given defect concentration revealed by this diagnostic technique. The defect detection and characterization knowledge obtained with the XCT diagnostic approach is useful for quality assurance and casting process monitoring, especially with the early prototype design and first items manufactured. It can also be used to establish a high quality defect baseline status against which more conventional NDE modalities could be calibrated.

Summary

The results attained in this study have been presented to demonstrate the remarkable engineering diagnostic capabilities of 3D defect characterization, analysis and visualization with the XCT modality using advanced VGSM voxel analysis and visualization software. In this example, some 4,346 individual unconnected internal defects were spatially located and the volume of each was determined. Defects encountered consisted primarily of porosity and some inclusions. The total defect level (with each pore volume $\leq 50mm^3$) was 1.11 % of the total casting volume.

Acknowledgements

Grateful acknowledgements are extended to Dr. Christof Reinhart of Volume Graphics, GmbH for providing guidance with the StudioMax software.

References

1. R. Monroe, "Porosity in castings," American Foundry Transactions, Paper 05-245(04), (2005).
2. J.H. Stanley,"Physical and Mathematical Basis of CT Imaging," ASTM, (1985).
3. A.C. Kak and Malcolm Slaney, *Principles of Computerized Tomographic Imaging*, IEEE Press, 1988.
4. ASTM E 1570-00, "Standard Practice for Computed Tomographic (CT) Examination, ASTM International and ASTM E 1441 -00, "Guide for Computed Tomography (CT) Imaging," ASTM International
5. M. Simon and C. Sauerwein, "Quality Control of Light metal Castings by 3D Computed Tomography,", http://www.ndt.net/article/wcndt00/papers/idn730/idn730.htm
6. Volume Graphics StudioMax v1.2.1 , www.volumegraphics.com
7. J.M. Wells, "On the Role of Impact Damage in Armor Ceramic Performance," Ceramic Engineering and Science Proceedings, v27, (2006) *In Press.*
8. J.M. Wells, "Considerations on Incorporating XCT into Predictive Modeling of Impact Damage in Armor Ceramics," Ceramic Engineering and Science Proceedings, v26, Issue 7, 51-58, (2005).
9. J. M. Wells, "Progress on the NDE Characterization of Impact Damage in Armor Materials," Proceedings of 22nd International Ballistic Symposium, ADPA, v2, 793-800, (2005).
10. H.T. Miller, W.H. Green, N. L. Rupert, and J.M. Wells, "Quantitative Evaluation of Damage and Residual Penetrator Material in Impacted TiB_2 Targets Using X-Ray Computed Tomography," 21st International Symposium on Ballistics, Adelaide, Au, ADPA, v1, 153-159 (2004).
11. J. M. Wells, N. L. Rupert, and W. H. Green, "Progress in the 3-D Visualization of Interior Ballistic Damage in Armor Ceramics," Ceramic Armor Materials by Design, Ed. J.W. McCauley et. al., Ceramic Transactions, v134, ACERS, 441-448 (2002).
12. J. M. Wells, N. L. Rupert, and W. H. Green, "Visualization of Ballistic Damage in Encapsulated Ceramic Armor Targets," 20th International Symposium on Ballistics, Orlando, FL, ADPA, v1, 729-738 (2002).

NATURALLY PRESSURISED FILLING SYSTEM DESIGN

John Campbell
Emeritus Professor
University of Birmingham, UK

Key words: castings, counter-gravity filling, turbulence, entrainment defects, bifilms.

Abstract

The conventional unpressurised and pressurised filling system designs employed to fill molds under gravity perform poorly, creating generous quantities of entrainment defects (bubbles, bifilms and sand inclusions), and thereby efficiently degrading castings and their properties. The new concept of the naturally pressurised system is described, and its successes and limitations are examined. It seems that contact pouring may be the only good solution at the present time to the gravity pouring of heavy steel castings using a bottom-teemed ladle. It is clear that ultimately, foundries should be designed to avoid the pouring of melts at any point during the melting or casting process. Thus new melt handling technology is required for the designs of foundries, and only counter-gravity filling of molds is seen to be capable of routinely producing substantially defect-free castings.

Introduction

The casting industry has never properly researched its filling system designs for castings. Thus current knowledge and practice is poor. This lamentable situation has cost the industry dearly, in terms of lost profits and lost customers. This short review supplements the more detailed accounts by the author [1], and attempts to make a critical overview and update of the current situation. Recent concepts by the author summarised elsewhere [1-4] have assisted to clarify and quantify the targets to be achieved in a filling system design. These included the so-called '*critical ingate velocity*' at which the melt has sufficient energy to entrain its own surface in a phenomenon called '*surface turbulence*'. This concept enshrines a further concept called '*entrainment*' that in turn is a mechanical folding action, so that '*entrainment defects*', bubbles and '*bifilms*', can be introduced into the melt. The major difference between a bubble and a bifilm is the amount of air that each entraps. However, both can be exceedingly damaging to the quality of a casting. Because the '*critical fall distance*' (the fall distance at which the critical velocity is exceeded for that particular metal) is only about 12 mm for liquid aluminum, the melts in the running systems of all gravity poured castings exceed this velocity by a large margin. Thus entrainment defects are extremely difficult to eliminate from gravity poured castings. The task for the designer of the filling system amounts in fact to a damage limitation exercise.

The problem of the filling of very thin sections involving flow by microjetting [2] a kind of naturally occurring micro-turbulence, is a key concern for very thin section castings, but space does not allow its inclusion here.

The design of improved systems in sections above a few millimeters thickness is now significantly facilitated by research tools such as the video X-ray radiography technique in which real liquid metals can be studied in detail in real molds [5]. In addition, of course, computer simulation has now come of age in this field, so that its intelligent use can be invaluable.

Traditional filling designs

The traditional methoding systems have included the so-called unpressurised filling design, in which an attempt is made to reduce the velocity of the metal prior to it entering the mold cavity. This worthy aim is attempted by expanding the flow channels. In terms of the areas of the sprue exit/runner/gate the relative areas chosen are often 1:2:2 or 1:2:4 etc. The assumption has been that the volume flow rate Q (m^3/s) and velocity V (m/s) relates to the area A (m^2) of the stream by the equation

$$Q = A.V \qquad (1)$$

The problem with this approach is that equation 1 is true for the melt stream itself, but not necessarily for the channel. Thus an expansion of the channel to reduce the velocity has practically no effect on velocity; the melt flow progresses at its original speed and cross section area, so the channel simply runs only part full. This is a potential disaster for melt quality, since the velocity is well above the critical velocity, so that air and oxides (i.e. more precisely, bubbles and bifilms) can now be entrained. Thus unpressurised systems are typically poor at reducing velocity, and deliver poor quality metal into the mold.

The design widely employed in the cast iron industry is the pressurised system. Here the so-called gating ratio is something like 1:0.9:0.9. The channels now can run full, providing air is not introduced in the basin. However, poor conical basin designs and wrongly tapered down-sprues as a result of the poor design of our automated greensand molding plants usually eliminates these advantages, with the result that air is unfortunately introduced at the entrance to the filling system. Thus although the potential of the pressurised system was promising, in practice the benefits were eliminated by poor front-end design. Finally, of course, the pressurisation effect resulted in very fast delivery of melt into the mold cavity, where the high velocity had the potential to create significant additional damage. In addition, damage generated in the mold is the worst kind of damage since it is necessarily incorporated into the casting. (In contrast to the damage created in the filling system, much of which appears to 'hang up' in the filling system and does not find its way into the casting).

This problem of good general principles, but poor front ends to the system, or other features that introduce defects, illustrates a significant aspect of filling system design: the whole of the system has to work well. Any small detail that is incorrect can introduce destructive amounts of irreversible damage, a significant proportion of which is likely to arrive in the casting. Gravity filling systems are hyper-sensitive to small errors as a result of the high velocities involved. This regrettable fact has been the source of major problems to the casting industry.

This situation contrasts with counter-gravity filling systems where the melt velocity can be controlled at all stages below the critical velocity, so that zero damage is introduced during filling, despite the presence of errors such as the ledges arising from the mismatch of channels etc. Counter-gravity systems, when properly operated, and despite modest errors in the filling system, represent robust technology that can routinely deliver good products.

The poor front-end designs used for many systems have effectively undermined many otherwise reasonable designs. Similarly, good front end features such as the off-set step pouring basin (the only basin to have been researched, and shown to be capable of avoiding air entrainment [3]) have often been mistakenly criticised for poor performance because of the deleterious actions of other features in the system, thus unfortunately adding to the overall confusion relating to filling system design.

One deleterious feature that has a long pedigree is the 'well' at the base of the sprue. Recent work has illustrated that the high velocity at this location, combined with the generous volume provided by the well, maximises the damaging churning action of the melt. This turbulence entrains clouds of bubbles, and probably, large quantities of oxide. The sprue/runner junction for the naturally pressurised filling design described below is simple and basic, and when properly designed and made, contributes not a single bubble to the downstream flow.

The Naturally Pressurised Filling System

The naturally pressurised system is characterised by the narrowness of its flow channels. Its central concept, fundamental to the whole approach of the naturally pressurised system, is that of maximising constraint on the flowing liquid. The concept of a 'choke' in the system becomes redundant, because the whole system, along its complete length, all acts as a choke. The system is a kind of 'continuously choked' design.

1. The Offset Stepped Pouring Basin

It is assumed that the offset stepped basin design is used in this system. Whilst not being exclusive to the naturally pressurised system, it appears to be the best basin design so far, and capable in some cases of delivering 100 per cent metal (i.e. zero content of entrained air bubbles corresponding to a discharge coefficient of 1.00 [7]). It seems to work adequately well for Al-Si alloys, particularly when a basin response time (i.e. its draining time if allowed to empty) of one second is selected. This usually gives adequately slow response so that the pourer can keep the basin filled, and appears to give adequate detrainment time for most of the air entrained by the pour into the basin.

Even so, the problem of basin design cannot be claimed to be completely solved, and in fact may not be soluble for some metals. Those metals that generate strong oxide films, such as aluminium bronzes and some stainless steels, may overload the system with oxides that cannot be detrained. In addition, the mass of strong films waving about in the flow like so much waterweed in a river will be expected to prevent detrainment, funnelling bubbles down the sprue.

Similarly, the pouring of steel from bottom-teemed ladles may also represent an insoluble situation, in which the extremely high velocity melt entering the basin overwhelms the system with oxides and bubbles that have insufficient time to detrain. The use of an argon shroud around the pouring stream is clearly an attempt to address this problem, and is somewhat beneficial. However, for a more rigorous solution a sophisticated version of 'contact pour' may be required in which the pouring basin is effectively exchanged for a bottom-poured ladle in which the outlet nozzle from the ladle makes direct contact with, and seals against, the entrance to the sprue (Figure 1). The technique has been demonstrated successfully for the first time for a 50 tonne steel roll casting [8] and represents a potential way forward for large steel castings. A system using small intermediate ladles sitting on the top of molds, and able to contain the complete volume of melt required for the pour, can be envisaged. The intermediate ladle is filled from the large ladle transferring metal from the melting furnace. A few seconds after the intermediate ladle is filled its stopper is raised and the mold is filled. A series of intermediate ladles on a series of molds might be workable with the early ladles being transferred forwards to be used a second time for later castings in the pour series.

2. The Sprue

The basis of the sprue design is the calculation of the form of a freely falling liquid stream. This frictionless flow is the starting point for the calculation of the areas for the sprue and the runner. For a volume per second flow rate Q, the area of the cross section area of the stream A after a fall of distance h from the upper surface of the melt is

$$A^2 = Q^2/(2gh) \qquad (2)$$

This area/height relation is the famous result describing a rectangular hyperbola (not a parabola as has often been assumed in error). It follows that the really important relation giving the diameter D of a rotationally symmetrical stream at height h is

$$D^4 = 8Q^2/(\pi^2 gh) \qquad (3)$$

This relation of h versus $D^{-1/4}$ is a significantly more complex curve, shown schematically in Figure 2 (with height in dimensionless units of $8Q^2/(\pi^2 g)$) emphasising the non-straightness of its taper, particularly near the top of the sprue, and the approach to asymptotes at each limit. The author now uses this theoretical sprue form for all castings above about 300 mm high.

Taking this theoretical shape, and introducing a mold around it, so as just to touch and contain the flowing stream changes the nature of the flow. Friction is now introduced for the first time. The natural frictional back-pressure now builds up along the length of the sprue (and all the subsequent channels if contact with the walls is maintained) causing the liquid to gently pressurise the walls of the channel. This *natural* pressurisation is of enormous value in assisting to keep air out of the system and is a fundamentally important feature of the naturally pressurised design.

3. The Sprue/Runner Junction

At bends in the system, particularly the sprue/runner junction, it has been widely assumed that significant friction is involved at such locations, causing up to a 20 % loss of velocity. If this is true the runner may now be expanded by 20 % to take advantage of the 20 % reduction in velocity giving a 'gating ratio' R = 1/1.2/1.4. A succession of right angle bends, when used in this way, can be successful in making significant reductions in final velocity at the end of the system. The succession of reductions in speed has been confirmed experimentally to be close to 1.0, 1.2, 1.4, 1.7, 2.1, 2.5 etc. [9]. Traditionally the right angle was feared because of the turbulence that it would create. This was true for oversize sections. However, for the slim sections in the naturally pressurised system the melt is constrained, so that no room is available to allow the metal to damage itself. Therefore instead of being a threat, the right angle may be used as a valuable speed-reducing feature in the new system. This L-Junction has been studied in detail by Hsu and others in a contribution to this conference [10].

4. The Runner

Alternatively, the runner need not be expanded at all, and may have the same area as the sprue giving R = 1/1/1. This will significantly assist to pressurise the melt in the runner and gates so that there will be little danger of the entrainment of air. If this option is taken, the choking action of the melt entering the runner will lead to a 20 % increase in filling time and a further 20 % at the gates, giving a total extension of approximately 40 % to the filling time. These losses at right angle turns are often overlooked, perhaps at least partly explaining why many calculated filling systems appear to run more slowly than predicted.

Alternatively again, the full 20 % increase in area for the runner need not be taken. If only 10 % were selected, giving R = 1/1.1/1.2, the runner would benefit from an extra 10 % pressurisation at each turn that would suppress entrainment defects to some degree, but filling time would be extended by approximately 10 + 10 = 20 %.

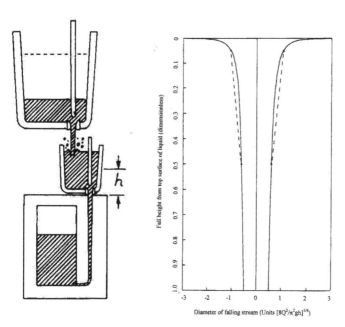

Figure 1. Teeming of steel into an intermediate ladle, filled to a minimum depth h.

Figure 2. The diameter of the falling liquid (the example of a straight taper from 50 to 500 mm for a 1m total height is seen to be a poor approximation).

After this initial adjustment of the area of the runner to allow for losses at the sprue/runner junction, its area can remain constant up to the first division or gate. This constant area will in any case contribute to the gentle pressurisation along the length because of the frictional loss.

In the early results of the detailed study of the L-Junction by Hsu and co-workers [10] the rule of 20% loss per right-angle junction is seen, unfortunately, to be grossly simplistic. Hsu's results suggest that the accumulating frictional losses along the lengths of channels would permit their gentle expansion, allowing a progressive reduction in speed without the penalty of reduced volume flow rate. This situation clearly requires more research.

5. The Gates
The placing of a series of equal gates along the length of the runner will favor delivery through the far gates if the runner retains its area. The delivery through the gates should be balanced so far as

possible by systematically reducing the area of the runner. The traditional practice of stepping down the area of the runner as each gate is passed is definitely not recommended, as a result of the deflection and turbulence caused at the steps. Far better is a linear taper, gradually reducing the area of the runner along its length. For many Al-Si alloy castings in the 10 to 50 kg weight range, and up to 0.5 m or so long, approximately 50 % reduction in area is often closely correct. (Tapering to zero area is an overkill that causes the early gates to be favored.)

Because many casting only have a single mold joint, and have to be gated on the joint, the gates have to be horizontal. If the gates now enter the mold cavity at more than the critical distance above the base of the casting, allowing the melt to fall inside the mold cavity, significant damage is to be expected to the properties of the casting.

Bottom gating, in which the gates enter at the lowest possible level of the casting are necessary to achieve best results. However, horizontal gates often cause the metal to jet over the base of the mold. Such uncontrolled directional propagation over a silica greensand mold surface can lead to metal damage in the form of oxide flow-tube defects, or mold damage at the edges of the jetting stream such rat-tails.

Vertical gates, by contrast, in which gravity assists both the filling of the gate and the slowing of the melt as the gates prime, work extremely well. It is a luxury therefore to have an additional joint beneath the casting, formed by a third mold part or cunningly designed core. Such features usually quickly recover their cost.

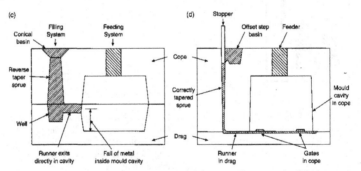

Figure 3. A poor traditional system and the naturally pressurised design.

The area of the gates should, if possible, be sufficient to ensure that the velocity of melt entering the mold is less than 0.5 m/s, although in practice, up to 1.0 m/s is often tolerable. However, in addition, it is essential to ensure that the gates do not form a hot-spot at the junction with the casting [1]. The use of narrow slot gates that become hot during filling, but rapidly lose their heat to act as cooling fins after the filling is complete, can be used to create a valuable temperature gradient for directional solidification and enhanced feeding efficiency.

6. Feeding
The author has to admit he is generally not in favor of feeding castings unless absolutely necessary (my First Rule of Feeding: "*Do not feed* (unless necessary)"[1]). Most castings that are to be seen in foundries are hugely overfed. This almost certainly is the result of a number of factors:
(i)   It is commendable that the founder should 'play safe'.

(ii)   The damage to the melt by poor running systems is usually so bad that most of the metal used to prime the system is no longer fit for contributing to the casting. Thus it is 'dumped' into the feeder, making the feeder oversize for its purpose of feeding.

(iii)   The computer simulation packages that check for solidification shrinkage in the casting are usually programmed with significantly inflated shrinkage factors, resulting in oversize feeders. Because of feature (ii) above, the oversize feeders are currently required, so that this problem has remained obscured.

When good filling systems are provided for the casting, the oversize feeders can usually be greatly reduced, greatly increasing the yield of the casting. It follows also that for good metal and a good filling system simulation packages will have to reduce their shrinkage factors to more realistic values. However, when reducing the volume of feeders I usually do not reduce their height. The height is valuable to pressurised the melt in the freezing casting, and thus suppress the opening of bifilms. If bifilms are allowed to open, properties are reduced, even though the bifilms may not have opened sufficiently to create visible porosity (the casting may still appear perfectly sound). Only if the bifilms open by some fraction of a millimeter will they start to become visible as microporosity (and thereby cause the properties to fall even further).

Filters

Since the beneficial action of a ceramic foam filter derives perhaps 5 % from filtration (depending on the quality of the metal) and 95 % from improved flow [11] the use of filters in filling systems is seen to be mainly that of the reduction in speed of the flow. In this way the melt downstream can remain cleaner simply by being less turbulent. To retain the benefits of the reduction in speed the filling channels after the filter require to be expanded. Thus if the filter can reduce speed by a factor of 5, an expansion of downstream area by a factor of 4 will retain some pressurisation but gain the benefit of speed reduction to one quarter of the speed entering the filter. The correct sizing of the post-filter part of the runner is usually overlooked. The author finds the filter most valuable in those systems entraining air because the front-end design is poor but cannot easily be rectified. The filter assists the back-filling of this part of the system. Naturally, the provision of a bubble trap to divert buoyant phases such as air (and in the case of ferrous castings, slag) away from the main flow can be useful in this case (Figure 4) [12].

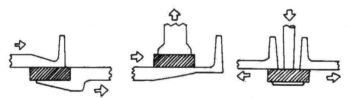

Figure 4. Examples of the combined use of filters with bubble traps.

Discussion

With combined reductions in filling and feeding systems metallic yields can easily rise from 45 to 70 per cent or more. Provided the metal that is poured is of adequate quality (which is not to be assumed) defects such as porosity and hot tears can be reduced if not eliminated [4]. In addition, the mechanical and corrosion properties of the casting can be significantly enhanced [13-15].

Although the author finds that the naturally pressurised filling system gives a significantly superior performance to current unpressurised and pressurised systems, the approach remains yet far from optimum. More research is needed to further refine the concept to take advantage of the progressive accumulation of frictional back-pressure to progressively expand sections and so reduce speeds without loss of volume flow rate. For this reason, computer simulations of pressure distribution against the walls of the filling system, to ensure the absence of unpressurised areas, will be more important than current simulations of velocity of flow.

Recommendations

1. Every foundry should, wherever possible and appropriate, convert exclusively to a good counter-gravity filling system to achieve filling at velocities below the critical velocity to achieve sound castings on a routine basis.
2. Processes that include unconstrained pouring by gravity (such most as low-pressure permanent mold techniques) have problems to deliver reliable quality.
3. If counter-gravity cannot be implemented, and gravity has to be used, then the next-best option is the introduction of the naturally pressurised filling principles. Thus
   (i)   Control air entrainment at entrance to system by offset basin or contact pour.
   (ii)   Fill and pressurise whole system with melt (simulate pressure distribution if possible)
   (iii) Control entrainment events at entry to mold, limiting jetting by either reducing velocity to less than 1 m/s by such devices as surge control or 2-stage filling [1].
4. Pressurise the casting during freezing with high top feeders to suppress the opening of bifilms, thereby suppressing porosity and tears, and reducing further loss of properties.

References

1. J Campbell; "Casting Practice" Butterworth-Heinemann 2004
2. J Campbell; "Castings 2nd Edition" Butterworth-Heinemann 2003
3. J Campbell; "Shape Casting: The John Campbell Symposium" Edited by M Tiryakioglu and P N Crepeau. TMS 2005 pp 3 – 12
4. J Campbell; Materials Science and Technology 2006 vol 22 pp 127 – 143 and correspondence pp 999 – 1008.
5. B Sirrell, M Holliday and J Campbell; The 7th Conf on the Modeling of Casting, Welding and Advanced Solidification Processes, London, 1995 Edited by M Cross and J Campbell.
6. X Yang and J Campbell; Internat Journal of Cast Metals Research 1998 vol 10 pp 239-253.
7. T Isawa and J Campbell; Trans Japan Foundrymens Soc 1994 vol 13 (Nov) pp 38-49
8. X Kang, D Li, L Xia, J Campbell and Y Li; "Shape Casting: The John Campbell Symposium" Edited by M Tiryakioglu and P N Crepeau. TMS 2005 pp 377-384
9. H Nieswaag and H J J Deen; 57th World Foundry Congress, Osaka, 1990 part 10 pp 2-9.
10.   F-Y Hsu, M Jolly and J Campbell; TMS Shape Casting Symposium 2007.
11.   T Din, R Kendrick and J Campbell; Trans AFS 2003 vol 111 paper 03-017
12.   A Habibollah Zadeh and J Campbell; Trans Amer Found Soc 2002 vol 110 19-35.
13.   N R Green and J Campbell; Trans AFS 1994 vol 102 pp 341-347
14.   C Nyahumwa, N R Green and J Campbell; Trans AFS 1998 vol 106 215-223.
15.   A. Keyvani, M. Emamy, M. Mahta and J. Campbell; The effect of bifilms on the corrosion of an Al alloy. To be published 2006-7.

Shape Casting: The 2nd International Symposium *Edited by Paul N. Crepeau, Murat Tiryakioğlu and John Campbell*
TMS (The Minerals, Metals & Materials Society), 2007

IMPROVED SOUNDNESS AND MECHANICAL PROPERTIES OBTAINED BY SOLIDIFICATION UNDER PRESSURE IN A206

Pavan Chintalapati [1], John Griffin [1]

[1]University of Alabama at Birmingham
1150 10th Avenue South, Birmingham, AL, 35294

Keywords: A206, Porosity, Solidification under pressure

Abstract

The goal of this study is to achieve forging properties with the benefits of cast structural components. The effect of applying 10 atmospheres of pressure during solidification on the soundness and mechanical properties of A206 aluminum was studied. Experimental wedge shaped castings were solidified in no-bake molds at 1 and 10 atmospheres pressures. A wedge shaped casting provided a range of porosity and controlled solidification front. Radiography verified that pressure increased soundness by an order of magnitude. Tensile coupons were removed from areas with selected cooling rates. Elongation improved by ~50 to 550% and tensile strength increased ~11 to 60% when the casting was pressurized during solidification. Similar improvements in cast properties of other alloys are possible under the correct metallurgical conditions. This paper discusses how to achieve these metallurgical conditions for complex three-dimensional production castings.

Introduction

Precipitation hardened A206 (Al 4.5 Cu 0.25 Mg) alloy is widely used in aerospace and military applications. Due to its high strength, approaching some grades of ductile iron, it is also used in some automotive applications [1]. Similar to many other commercial aluminum casting alloys, A206 is prone to porosity caused by dissolved hydrogen and/or shrinkage. As far as the authors were able to ascertain, very few publications besides one by Griffin et al. [2], were published relating A206 properties to porosity. Current research was done to relate the mechanical properties of A206 with volume percent porosity and the effect of application of hydrostatic pressure.

Safety critical components are usually subjected to Hot Isostatic Pressing (HIP) to heal the pores in the casting. As it is an additional step and involves high temperature and pressure, HIPping adds to cost and time. Application of pressure during solidification was studied as an alternative to obtain sound material and improved casting properties.

Experimental

The experimental casting chosen for these trials was wedge shaped. The wedge shape was selected to obtain a variety of cooling rates and porosity concentrations at preferred locations. A chill was placed at the thinnest section to obtain directional solidification. Metal was melted in a SiC crucible and degassed with a mixture of Ar and 2% Cl_2 gas. Reduced pressure test was done after degassing to ensure adequate degassing of the aluminum melt. A pouring basin with a stopper was used to minimize air entrainment and a ceramic filter was placed in the runner system for entrapment of any remaining oxides. TiBor was added as a grain refiner. The dimensions of the casting along with runner system are shown in Figure 1. A plate machined from the center of the casting was heat treated (T4 condition, described below) and samples were machined (as shown in Figure 2) for mechanical testing. For castings solidified under pressure, the assembly of mold and basin was placed inside the pressure vessel (shown in Figure 3) and the vessel sealed and pressure applied as soon as the mold cavity was filled. Density measurements were performed on the blanks using mensuration, by measuring their mass and dimensions accurately prior to machining tensile samples. Tensile samples were machined according to ASTM E8/B557 for tensile testing.

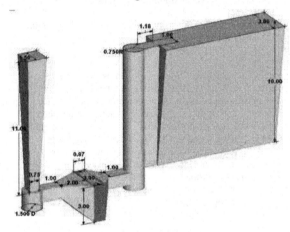

Figure 1. Schematic showing the wedge casting along with dimensions. Dimensions are in inches.

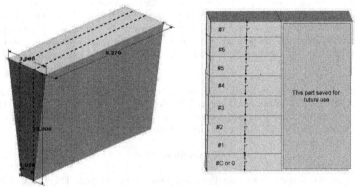

Figure 2. Schematic showing the wedge casting. Dotted lines on left figure indicate the location of plate for testing. Figure on the right shows the location of the samples tested. Dimensions are in inches.

Figure 3. Pressure vessel used for application of pressure during solidification.

<u>Heat Treatment</u>

Plates obtained from the castings were heat treated to T4 condition. Specifications: solutionize at 930 °F (~500 °C) for two hours at temperature – increase temperature to 985 °F (~530 °C) and hold at temperature for 12 hours.

Results

The effect of pressure on the density at different casting thicknesses is shown in Figure 4. The chart is plotted for densities at different distances from the chill. As the distance increased, casting wall thickness increased. It could be seen from this figure that with application of 10 atm pressure, constant densities of about 2.77 g/cm^3 are obtained at different locations, while densities gradually decreased for casting solidified at atmospheric pressure.

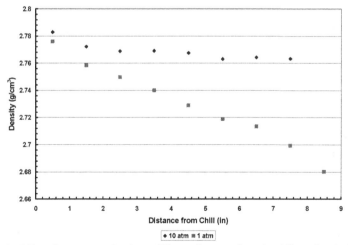

Figure 4. Effect of pressure on density at different distances from the chill. Notice the higher densities achieved by pressurization.

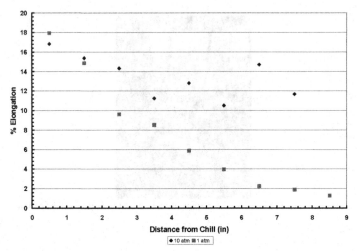

Figure 5. Effect of pressure on % elongation at different distances from the chill. Notice the significant increase in % elongation with distance from the chill. Each point represents data from one tensile test.

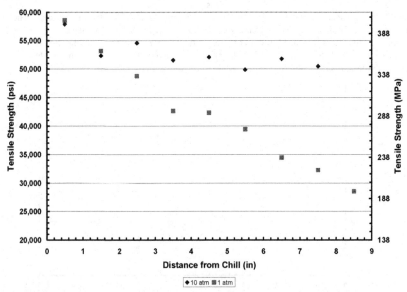

Figure 6. Effect of pressure on tensile strength at different distances from the chill. Notice the significant increase in UTS with distance from the chill. Each point represents data from one tensile test.

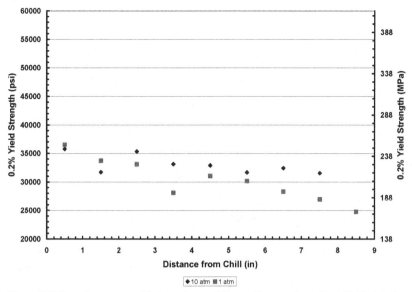

Figure 7. Effect of pressure on % elongation at different distances from the chill. Each point represents data from one tensile test.

The effect of application of pressure on the static mechanical properties is shown in Figures 5-7. The effect of pressure on porosity is shown in micrographs in Figure 8. The figure compares montages of samples from 1 and 10 atm pressures that are 1.5" from the chill. Significant decrease in porosity could be witnessed.

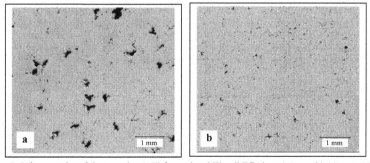

Figure 8. Micrographs of the samples 1.5" from the chill solidified at a) 1 atm b) 10 atm. Notice significant changes in size and volume of porosity.

Discussion
Mechanical Properties

Application of pressure during solidification significantly reduced the porosity and increased the density at different thicknesses. Porosity is known to have a detrimental effect on the elongation of aluminum. Decrease in porosity should increase elongation. By application of 10 atm pressure, densities at different casting thicknesses were increased by 0.25 to 2.5 %. This range in variation could be ascribed to variation in thickness of the casting and effectiveness of the pressure as described later. Figure 5 shows this effect of pressure on % elongation. It could be seen that at the same distance from the chill, application of pressure improved % elongation by ~ 30 to 550 %.

The effect of decrease in porosity by application of pressure on tensile strength is shown in Figure 6. It could be observed from this figure that at the same distance from the chill, application of pressure resulted in increase of tensile strengths by ~11 to 60 %. Yield strength is a material property and is mostly affected by chemical composition and heat treatment. It is not expected to change with change in porosity. Results concurrent with this theory are obtained in present research as shown in Figure 7.

Hot iso-static pressing is usually performed on the critical components, in which porosity is healed. This healing has not always resulted in improvement of static mechanical properties [3]. The effect of application of pressure during solidification was studied by few earlier researchers [4-6]. However, the static mechanical properties reported did not show significant improvement and were mixed. The mixed results reported by earlier researchers, despite reducing porosity significantly, suggests one other important determining parameter in aluminum casting: metal quality. As suggested by Campbell [7], the presence of bi-films in the melt could significantly reduce the mechanical properties, as they behave as pre-existing cracks in the material. The quality of the melt was not reported in these reports, so the poor observed properties may be ascribed to poor melt quality.

In the present research, proper care was taken to obtain a good quality melt and casting. As mentioned before the melt was degassed with a mixture of argon and chlorine. Chlorides formed at the gas bubble and metal interface (Al reacts with Cl_2 to give $AlCl_3$) change the surface tension as a result of which, oxide inclusions in the melt gets attached to the bubbles and float to the surface [8]. A ceramic filter was placed in the metal flow path as shown in Figure 1 to entrap any remaining oxide films. The design of the casting was made to minimize the turbulence during the flow. All these factors contributed to good casting quality.

Solidification under Pressure

Application of pressure during solidification reduces both the shrinkage and gas porosity, as the applied pressure suppresses the growth of gas pore and increases the feeding capability of the metal through interdendritic channels.

Pressure could reduce the gas porosity in aluminum castings as explained by Sievert's and Boyle's laws.

Sievert's law states that
$$C_s = K\sqrt{P}$$

where, C_s is solubility of hydrogen in solid, K is a constant and P is the pressure of hydrogen inside the pore, which would be equal to applied pressure. By increasing the applied pressure, solid solubility of hydrogen in Al increases and hence decreases the amount present for pore formation.

By Boyle's law:

$$V_v = C(\frac{1}{P})$$

where V_v is the volume of the pore, C is a constant and P is the applied pressure. So the volume of gas pore decreases with increasing pressure.

Shrinkage porosity is caused by local insufficient feeding during late stages of solidification. By the application of pressure during solidification, liquid metal could be pushed interdendritically to feed the regions that are starved of liquid. This could reduce (or eliminate) shrinkage porosity.

This effect was seen at various casting thicknesses. As observed by earlier researchers Boileau et al [3], Murthy et al [5], and Eady et al [9], porosity gas a significant effect on elongation followed by tensile strength. The results obtained in the present research are concurrent with these observations. However, these results contradict the observations by Radhakrishna et al [10]. With reduction in porosity, fatigue life of the component increases significantly as observed by Boileau et al and Wang et al [11]. Hence by application of pressure during solidification, fatigue life is expected to increase, besides the static mechanical properties. This could be a very useful technique in casting the components demanding improved properties.

The design of the casting as well as the gating and runner system plays a crucial role in extending this technique for complex castings. In a casting, the applied pressure will produce the desired results only when it could reach the required areas in the casting such as the areas prone to porosity. Proper design of the casting and runner system resulting in directional solidification is required. One obvious route for this application would be through the riser. However, more 'provisions' could be made in the casting through which the applied pressure could reach the required areas.

In the current research, porosity was reduced significantly but was not completely eliminated. This could also be ascribed to the casting design, which as previously mentioned demands the directional solidification. The physical phenomenon resulting in increased casting soundness with application of pressure is not completely understood and requires further research.

References

1. G.K. Sigworth, F. DeHart, and S. Millhollen, "Use of High Strength Aluminum Castings in Automotive", Journal of Light Metals, (2001), 313-322
2. J. A. Griffin, J. Church, and D. Weiss, "Fatigue, Tensile and NDE Relationships in Cast A206-T4 Aluminum", Transactions of American Foundry Society, 111 (2003), 289-301.
3. J.M. Boileau, J.W Zindel, and J.E. Allison, "The Effect of Solidification Time on the Mechanical Properties in a Cast A356-T6 Aluminum Alloy", SAE Special Publications, 1251 (1997), 61-72.
4. S.Z. Uram, M.C. Flemings, and H.F. Taylor, "Effect of Pressure during Solidification on Microporosity in Aluminum Alloys", Transactions of American Foundry Society, 66 (1958), 129-134.

5. K.S.S. Murthy, and E.O. Edwards, "Effect of Pressure Applied During Solidification on the Soundness of Al-7% Si-0.3% Mg (SG70-British Equivalent, BS 1490; LM 25) Alloy Sand Castings", The British Foundryman, 68 (1975), 294-304.

6. D. Hanson, and I.G. Slater, "Unsoundness in Aluminum Sand Castings. Part-III.- Solidification in Sand Moulds Under Pressure", Journal of Institute of Metals, 56 (1935), 103-123.

7. J. Campbell, "New Metallurgy of Cast Metals", Conf. Proc. Advances in Aluminum Casting Technology II, (2002) 11-17.

8. J.E. Gruzleski, and B.M. Closset, "The Treatment of Liquid Aluminum-Silicon Alloys, Brown, J.E., Heat Treatment of Aluminum Alloys", American Foundrymen's Society Inc, Des Plains, IL (1990).

9. J.A. Eady, and D.M. Smith, "Effect of Porosity on the Tensile Properties of Aluminum Castings", Materials Forum, No. 4, 9 (1986) 217-223.

10. K. Radhakrishna, S. Seshan, and M.R. Seshadri, "Effect of Porosity on Mechanical Properties of Aluminum Alloy Castings", Transactions of the Indian Institute of Metals, No. 2, 34 (1981) 169-171.

11. Q.C. Wang, P. Jones, and M. Osborne, "Effects of Applied Pressure during Solidification on the Microstructure and Mechanical Properties of Lost Foam A356 Castings", Conf. Proc. Advances in Aluminum Casting Technology II, (2002) 75-84.

Shape Casting: The 2nd International Symposium *Edited by Paul N. Crepeau, Murat Tiryakioğlu and John Campbell*
TMS (The Minerals, Metals & Materials Society), 2007

THE DESIGN OF L-SHAPED RUNNERS
FOR GRAVITY CASTING

Fu-Yuan Hsu[1], Mark R. Jolly[2] and John Campbell[2]

[1]Auspicium Co., Ltd., 2F-3, No.4, Sec.1, Jen-Ai Road, Taipei, Taiwan, R.O.C.
[2]University of Birmingham, Edgbaston, Birmingham B15 2TT, UK

Keywords: Simulation, Gravity Casting, Runner design, Junction

Abstract

The purpose of this research is the development of guiding principles and rules for the design of filling systems for aluminium gravity castings. The approach employs computational modeling and real-time X-ray radiography. Out of five geometries of L-junctions that occur in filling systems and which have been studied, space dictates that only the sprue/runner junction is presented. Progressive filling along the L shaped geometry has been achieved without surface turbulence by eliminating excess cross sectional area of the channels, and in particular reducing the "dead zone" in the corner of the junction. Higher flow rate and less turbulent filling of the mould cavity can be achieved by the new L shaped junction design.

Introduction

Rules for designing the pouring basin and sprue were well established in previous works [1, 2] leaving this research to focus on the development of guidelines for the design of the next feature of the running system, the sprue/runner junction.

The requirements for a good running system can be summarized as the prevention of the entrainment defects into the casting in the early stage of mould filling [2-4]. The liquid metal should fill progressively along the running system without developing surface turbulence that would risk the entrapment of defects such as bubbles and bifilms into castings.

However, at the base of the sprue, the high velocity of the liquid metal is inevitable, well above the critical velocity [5] introducing the significant risk of entrainment problems. To avoid the formation of defects, the design of the L-shaped junction, between the vertical runner (the sprue) and the horizontal runner, becomes a critical task. This junction does not appear to be well understood in terms of fluid dynamics; hence the reason for this study.

Method

Computational modeling
A computational fluid dynamics (CFD) code, Flow-3D™ [6], has been used for these studies. Flow-3D™ is based on the finite-volume-method, which is originally developed as a special finite difference formulation. Since the flow phenomena in running systems were mainly considered here, one fluid (i.e. liquid metal) with sharp interface tracking using the volume of fluid (VOF) algorithm was employed in the modeling. The sand mold was assumed to have a high permeability, so that the generation of back-pressure from air and/or gas in the cavity during

filling was ignored. Empty cells within a domain were therefore present as voids at no pressure, equivalent to the assumption that atmosphere pressure exists throughout the filling process. Isothermal conditions were assumed allowing any change of viscosity of the liquid to be ignored as approximately only 1 per cent of heat loss was expected (c.f., Richins and Wetmore [7]).

The simulations were carried out on a Silicon Graphics Octane (R10000-IP30) machine, running a 175 MHz CPU and 640 MB memory. The operating system was IRIX version 6.5 (Silicon Graphics). The input parameters, which were selected for the modeling of liquid alloy Al-7Si-0.4Mg in the options of Flow-3D™, have been listed previously [8, 9].

<u>L shaped junctions in 2-dimensions</u>
The 2-D condition is a theoretical approximation to reality that has the benefit of simplicity. It may be considered when the flow in the third direction is non-existent and when the effect of friction forces resulting from the boundary walls in that direction can be ignored.

A series of 2-dimensional geometries were constructed having one-directional flow with a uniform velocity impacted into the right angle of different L-shaped junctions. The geometry of flow issuing from the junction depended on the space available. Effectively, the approach constrains one side of the flow stream and measures the profile of the other free side in a 2-dimensional condition. With sufficient space the unconstrained flow profile in this L-junction could be characterized by a number of measurements, but in this report limitations of space dictates a focus on only one dimension, probably that of greatest interest to the casting engineer: the inner radius R_i. Fitting an appropriate inner-radius R_i to the curve of the transition flow, a three-point method was used that involved a potential error of about ± 2.5 mm.

<u>The computational approach for 2D an L-shape junction</u>
Figure 1 presents a schematic illustration of an L-shape junction. The basic variables, the "Inlet-opening I_o" and "inlet-velocity V_i" were systematically studied to characterize their influence on the design of the junction.

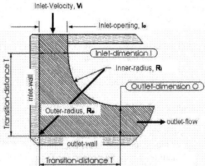

Figure 1 schematic of a L-junction and the profile of transition flow determined by the five dimensions (i.e., I, O, R_i, R_o, and T)

All the outputs have been reported in terms of the "Inlet-opening I_o". In the 2-dimension case, the Inlet-opening I_o can, of course, be simply characterized by a length. The variables geometrical assumptions were
1. Various inlet-opening I_o (1, 4, 8, 12, 16, 20, 24, 28, and 32 mm for L-shape wall) were built.

102

2.   Various unidirectional inlet-velocities V_i (1, 2, 4, and 8 m·s^{-1}) were applied as initial boundary conditions.

Figure 2 demonstrates an example of the 2D simulation with the inlet-velocity $Vi = 2.0$ m·s^{-1} and inlet-opening $I_o = 28$ mm (L-shape junction V_i2I_o28) at various time frames.

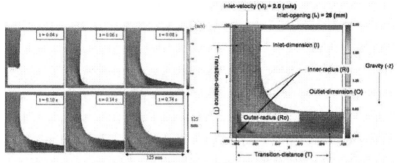

Figure 2 a series of frames for the 2-D modelling result of the L-shape junction V_i2I_o28.

L shaped junctions in 3-dimensions

To explore the appropriateness and validity of the 2-D design a 3-dimensional computational model was constructed (Figure 3(a)). Because of the velocity boundary condition of uniform one-direction flow entering the top of the domain, the geometry of the hyperbolically tapered sprue was constructed assuming conservation of mass. An entrance velocity into the sprue of 0.891 m·s^{-1}, equivalent to a head height of 40 mm in the pouring basin, was assumed. Thereafter, a 260 mm height of hyperbolic shape of sprue was designed, giving a total head height of 300 mm.

A Radius-Bend geometry (Figure 3(b)) was modeled to compare with the former sharp L-junction. An appropriate geometrical profile of the channel is not known a priori. Thus an outer-radius, tangential to the inlet-wall and the outlet-wall, was constructed assuming the depth in the middle of the bend corresponded approximately to the average of the inlet- and outlet-dimensions of the flow in the original right angle L-shape.

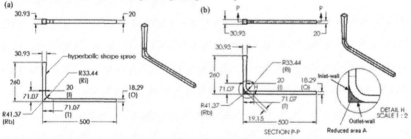

Figure 1 The 3-dimensional (a) "L-shape" and (b) "Radius Bend" of the sprue-runner junction.

The main role of the real-time X-ray radiography was to verify the results of computational modeling. A 'real' casting incorporating the sprue-runner junction was investigated using a square sprue of total head height 300 mm. The technique has been described previously [9]. Radiographic images do not reproduce clearly, and so are not presented here. Radiographic results are available on request from the corresponding author.

Results

L shaped junctions in 2-dimensions.
The effect of various inlet-velocities V_i on the profile of flow for the L-shape junction are plotted in Figure 4. The dimensions of the outer-radius R_o, inner-radius R_i, and outlet-dimension O are normalized by inlet-dimension I.

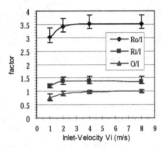

Figure 4 In the 2-D L-shape junction, the ratio of the dimensions, which are the outer-radius R_o, the inner-radius R_i, outlet-dimension O, over the inlet-dimension I, against to various inlet-velocity V_i.

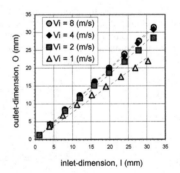

Figure 5 The outlet-dimension O versus to the inlet-dimension I for the 2-D L-shape junction with various inlet-velocity V_i.

L shaped junctions in 3-dimensions
Three experiments for sprue-runner junction designs are investigated in this section. They were a 3-D L-shape simulation, a 3-D Radius-Bend simulation, and an L-shape in real casting.

104

3-D L-Shape Simulation

The simulation results of the L-shape in the center plane are shown in Figure 6(a). Before the time frame of 0.13s, the liquid metal filled completely the hyperbolic square sprue. Also, the flow front advanced progressively without leaving any empty region in its wake. At 0.13s, the liquid metal starts to fill the L-shape of the sprue-runner junction. The magnitude of velocity at that time was 2.27 m·s^{-1}. At 0.15s, the flow impacted on the base of the runner. At 0.24s, the flow filled the L-shape completely. At 0.31s, the running system was completely filled and the maximum velocity of the flow at the exit of the runner was 2.56 m·s^{-1}.

3-D Radius Bend Simulation

Figure 6(b) shows the results of the radius bend in the center plane. Again, before 0.13s, the flow gradually filled the square-sprue down its hyperbolic shape. At 0.13s, the flow began to fill the sprue-runner junction of the Bend-shape. The velocity was 2.22 m·s^{-1} at that time. After 0.13s, the flow filled the outer side of the Bend-shape and the base of the runner. The flow moved toward the inner side of the Bend-shape progressively without creating empty regions. At 0.18s, the bend was completely filled. At 0.31s, the whole running system was filled and the maximum velocity of the flow at the exit of the Bend-shape runner was 2.65 m·s^{-1}.

(a)   (b)

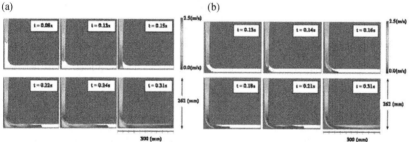

Figure 6 The modeling result of (a) the 3-D L-shape and (b) the 3-D Radius-Bend shape on the centre plane.

L-shape in real casting

At the time frame of 0.16s, the flow entered the L-shape junction. At 0.20s, the flow impacted on the base of the junction. At 0.24s, the flow direction changed from vertical to horizontal direction. At 0.28s, the junction was completely filled. The velocity at the end of the runner was measured by the trajectory method in which the melt was observed to create a free-fall parabola shape during which was captured by video and measured (air resistance to the trajectory was neglected). The velocity of the flow was highest, 2.23 m·s^{-1}, at 0.08s. After this peak velocity, the flow settled to a constant velocity of 1.69 m·s^{-1} from 0.4s to 3.16s, after which it declined until the end of flow at 4.5s. Due to parallax and other errors, the total potential error of measurement could be 2.63 mm corresponding to a potential velocity error 0.11 m·s^{-1}.

Discussion

L shaped junctions in 2-D

The following discussion will be based on the example of L-shape junction (Figure 2), which has inlet-opening $I_o = 28$ mm and inlet-velocity $V_i = 2$ m·s^{-1}. In Figure 5 the outlet-dimension versus the inlet-dimension is plotted. It shows that the inlet and outlet dimensions tend to become identical as the inlet-velocity approaches 8 m·s^{-1} implying that the influence of gravity reduces at high velocity.

In the exercise described in this work, only the walls on the outside of the bend constrain the flow, whilst the inner side of the flow remains open to the free space of the runner. However, by tailoring a channel to the exact predicted shape of the junction so that the channel walls would contain the flow completely, confining the flow by walls on all sides raises the issue that the additional wall naturally provides friction, locally reducing its velocity. However, of course, the velocity profile of the flow will change relatively little since the majority of the flow will be concentrated in the centre of the runner. The additional, relatively small, influence of friction will act to assist the channel to remain fully filled by a slight overcompensation, conferring a reassuring factor of safety for the complete filling of the channel.

Even so, the 2D model is probably too oversimplified to be helpful in the design of real 3D systems mainly because as it stands only the outer surfaces contribute to frictional drag. This minimal drag is corroborated by the inlet and outlet velocities being nearly equal. For the 3D model the area of surfaces contributing to frictional drag is four times greater.

Junctions in 3-D
Simulated L-shape and Radius Bend in modeling
In the modeling of the L-shape of sprue-runner junction, the flow enters the geometry at 0.13s. At 0.24s, this geometry is completely filled. Thus, the duration of filling this geometry, the "clean-up" time, is 0.11s. For the Bend-shape, the "clean up" time can be taken to be merely 0.05s, being the time from 0.13s to 0.18s taken for the flow to travel around the bend. This shorter time corresponds to the reduced volume of the junction by the removal of the 'dead zone' in the corner of the junction. It is seen therefore that the Radius Bend approach is not only economical but has an improved flow behavior.

Considering the 'clean up' period more strictly, since no entrainment or air or surface appear to occur at any time, this time can be said to be zero for both of these junctions. Even if not considered zero, these times are far less than those required in the traditional sprue-runner junction designs by Grube et al [10], in which their optimum 'clean-up' times are from 1 to 3 seconds.

L-shape in a real casting
In the filling of the real casting observed by video X-ray radiography, it is clear that there are some empty regions in the base of the L-shape junction. These empty regions are probably the result of the rather 'ragged' fall of the melt down the sprue for the first fractions of a second. After the flow impacts on the floor of the system the flow is deflected as a jet toward the exit of the runner. As a free stream emerges from the end of the runner, contact with the walls is minimal, so that frictional resistance is negligible. In the results of the real L-shape casting experiment the exit velocity from the junction at the beginning of the flow was 2.23 m·s^{-1}; quite close to the theoretical velocity 2.43 m·s^{-1} for a free fall from 300 mm, confirming that the jet does not totally contact the surfaces of the walls of the running system in the early phase of the filling.

The high-speed jet began to slow down as the flow completely filled the whole system. The "clear-up" time of the L-shape junction is 0.12s (from the beginning of entering at 0.16s to the fully filled condition at 0.28s). The velocity measured by the trajectory experiment at this stage was 1.69±0.11 m·s^{-1}; a 30% reduction in the speed at the sprue base. In Table 1 the head losses on each part of the real casting and the loss coefficient of the L-shape junction were estimated. It shows that the head loss of the L-shape junction (0.016m) is a minor loss in comparison to the major loss of sprue (0.082m). And, in the calculation, the loss coefficient, K_L, of the L-shape junction is 0.110. The detail calculation was shown elsewhere [8].

Generally, the modeling appeared to give substantially similar profiles of the flow to the real casting at the various stages. However, some details of the flow appear different as discussed below.

The filling time in the model is always less than that in the real casting. This feature was also found in the benchmark test 1995 of Flow-3D modeling. In an attempt to counter this Barkhudarov and Hirt [11] used an effective viscosity 5 times higher than the molecular viscosity of pure aluminium to prolong the filling time from the original modeling prediction of 1.5s to the experimental time of 2.2s. In this work also the original molecular viscosity was employed, corroborating once again shorter predicted filling times. The assumptions of isothermal and one-fluid (i.e. no air) are thought to have negligible influence on this problem. It seems more likely that bifilm defects in suspension in the melt are likely to significantly increase the effective viscosity; a factor of 5 does not seem unreasonable for typical concentrations of bifilms. (High concentrations of bifilms can increase the viscosity by a factor of 10^3 or more, and are not uncommonly seen as melts of porridge-like consistency). In addition, the presence of the oxide film on the advancing meniscus, having to be continuously broken (but of course continuously re-forming) as a result of its being dragged back against the walls of the channels, is likely to contribute a real and observable drag on the melt.

Table 1 head and head losses for each parts in the real casting of the L-shape junction.

Parts		Head height (m)	Loss coefficient, K or friction factor, f	Source of coefficients	Others data for calculations
Head losses	Sprue entrance	0.014	$K_E = 0.30$	Richins and Wetmore[7]	$VE = 0.95 \text{ m·s}^{-1}$
	Sprue	0.082	$K_S = 0.563$	Srinivasan [12]	$\overline{V}_2 = 1.69 \text{ m·s}^{-1}$
	L-shape junction	0.016	$K_L = 0.110$	Present author calculated	$\overline{V}_2 = 1.69 \text{ m·s}^{-1}$
	Runner	0.032	$f_R = 0.04$	Richins and Wetmore [7]	$\overline{V}_2 = 1.69 \text{ m·s}^{-1}$ $l_R = 0.070 \text{ m}$ $D_R = 0.01277 \text{ m}$
Velocity head at exit of the runner		0.156	The kinetic energy coefficient, $\alpha = 1.07$	Present author estimated (c.f., Fox and McDonald [13])	$\overline{V}_2 = 1.69 \text{ m·s}^{-1}$ (Trajectory method)
Total head height		0.300			

The breaking away of the flow from the surface of the channel contrasts with the modeling that indicates that the flow is always against the surface of the wall. The relatively coarse mesh selected for the simulation to reduce the computation to reasonable times of the order of 48 hours may have contributed somewhat to this small deviation from realism. However, the major contributor in practice is more likely to reside with the experiment rather than the simulation as a result of the imperfect initial priming of the sprue in real casting conditions.

Conclusions

1. Guidelines for the designing of L-junctions have been developed.

2. The 2D models are relatively poor simulations compared to the 3D models of junctions.

3. Progressive filling along the L-junction geometry can be improved by reducing the area of the "dead zone".

4. L-Junctions, if designed as in this study, have a relatively small frictional loss compared to the longer length channels of the filling system such as the sprue and the runner.

Acknowledgments

F-Y Hsu acknowledges the help of Jean-Christophe Gebelin, Nan-Woei Lai, H.S.H. Lo, Masood Turan, A. Caden and Jörg Pfeiffer, and the sponsorship of the Overseas Research Students Awards Scheme (ORS) of Universities UK from 2000 to 2002.

References

1.   Yang X., Campbell J., liquid metal flow in a pouring basin, Int. J. Cast Metals Res., (1998), 10, pp.239-253
2.   Campbell J., (1991), Castings, Butterworth-Heinemann, Oxford, UK.
3.   Green N.R., Campbell J., Trans. AFS, (1994), pp.341-347
4.   Campbell J., Invisible Macrodefects in Castings, Journal de Physique IV, (1993), 861-872, Coliogue C7, supplement au Journal de Physique III, volume 3, Nov.
5.   Runyoro J., Boutorabi S.M.A., Campbell J., Trans. AFS, (1992),37, 225-234
6.   Flow Science Inc., 683 Harkle Rd Suite A, Santa Fe, NM 87505m U.S. http://www.flow3d.com
7.   Richins D.S., Wetmore W.O., Hydraulics applied to molten aluminum, Trans. ASME, (1952), July, pp.725-732
8.   Hsu F.-Y., "Further Development of Running System for Aluminium Castings" (PhD thesis, The University of Birmingham, U.K., 2003)
9.   Yang X, Jolly M.R., Campbell J., Minimizations of surface turbulence during filling using a vortex-flow runner, Aluminium Transactions, (2000), 2 (1), 67-80
10.   Grube K., Jura J.G., Jackson J.H., A study of the principles of gating as applied to sprue-base design, Trans. AFS, (1952), v.60, pp.125-136
11.   Barkhudarov M.R., Hirt C.W., Casting simulation: mold filling and solidification-benchmark calculations using Flow-3D®, Modeling of Casting, Welding and Advanced Solidification Processes VII, (1995), pp.935-946
12.   Srinivasan M.N., "Applied hydraulics to gating systems" (Ph.D. thesis, The University of Birmingham, U.K.,1962)
13.   Fox R.W., McDonald A.T., Introduction to fluid mechanics, Fourth edition (John Wiley & Sons, 1994)

ANALYSIS OF A CONFLUENCE WELD DEFECT IN AN ALUMINUM CASTING

O. García-García[1], M. Sánchez-Araiza[2], M. Castro-Román[1], J. C. Escobedo B.[1].

[1]Cinvestav-Saltillo; Apartado Postal 663, 25000-Saltillo, Coahuila, México
Email: oscar.garcia@cinvestav.edu.mx; jose.escobedo@cinvestav.edu.mx;
manuel.castro@cinvestav.edu.mx
[2] Industrias Castech, S.A. de C.V., Blvd. Industria de la Transformación 3140, Parque Industrial
Ramos Arizpe sector 2, Ramos Arizpe Coah., México C.P. 25900
Email: sanchez.miguel@castech.com.mx

Keywords: Confluence weld, oxide lap, gas back-pressure

Abstract

A leakage problem in an aluminum cast cylinder head was studied. Metallographic characterization showed the presence of a confluence weld defect caused by the meeting of two different metal streams not able to join together. Process key variables such as pouring temperature, mould temperature, etc., were analyzed to determine their influence on the occurrence of the problem. Computer simulation of the filling process suggested that air entrapment could be related to the origin of the defect. Additionally, core outgassing was found to be an important factor increasing the local pressure at the area of interest. Optimization of process key variables contributed significantly to reduce the scrap rate. However, the complete elimination of the defect required a major change in casting design, which was made by considering both, the formation mechanism of the confluence weld defect and the increase of local pressure due to core outgassing during the filling operation.

Introduction

The reduction of scrap rates is important for foundries in order to reduce costs and improve quality. While there is a lot of information regarding casting defects, the solution to a specific foundry problem requires finding specific causes, as castings and processes are different among each other. The present work is intended to understand and to find a solution to what was designated as a confluence weld. This type of defect has been described by Campbell [1] as an oxide lap problem, where the formation of films on the advancing fronts of two converging metal streams prevents them from joining successfully. Contrary to the continuous advancing of the metal fronts where a joint with full strength would be expected, if a front comes to a stop while the other continues advancing, the stationary front would build up a thicker oxide layer. When the live front meets the just built oxide layer, a discontinuity would be created completely across the casting. This paper presents a case study carried out during the production of an aluminum cast cylinder head in a semi permanent mold, where high scrap rates caused by a leakage problem resulted in a critical issue. The cause of the leakage was found to be a confluence weld defect.

Characterization of Confluence Weld

In order to observe the filling sequence of the problem zone some cores were removed from the assembly set in the semipermanent mold. Then, a casting was poured and video recorded. Two metal bars were utilized to keep the remaining cores present in the assembly in position, and prevent them from floating during filling of the mould.

Figure 1 shows the filling sequence of the section were the problem occurred. As can be seen, more than one metal stream came to a stop. It is worth mentioning that the problem section had a thin 5 mm wall thickness at its narrower section, so the stopping effect could be increased by the effect of surface tension.

Non advancing metal fronts

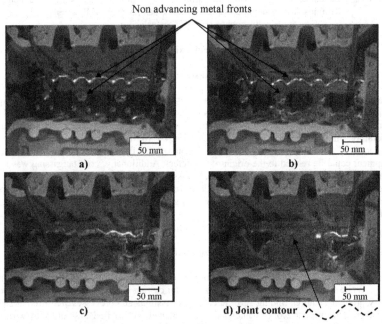

Figure 1. Filling sequence of the problem zone: (a) Arriving of metal streams. (b) 0.3 seconds after arriving of metal streams, various metal fronts stop their advance. (c) 1 second after arriving of metal streams. (d) 1.2 seconds after arriving of metal streams, the welding contour of the different metal fronts is observed.

The appearance of the confluence weld defect is displayed in Figure 2. It is clear that the contour of the defect corresponds to that of the joint of the metal streams showed in Figure 1d. Figure 3 illustrates the detached surfaces of the confluence weld, which present horizontal wrinkles created by the discontinuous advance of the liquid metal, Figure 3c.

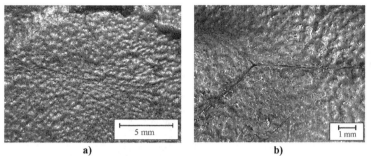

a)   b)

Figure 2. Appearance of the confluence weld defect: (a) Formed by two metal fronts, (b) Formed by three metal fronts.

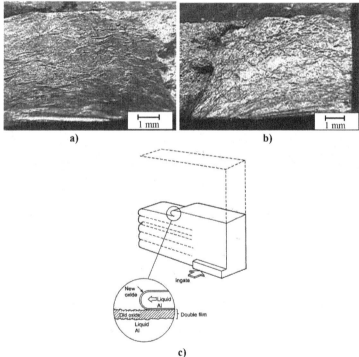

a)   b)

c)

Figure 3. (a) and (b), Illustration of detached surfaces of the confluence weld defect; (c) Formation of laps due to unstable advancing of a film forming liquid metal. [1].

SEM characterization of both metal fronts was performed in order to obtain physical evidence of the oxide film. This was not possible, probably because the thickness of the oxide lap was too thin. The oxide thickness depends on the time available for its formation, ranging from 1nm to

1μm for times of up to 1 second [1]. Another possibility is that the oxide film had been torn away from the wall during sample preparation. Figure 4 shows the presence of an oxide film with primary α–Fe (Al$_{15}$(MnFe)$_3$Si$_2$) intermetallics on the tip of one of the non-advancing fronts. Primary α-Fe particles tend to form on the wetted surface of several oxides encountered in aluminum [2]. The nucleation of precipitates on only one side indicates that the defect corresponds to an asymmetrical thick/thin double oxide film [1].

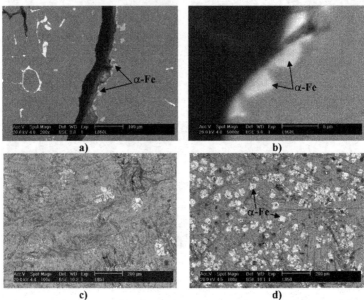

Figure 4. SEM backscattered electron images showing; (a) Cross section of the confluence weld with α–Fe (Al$_{15}$(MnFe)$_3$Si$_2$) particles at the surface of the stopped front. (b) Tip of the stopped front viewed at a larger magnification, (c) and (d) the two opposing detached surfaces of the confluence weld defect, showing primary α-Fe particles nucleated on the tip of the stopped front.

Process Variables Affecting Confluence Weld Formation

The oxide lap can be associated with a fluidity problem, as the formation of surface films in liquid aluminum increases the surface tension by a factor of three, becoming an important factor in narrow channels [3]. As process parameters have an important effect on fluidity, key variables influencing the confluence weld formation were studied. A statistical control analysis based on p control charts of several process variables known to affect fluidity was performed.

Venting

The venting practice of mold cavities and its influence over fluidity has been a subject of much debate [4]. Usually, the venting practice has been focused on cores and related to blowhole defects where it is necessary that the gas pressure in the core exceed that of the metal head. Another important effect of the gases evolved from a core, is that related to a premature evolution before the metal fills a cavity. The presence of gas in a closed section would increase

112

the pressure in the mold cavity hindering the advance of the metal fronts. In order to understand the influence of core outgassing on the filling process, a simulation of the distribution of air pressure was performed using Magmasoft® software. This was made possible by setting the sand permeability value to zero, which increases the effect of the air present in the cavity. Figure 5 illustrates the results obtained from the simulation, it can be seen that the air pressure would concentrate in the problem zone causing an empty space.

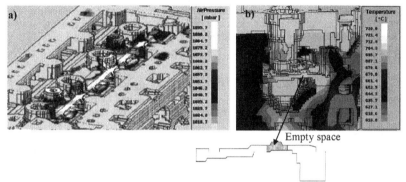

Figure 5. Computer simulation of filling and solidification processes: (a) Showing the zone prone to elevated air pressure effects, (b) Filling temperature profiles, showing the empty space in the defect zone due to the elevated gas pressure.

Metal Temperature

It is important to differentiate an oxide lap-confluence weld defect from a cold lap. The latter is created because of a premature freezing of metal during pouring, while the former is a consequence of metal oxidation. A cold lap can be prevented by increasing the pouring or die temperatures. The metal temperature at the problem section was measured; it was found that the metal arrived at nearly 620°C, Figure 6, which corresponds to a superheat of 50°C. The cooling rate obtained from measurements was 4.5°C/s; this indicated that the metal had over ten seconds to fill the problem zone before solidification started. A process survey carried out on the incidence of the confluence weld indicated that it occurred independently of the metal and mould temperatures, discarding a cold lap as the resultant defect.

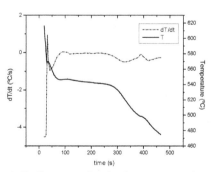

Figure 6. Cooling curve for aluminum, measured at the zone of the confluence weld defect.

113

Pouring Rate

It has been proposed that an oxide lap problem can be eliminated by keeping the liquid fronts moving. This can be accomplished by casting at a higher rate [1]. The rate at which the mould is filled can be increased either by increasing fluid velocity or by increasing the cross-sectional areas of the runner, gates or sprue. Care must be taken in order to avoid turbulence during filling, which limits the use of high pouring rates. Figure 7 shows that by increasing the pouring rate by about 20%, a decrease of 40% of the mean scrap value was obtained. Higher pouring rates were not satisfactory because of turbulence problems.

Sand Permeability

In order to increase the venting capacity of the mould cavity, the permeability was increased by employing a coarser grain sand distribution, with a lower quantity of fines. As illustrated in Figure 7 the defect incidence decreased 25% with respect to the value obtained after increasing the pouring rate.

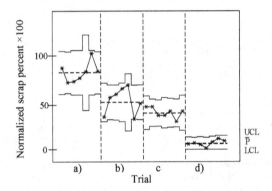

Figure 7. Normalized scrap for confluence weld defect after different trials: (a) Normal production, (b) After a 20% increase in pouring rate, (c) After a 50% increase in sand permeability, (d) After design change.

Discussion

A confluence weld defect was characterized and various process variables optimized in order to prevent its occurrence. The optimization of process variables, as pouring temperature, pouring rate and sand permeability yield a decrease of 50% on the incidence of the defect; the latter is shown in figure 7. Other conditions mainly related with the venting of sand cores were found to have a considerable impact on the generation of the defect. It was observed that those dies presenting high scrap rate had partially blocked vents, either because of carbon and volatiles build up, or by aluminum infiltration. The casting design, which originally had a section of 5mm, also contributed to the occurrence of the defect. The venting capacity of cores was reduced by the clogging of venting orifices located on the base of the permanent part of the mould; this in turn increased the gas pressure at the interior of cores [5] and mould cavity. The augment of

114

pressure in the system works against the metal flow, stopping it and thus promoting the formation of the defect. This effect can be explained using Equation 1.

$$\rho g h = \gamma/r + P_{gas} \qquad (1)$$

where ρ is the metal density, g is the gravity's constant, h is the height of the metal head, γ is the surface tension of the liquid metal and r is the radius which characterize the local shape of the surface. In the present case, r corresponds to the metal front radius and P_{gas} is the backpressure exerted by the gas in the cavity.

The term at the left side of Equation 1 represents the hydrostatic pressure, while the first term at the right represents the backpressure exerted by capillarity repulsion in a thin wide strip and the second, the gas pressure accumulated in the system because of insufficient venting. In order to keep the metal front moving, the hydrostatic pressure must exceed the backpressure created by the capillarity repulsion and the pressure of gas in the cavity. An increase in the metal head, by increasing h, was discarded as a possible solution because it could result in a raise of the metal velocity, causing turbulence problems. This consideration led us to focus on reducing the surface tension effects by increasing the radius r, and reducing P_{gas} by improving the venting capacity. Figure 8 shows the modification applied to the casting. It consisted in connecting the cores forming the cavity to each other to reduce the internal pressure by providing an escape path to the gas generated by the pirolysis of the binder and vaporization of humidity. On the other hand, the increase in the cross section, shown in Figure 7 as "design change", reduced the effect of surface tension, resulting in a significant decrease on the incidence of the confluence weld defect.

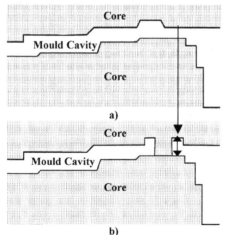

Figure 8. (a) Original casting design, (b) Casting design change considering the reduction of cavity pressure by cores connection and reduction in capillarity repulsion by increasing section thickness.

Conclusions

In the present work a confluence weld defect was analyzed, the following conclusions were obtained:

1. The nucleation of α-Fe particles on the surface of a stopped front of metal during filling indicates that a double asymmetric oxide film caused the confluence weld defect.
2. Pouring at a high rate can reduce the incidence of a confluence weld defect. However, this is limited by considerations of turbulence problems.
3. The gas evolved from cores and the air present in the mould cavity build a positive pressure that works against the metal flow. This backpressure of gases can be reduced by improving venting in the mold cavity.
4. The stopping of the metal front was related to surface tension effects and to the increase of gas pressure in the cavity, the influence of both variables can be visualized in commercial simulation software.

Acknowledgements

The authors acknowledge financial support from Consejo Nacional de Ciencia y Tecnología (CONACYT) and Castech S.A de C.V.

References

1. John Campbell, *Castings* (Oxford, UK: Butterworth-Heinemann, 1991), 40-44, 74.

2. X. Cao and J. Campbell, "The Nucleation of Fe-Rich Phases on Oxide Films in Al-11.5Si-0.4Mg Cast Alloys", *Metal. Trans. A*, Vol. 34A, (2003), 1409-1420.

3. Peter Beeley, *Foundry Technology* (Oxford, UK: Butterworth-Heinemann, 2nd Edition, 2001), 19.

4. Robert C. Voight, et al,"Fillability of Thin-Wall Steel Castings" (Report DOE/ID/13580, Pennsylvania State University US, 2002).

5. Y. Maeda, et al, "Numerical simulation of gas flow trough sand core", *Int. J. Cast Metals Research*, Vol. 15, (2002), 441-444.

STUDY OF MOULD TEMPERATURE EFFECT ON THE INCIDENCE OF POROSITY IN A CAST CYLINDER HEAD

J. Vargas-Orihuela[1], M. Castro-Roman[1], M. Herrera-Trejo[1], M. Sánchez-Araiza[2]

[1]Cinvestav Saltillo, Apartado Postal 663, 25000-Saltillo, Coahuila, México
Email: jacobo.vargas@cinvestav.edu.mx; manuel.castro@cinvestav.edu.mx;
martin.herrera@cinvestav.edu.mx
[2]Castech S.A. de C.V., Blvd. Industria de la transformación 3140, Parque Industrial Ramos
Arizpe sector 2, Ramos Arizpe Coah., México C.P. 25900
Email: sanchez.miguel@castech.com.mx

Keywords: porosity, heat transfer coefficient, simulation, semi-permanent mould

Abstract

Software simulation allows detecting casting zones with susceptibility to foundry defects. Some defects are only detected after castings are manufactured at high production rates. The changes needed to eliminate the problem imply economic losses and time waste. The present work analyzed the possibility of detecting low incidence defect zones by simulation. Simulation was calibrated with experimental data obtained from cooling curves, measured during the filling and solidification processes at the zones were potential defects appeared. The effect of changing the mould temperature was considered. Variations of mould temperature were made by delaying the time between castings. Cooling of the mould, before pouring a subsequent casting, was allowed both with and without the previous casting inside of the mould. The information obtained from simulations was compared with experimental results. It was found that after delaying the molding process, the number and extension of zones with porosity defects were increased.

Introduction

There are many defects prone to occur during the fabrication process of aluminum casting alloys. The most frequent are those associated with porosity. In this type of alloys, porosity forms during solidification, in the mushy zone, where two mechanisms take place:

- Hydrogen segregation and precipitation (hydrogen porosity)
- Poor interdendritic feeding (micro-shrinkage) [1]

Under normal conditions, both mechanisms act at the same time, so each pore is caused by hydrogen precipitation and shrinkage. Porosity defects reduce considerably the mechanical properties of castings and is a major cause of leak defects.

During the last years, simulation software allowed predicting and preventing porosity defects in castings. Thanks to the development of low cost and high capacity computational equipment, this type of software is now accessible to a greater number of foundries around the world. The precision in the prediction of porosity depends on the adequate use of physical and thermal properties of the alloy and mould under study, as well as the initial and boundary conditions that define the system. In addition to precision, the prediction of porosity defects depends on the mathematical model capacity to simulate correctly its formation.

The porosity prediction model based on the detection of isolated liquid or semi-solid metal pockets rounded by a more solidified metal is the most employed in foundries. This model needs the definition of a critical solid fraction to detect casting zones susceptible to porosity defects. These types of models require of a good knowledge of the thermal transfer between metal and its surroundings.

The present work presents a case of study where the data utilized in the simulation programs was optimized to make it representative of a semi-permanent mould process, in order to detect zones with high porosity susceptibility. Since the prediction of these zones is not 100% accurate, small solid fractions were selected to maximize the occurrence of defects during the simulation. With this strategy, a large number of susceptible zones were detected. However, other specific zones can still present low incidence porosity during normal production, due to specific process conditions.

The simulation of the mould temperature distribution and metal cooling curves was analyzed and compared with experimental values. For this purpose, critical sections of the cylinder head were instrumented with thermocouples to measure temperature during solidification and thus obtaining the cooling curves. Infrared thermography was employed to determine the distribution of temperature in the mould. The experimental data was utilized to improve the accuracy of the filling and solidification simulation, calculations were done using ProCast®. The results obtained after optimizing the simulation showed that variations in the mould temperature increased the susceptibility of the casting to present porosity.

Experimental

The castings studied in the present work were produced in a semi-permanent mould with cooling elements located in the metallic part of the die to produce a directional solidification. As one of the main control parameters in the process is the mould temperature. Pouring is not allowed if the temperature is outside specified control limits.

Experimental temperature measurements were made at specific points of the cylinder head in order to compare results with cooling curves obtained from simulation. The selected points were two massive sections limited by thin walls, named 1B and 2B. Thin walls tend to solidify faster, contributing to the formation of hot spots in the massive zones, Figure 1. During normal production, only section 1B presented porosity problems, even when both sections were geometrical similar. Measurements in section 2B were made for comparison purposes. Additionally, an infrared camera was employed to determine the temperature distribution of the base of the mould before the beginning of each measurement.

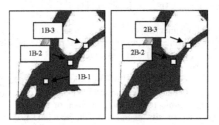

Figure 1. Selected sections of the casting 1B and 2B, blank squares represent the location of thermocouples.

Simulation

The simulations were made using ProCast®. This software solves the conservation equations using the finite element method. The mesh employed in the solution of the system is made of fine tetrahedral volume elements. A mesh of 359,364 nodes and 1,960,920 elements was employed to define the mould, cores and casting. Figure 2 displays the mesh at a section of the casting simulated.

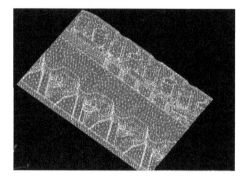

Figure 2. Example of the mesh utilized for the simulations.

The defect predictions were made in "Poros 4" module, which uses the thermal results generated during the simulation. An optimal solid fraction of 0.35 was employed for porosity calculations in order to display the largest number of zones prone to porosity problems.

Comparison of Simulation and Experimental Results

The thermograph of the base of the mould and that obtained with the simulation are presented in Figure 3. The thermograph showed that the hottest zone corresponded to the middle section of the mould base, known as combustion chambers. There are four chambers in the base of the mold. The one located farther from the pouring cup was the coldest one. This was opposite to the results obtained from simulation were the chamber located farther from the pouring cup was the hottest.

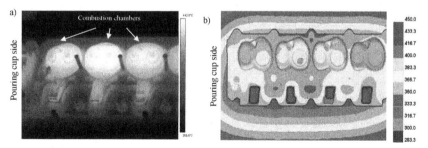

Figure 3. Temperature distribution at the base of the mould obtained by: a) Infrared thermography and b) simulation.

119

The difference in the temperature distributions obtained by thermography and simulation can be due to aspects related with the heat transfer coefficient (h). An important consideration is that a single value of h was considered for the entire contact surface between mould and metal in the case of the simulation. However, there exist differences in the values of h depending on the formation rate of the air gap between the mould and the casting at a given area.

The air gap formation is influenced by the thermo-mechanical movement of the casting during cooling due to metal contraction [2] and its growth depends on the mould temperature. A lower temperature can lead to a faster formation of the air gap, therefore, reducing h [3]. As a result, the amount of heat received by this part of the mould would be smaller. The mould temperature distributions showed by the thermograph indicated that the chamber located farther from the pouring cup was the cooler one, suggesting that the h value was minimum in this zone.

An indirect verification of this behavior was done by comparing the experimental cooling curves for point 1B-1, which was near to the pouring cup, and point 2B-1, located in the opposite side. Figure 4 displays the experimental cooling curves. Although point 2B-1 was located in the coldest zone of the mould, the cooling rate in this zone was lower compared to that corresponding to point 1B-1. This indicated that the heat transfer was lower at point 2B-1, suggesting that there exist different values of h between the mould base-metal interface.

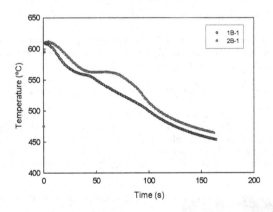

Figure 4. Experimental metal cooling curves for points 1B-1 and 2B-1.

Calibration of the Simulations

The lack of knowledge of h values at the different zones of the mould-metal interface, led us to look for an improvement of the set up of the simulation, to adjust their results to those obtained from the experimental cooling curves. This was done by changing the value of h between cooling elements and the base of the mould. The comparison of experimental and simulated cooling curves for points 1B-1 and 2B-1, after optimizing h, is shown in Figure 5.

120

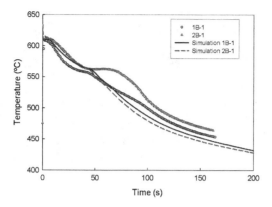

Figure 5. Experimental and simulated cooling curves for points 1B-1 and 2B-1 after optimizing heat transfer coefficient h.

A comparison of the experimental and simulated cooling curves was made considering the average cooling rate between the liquidus and eutectic temperature interval of the casting alloy. The average cooling rate was calculated using seven points between 600 and 570°C, for cooling curves obtained at points 1B-1, 1B-2 and 2B-1. For curves 1B-3 and 2B-3, five points in the interval of 590 to 570°C were used, because at these points no higher temperatures were registered. The simulated and experimental average cooling rate is presented in Table 1.

Table 1. Experimental and simulated average cooling rate obtained after optimizing h.

Simulation	Cooling rate (°C/s)				
	1B-1	1B-2	1B-3	2B-1	2B-3
Experimental	2.08	1.55	0.92	1.70	0.82
Simulation	1.46	1.24	0.72	1.67	0.83
% Difference	29.51	19.59	21.97	1.79	1.72

The simulated average cooling rate showed a good agreement with respect to the experimental one. The greater differences are present in section 1B, been nearly 30%. These differences could appear to be too big, but they are relatively low compared to those obtained in other investigations [4].

Effect of Mould Temperature Variations on Porosity Prediction

The incidence of porosity defects in the casting sections studied can be associated to discontinuities in the casting process. Simulation was used to analyze the effect of delaying cycle time between castings, with and without casting inside of the mould. These conditions influenced the mould temperature distribution. The porosity prediction results using three different process conditions are presented in Figure 6.

121

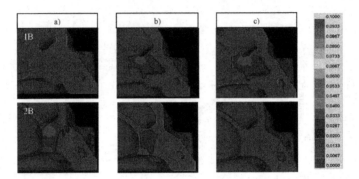

Figure 6. Porosity prediction in sections 1B and 2B for a) normal production, b) delayed cycle with casting inside the mould and c) delayed cycle with empty mould.

The cycle time utilized for normal production was 305 seconds, while the delayed cycle with casting inside of the mould was 905 seconds, and without casting in the mould was 600 seconds. The times for the delayed cycles were representative of variations observed in the process.

The simulation results of the delayed cycles showed a change in the mould temperature respect to the normal production cycle. The mould temperature was increased when the delay was made with casting inside of the mould and decreased when the casting was out of the mould. The changes in the mould temperature modified the solidification process, changing the susceptibility of the selected sections to present porosity. Simulation showed a porosity problem in section 1B for both delayed cycle conditions, while section 2B only display porosity problems with casting outside the mould.

During the delayed cycle with the casting inside the mould, the mould base surface became hotter, reducing heat transfer and in consequence, affecting the directionality of the solidification. Therefore, hot spots were created during the solidification of the casting, favoring the formation of porosity. In the case of the delayed cycle with the casting outside the mould, the mould became colder; then when a new casting was poured, the cooling rate was increased. The increase of the cooling rate was not homogeneous along the entire casting. The hot spot observed under this condition suggested that the thin walls of the casting that were in contact with the mould surface solidificated faster than the massive zones, causing a non-directional solidification profile.

Figure 7 shows the influence of changing the mould temperature on the directionality of solidification. Under normal conditions, solidification proceeds as indicated by the path described by "a-b-c" in Figure 7, while for the delayed cycles, solidification starts first at "b", creating a hot spot at "a".

These results showed that simulation could be very useful in identifying the cause of porosity problems; even with the limitations related to the heat transfer accuracy mentioned before. In the case of section 1B, a possible solution to eliminate porosity could be obtained by increasing the thickness of the thin wall located over the hot spot. In thin walls, 4 mm for instance, an increase of 0.5 mm could be enough to augment the solidification time in about 26%, according to Chvorinov's rule.

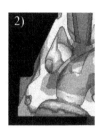

Figure 7. Solid fraction images from simulation at point 1B 1) normal condition, 2) delayed cycle with casting inside and 3) delayed cycle with casting outside of the mould.

Conclusions

The correlation between the experimental and simulated results obtained in the present work for an aluminum cast cylinder head shows that:

1. The simulated and experimental cooling curves are similar.
2. In order to obtain accurate porosity predictions is necessary to adjust the boundary conditions, which require knowing the variations of the heat transfer coefficient along the mould-metal interface.
3. A delay in the cycle time, either with or without a casting inside the mould, increases the susceptibility to porosity problems in the sections studied.

Acknowledgments

The authors acknowledge financial support from Consejo Nacional de Ciencia y Tecnología (CONACYT) and Castech S.A de C.V.

References

1. R. Fuoco, E.R. Correa, M. de Andrade Bastos, "Microporosity Morphology in A356 Aluminum Alloy in Unmodified and in Sr-Modified Conditions", *AFS Transactions*, 01 (168) (2001), 659-678.

2. K. Ho, R.D. Pehlke, "Mechanism of Heat Transfer at a Metal-Mould Interface", *AFS Transactions*, 61 (1984), 587-598.

3. F. Michel, P.R. Louchez, F.H. Samuel, "Heat Transfer Coefficient During Solidification of Aluminum-Silicon Alloys", *AFS Transactions*, 103 (1995), 275-283.

4. P.D. Lee, R.C. Atwood, R.J. Dashwood, H. Nagaumi, "Modeling of Porosity Formation in Direct Chill Cast Aluminum-Magnesium Alloys", *Materials Science & Engineering A*, A328 (2002), 213-222.

5. ProCast® User Manual, Version 2005, (2005).

6. Robert D. Pehlke, "Computer Simulation of Solidification Processes, the Evolution of a Technology", *Metallurgical and Materials Transactions A*, 33A (2002), 2251-2273.

7. George E. Totten and D. Scott Mackenzie, *Handbook of Aluminum, Vol.1. Physical Metallurgy and Processes* (New York, NY: Marcel Dekker, Inc. 2003),573-589.

8. D. M. Stefanescu, "Computer Simulation of Shrinkage-Related Defects in Castings, a Review", FTI, July/August (2005), 189-194.

9. K. Ho, R.D. Pehlke, "Mechanism of Heat Transfer at a Metal-Mould Interface", *AFS Transactions*, 61 (1984), 587-598.

10. W.D. Griffiths, "The Heat Transfer Coefficient During the Unidirectional Solidification of an Al-Si Alloy Casting", *Metallurgical and Materials Transactions B*, 30B (1999), 473-482.

SHAPE CASTING:
2nd International Symposium

Structure/Property

Session Chair:
Glenn Byczynski

Shape Casting: The 2nd International Symposium *Edited by Paul N. Crepeau, Murat Tiryakioğlu and John Campbell*
TMS (The Minerals, Metals & Materials Society), 2007

STRESSES IN THE EUTECTIC SILICON PARTICLES OF STRONTIUM-MODIFIED A356 CASTINGS LOADED IN TENSION

T.R. Finlayson[1], J.R. Griffiths[2], D.M. Viano[2], M.E. Fitzpatrick[3], E.C. Oliver[4] and Q.G. Wang[5]

[1] School of Physics, Monash University, P O Box27, Clayton, VIC 3800, Australia
[2] CSIRO Materials & Manufacturing Technology, P O Box 883, Kenmore, QLD 4069, Australia
[3] Department of Materials Engineering, The Open University, Milton Keynes, MK7 6AA, UK
[4] CCLRC Rutherford Appleton Laboratory, Chilton, Didcot, OX11 0QX, UK
[5] General Motors, Powertrain Engineering, 823 Joslyn Avenue, Pontiac, MI 48340-2920, USA

Keywords: Composites, Aluminum Castings, Particle Stresses, Neutron Diffraction

Abstract

Neutron diffraction methods have been successfully used to measure the stress in the eutectic silicon particles in a casting as a function of the applied tensile strain at all strains up to fracture. These measurements have been made for four microstructural conditions – fine and coarse secondary dendrite arm spacings and low and high yield stresses. We have identified and characterized the three classic components of particle stress (i) the thermal misfit stress, resulting from mismatch in the coefficients of thermal expansion, (ii) the elastic misfit stress, resulting from differences in the elastic constants, and (iii) the plasticity misfit stress, resulting from the plastically inhomogeneous nature of the silicon and aluminum constituents of the casting. We have estimated a lower limit to the fracture strength of the silicon particles and suggest possible explanations for its value.

Introduction

During tensile tests of A356 castings, the eutectic silicon particles begin to cleave at a macroscopic applied plastic strain of about 1%, this value depending on particle size and aspect ratio. The number of broken particles then increases with the applied strain until complete fracture of the alloy occurs [1,2,3]. Controversy exists (a) about the fracture stress of the particles [4,5] and (b) about how the particles contribute to the yield stress and work-hardening rate of the alloy [6,7]. These controversies centre on the partitioning of stress between the aluminum matrix and the silicon particles during a tensile (or compression) test. Surprisingly, there have only been three direct measurements of the stresses in the silicon particles. In two of these cases observations could only be made on particles lying in the surfaces of the specimens because of the techniques used (X-ray diffraction [8] and Raman spectral analysis [9]). In the third case high energy synchrotron radiation was used to sample the entire volume of the specimen but measurements were limited to small applied strains [10]. Discussion about how the particles affect the plastic flow and fracture properties has, therefore, been speculative and this present work was undertaken to remedy the situation.

The ductility of Al-7Si-0.4Mg castings is, in the absence of casting defects, determined by (a) the heat-treatment condition, (b) the secondary dendrite arm spacing, SDAS, and (c) the size and shape of the silicon particles (as influenced by chemical and thermal modification) [3,7,11]. In this paper we only report the effects of the first two of these.

Materials and Methods

<u>Materials</u>

Five sand cast plates (140 x 160 x 25 mm^3) were made from one melt of A356 ingot with added strontium modifier. The molds had a chill at one end to aid in quasi-directional solidification and to produce a range of secondary dendrite arm spacings from 20 μm at the chill end to 70 μm at the other end. The plates were hot isostatically processed at 525°C and 103 MPa for 2 hours to eliminate essentially all the casting porosity. Slices 18 mm wide were cut from either end of the plates (i.e., with the extremes of SDAS) and these were solution heat-treated at 540°C for 6 hours followed by a cold water quench. Some bars were left in this as-quenched state, denoted as T4, and some were aged for 6 hours at 170°C, denoted as T6. Cylindrical tensile test specimens were then machined from the bars with a parallel gauge length of 20 mm and a diameter of 8 mm. The time between heat-treatment and the tensile tests was about six weeks.

The chemical composition of the plates was (wt%) 6.6Si, 0.4Mg, 0.05Fe, 0.18Ti, 0.019Sr with Cu, Mn, Zn all <0.01. The tensile properties are shown in Table 1 (means of three tests with the maximum observed differences from the means recorded as ±).

Table 1. Tensile properties of casting

heat-treatment/SDAS	0.2% proof stress, MPa	tensile strength, MPa	elongation, %
T4 / 70 μm	148±1	257±8	12±3
T6 / 20 μm	279±3	330±4	8±2
T6 / 70 μm	268±5	316±5	4±1

We have assumed various physical constants and these are shown in Table 2. We assume that aluminum and silicon are elastically isotropic with elastic constants given by Lubarda [12]. Young's modulus for the A356 casting was measured as 73 ±2 GPa, a value that is close to the 75 GPa reported in [8] and that is consistent with the theory of two-phase materials.

Table 2. Physical constants for Al and Si (lattice parameters, a, are described below "Calibration")

	Young's modulus GPa	Poisson's ratio	coefficient. of thermal expansion °C^{-1} x 10^6	surface energy J m^{-2}	lattice parameter, a nm
Al	70	0.34	23		0.404921
Si	162	0.22	2	1.2	0.543088

<u>Neutron diffraction</u>

In situ neutron diffraction lattice strain measurements were made during tensile testing using the ENGIN-X instrument at the ISIS pulsed spallation neutron source [13]. The instrument has two fixed-angle detector banks centered on scattering angles of ±90°. The detectors measure time-resolved spectra corresponding to scattering vectors aligned at ±45° to the incident beam. The load axis was aligned at 45° to the incident beam, parallel to the scattering vector of one bank, allowing simultaneous measurements of lattice strains in directions both parallel and perpendicular to the applied load. The volume of material irradiated by the neutron beam was 4 x 4 x 8 mm^3. Measurements were made at a series of increasing strains using count times of about 20 minutes. This meant that some load relaxation occurred during the count time. The diffraction spectra were analyzed by Rietveld refinement of the complete spectrum over a range of d-spacings from 0.076 nm to 0.24 nm using the software of Larson and Von Dreele [14].

Results and Discussion

Calibration

A diffraction spectrum was taken from a NIST standard silicon powder. The NIST standard value for silicon is 0.543119 nm and the value measured from the diffraction spectrum was 0.543088 nm. We did not have powder samples of silicon derived from our castings and so we take this experimental diffraction value as the "zero strain" lattice spacing[1]. A spectrum was also taken from a commercially pure aluminum powder sample, giving a value of 0.404935 nm. Spectra were also taken for powder samples made from the castings by manual filing followed by heat-treating, giving a value of 0.404921 nm for both T4 and T6 treatments. The similarity in lattice spacing between the pure aluminum and the castings suggests that the combined effect of the alloy solute elements in the aluminum matrix is small. This is expected [12] since the atomic radius of Mg is greater, and of Si is less, than that of Al. We take 0.404921 nm as the zero strain value for Al. These assumed lattice spacings are given in Table 2 above.

Thermal misfit stresses in the silicon particles and in the aluminum matrix at zero applied load

A thermal misfit stress exists in the silicon and the aluminum as a result of thermal expansion mismatch (Table 2) and the temperature history during heat-treatment of the alloy. To measure these stresses a diffraction spectrum is required for an unloaded sample. Operational constraints meant that spectra were only collected once specimens were in the tensile machine and the first spectrum was collected at a small positive stress (~10 MPa). To estimate the lattice strains at zero stress the plots of lattice strain *vs.* applied stress were extrapolated to zero applied stress. An example is shown in Figure 1.

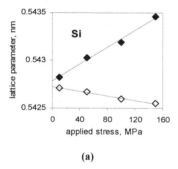 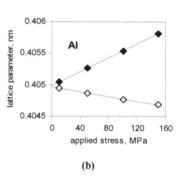

| (a) | (b) |

Figure 1. Lattice spacings as a function of applied stress; (a) silicon particles; (b) aluminum matrix. T6 heat-treatment and fine SDAS. (Solid symbols are axial strains and hollow symbols are transverse.)

The axial and transverse lattice spacings for the silicon particles extrapolated to zero applied stress are 0.542784 nm and 0.542722 nm respectively. The corresponding values for the aluminum matrix are 0.404992 nm and 0.404966 nm. It can be seen that the particles are, to a first approximation, under a hydrostatic compressive strain of -617×10^{-6} and the matrix is under a hydrostatic tensile strain of 143×10^{-6}. (Assuming a hydrostatic stress implies assuming that the particles are spheres, which is nearly true for our castings in which the particles are randomly

---

[1] We are presently collecting strain-free Si particles by dissolving the Al matrix but this is not complete at the time of writing.

oriented and have a mean aspect ratio of 1.6 [11].) Using the elastic constants in Table 2 the stress in the silicon is -180 MPa and that in the Al is +29 MPa. The volume fraction of particles is 0.0757 so the net stress is (-13.6 + 26.6) MPa, implying a violation of stress equilibrium. The effect was noted in all six experiments, suggesting a systematic error, but it is small enough that we neglect it in the present paper. Calculations for the thermal misfit stress for all experiments (silicon particles only) are summarized in Table 3.

Table 3. Thermal misfit stresses in the silicon particles

heat-treatment/SDAS	stress (as measured), MPa	stress imbalance, MPa
T6 / 20 μm	-180	13
T6 / 20 μm	-182	18
T6 / 70 μm	-173	19
T4 / 20 μm	-131	39
T4 / 20 μm	-192	11
T4 / 70 μm	-171	26

These residual stresses are near those reported in [8] but are higher than reported in [9]. The misfit stress in the particles predicted by Eshelby's method [15] is 1.5 MPa/°C so that a residual stress of -180 MPa is consistent with a temperature difference of 120°C. This temperature difference is smaller than that experienced by the material during its processing heat-treatments and it implies stress relaxation during cooling – it may be significant that the higher yield stress T6 samples have a slightly greater residual stress than the softer T4 samples.

Elastic misfit stresses in the silicon particles and in the Al matrix under elastic applied strains

Elastic misfit stresses are produced in the silicon particles and the aluminum matrix under an applied load because of differences in their elastic constants. To measure these misfit stresses, spectra were collected at several applied stresses below the yield stress and the lattice parameters plotted as in Figure 1. The data were then analyzed to give the stresses in the two phases. These calculations are summarized in Table 4 together with the stress imbalance at an applied stress of 1 MPa, $(V_f \sigma_{Si} + (1 - V_f)\sigma_{Al} - 1)$, where the σ terms refer to the elastic misfit stresses and V_f is the volume fraction of silicon particles. Table 4 shows that in the elastic loading regime stress equilibrium is satisfied to better than ~5%. The predicted elastic stress partitioning using Eshelby's method and assuming spherical inclusions is 1.27 in the particles and 0.98 in the matrix. It can be seen that the data are in good agreement with the Eshelby theory. This is despite the fact that the Eshelby method is for isolated and randomly distributed particles whereas the eutectic silicon particles are clustered in the inter-dendritic spaces. There is some suggestion in Table 4 that the elastic stress partitioning depends on the dendrite arm spacing. Such an effect might be due to the different distribution of the eutectic particles in the fine and coarse microstructures but far more experiments would be needed to verify this.

Table 4. Partition of stress to the silicon and aluminum phases under elastic loading

heat-treatment SDAS	experimental measurements (MPa per 1 MPa applied stress)		
	Si	Al	stress imbalance
T6 / 20 μm	1.28	0.93	-0.044
T6 / 20 μm	1.23	0.91	-0.069
T6 / 70 μm	1.07	0.94	-0.048
T4 / 20 μm	1.26	0.95	-0.031
T4 / 20 μm	1.28	0.98	+0.005
T4 / 70 μm	1.21	0.95	-0.030

Plasticity misfit stress in the silicon particles during plastic deformation of the aluminum matrix

When the matrix flows plastically around the particles a plasticity misfit stress is generated. Evidence for its existence is shown in Figure 2 in which the Si and Al strains and stresses are plotted as functions of the applied stress. The straight lines are the elastic response (i.e., from Figure 1) extrapolated to the fracture stress. Recalling that the 0.2% proof stress is 279 MPa for the T6 heat-treatment (Table 1), it can be seen how plastic flow in the matrix causes the stress in the particles to increase above the value that would result from purely elastic stressing. The matrix response to yield is less dramatic because its high volume fraction means that only small decreases in stress are needed to balance increases in the particle stress.

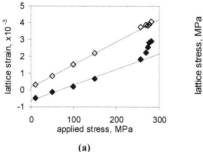
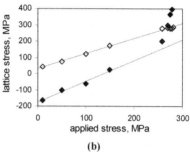

(a)   (b)

Figure 2. Axial strains (a) and stresses (b) in the Si and Al as the specimen is loaded to fracture. Solid symbols denote Si and hollow symbols denote Al. The lines are for elastic loading. T6 / fine SDAS.

It is tempting to use Figure 2(b) to calculate the plasticity misfit stress component in the silicon particles by measuring the difference between the total stress and the sum of the thermal and the elastic misfit stresses. That is, to assert that $\sigma_{total} = \sigma_{thermal} + \sigma_{elastic} + \sigma_{plastic}$ in which $\sigma_{thermal}$ is a constant, $\sigma_{elastic}$ is proportional to the applied stress and $\sigma_{plastic}$ is a function of the applied plastic strain [8]. However, two factors complicate the situation. The first is the interaction of plastic flow with the thermal residual stress. The second is that, as noted at the start of this paper, the silicon particles begin to crack at a plastic strain of ~0.01. We comment on each factor.

Plasticity and the thermal misfit stress

The compressive thermal misfit stress in the particles is balanced by tensile elastic strain gradients in the surrounding matrix. Once the matrix deforms plastically these strain gradients largely disappear and, with them, the residual stress. The decrease in the residual stress increases with plastic strain; in terms of the equation above, $\sigma_{thermal}$ is not constant but is a function of the applied plastic strain. The analysis to quantify this effect is outside the scope of this paper.

Particle cracking and stress relaxation

When a particle cracks the stress in it decreases by some large amount. A nice direct observation of this is reported by Harris et al. [9] who measured a decrease in the stress in a silicon particle in an A319 alloy from 640 MPa to 230 MPa following its fracture. The stress does not drop to zero because in these alloys (and in A356) the particle/matrix interface does not fail and so is able to transmit stress to the two halves of the cracked particle. Thus, once particles start to

131

crack the strains seen by the neutrons are a mix of fully-loaded and part-loaded particles so that the elastic misfit stress, $\sigma_{elastic}$, is not given by the extrapolated line drawn in Figure 2.

The plasticity misfit stress

Despite the foregoing reservations, the silicon particle data from Figure 2 are re-plotted in Figure 3 in terms of the stress difference compared to the extrapolated elastic values. Results for a replicate test on a T6/fine SDAS specimen are included to show the excellent reproducibility of the ENGIN-X data. Also included are T6/coarse SDAS data to show the effect of SDAS. Observations could not be made at the instant of final fracture and so extrapolated estimates for final fracture are included in Figure 3 (note the lower ductility of the coarse SDAS material). It can be seen that the fracture strains are lower than those for normal tensile tests (Table 1) and this may be a result of the periodic interruptions to take diffraction data. The plasticity misfit stress is expected to be proportional to the plastic strain, ε_{pl}, at low strains and to $\varepsilon_{pl}^{1/2}$ at high strains [7]. Although the data do not allow the limits of the linear behavior to be estimated with precision it appears that the prediction is qualitatively obeyed.

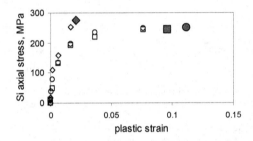

Figure 3. The plasticity misfit stress in the silicon particles as a function of the applied plastic strain. ◊ denotes T6/coarse SDAS; □ and ○ denote T6/fine SDAS. Filled symbols are extrapolations to the macroscopic fracture strain.

The development of the plasticity misfit stress at low strains ($\varepsilon < 0.01$) is similar in the coarse and fine SDAS material. Differences at higher strains may be associated with particle fracture and with relaxation of thermal misfit stresses and this will need to be tested in future research.

The fracture strength of the silicon particles

Silicon particles begin to crack at an applied plastic strain of ~0.01, at least in T6 material. The first particles to fracture are the larger ones and those with higher aspect ratios, and the volume fraction of broken silicon particles when the specimen breaks (i.e., the volume of broken particles divided by the total volume of particles) is ~0.45, regardless of SDAS (again, at least for T6 material) [11]. It follows that there is a range of fracture strengths[2] but in this present experiment we are unable to measure the (fracture strength/size/aspect ratio) details.

Because of the uncertainties associated with calculating particle stresses at plastic strains > 0.01 we estimate the stress in the particles at an applied plastic strain of 0.01 where the volume

---

[2] We assume that the development of stress in the particles is size-independent and note that for Sr-modified castings the aspect ratio of all particles is ~1.6, irrespective of SDAS [11].

132

fraction of cracked particles is expected to be $0.01 - 0.02$ [11]. This stress is, therefore, a lower limit to the fracture strength of the particles. That is, it is a measure of the weakest particles in the casting. The strengths calculated in the six experiments are summarized in Table 5.

Table 5. The stress, $\sigma_{thermal} + \sigma_{elastic} + \sigma_{plastic}$ in the silicon particles at an applied plastic strain of 0.01

heat-treatment/SDAS	stress, MPa
T6 / 20 μm	330
T6 / 20 μm	310
T6 / 70 μm	300
T4 / 20 μm	220
T4 / 20 μm	200
T4 / 70 μm	220

These results suggest that the strength of the particles is not constant – it appears to be smaller in the T4 material. Further analysis is required to verify this rather odd observation. Regardless of this, however, it appears that the lower limit to the strength of the particles is in the range 200 to 300 MPa. The lower limit value is, of course, less than the mean strength of the particles and so cannot be easily compared to the 1200 MPa suggested in earlier work [7].

This strength is far less than the ideal, or defect-free, strength of silicon which is >30 GPa [16,17,18] and to explain it we must postulate the existence of credible defects. The effect of cracks and crack-like defects on the fracture strength is described by fracture mechanics. For elastic fracture, as occurs in silicon, the fracture strength is given by $\sigma_f = \alpha\sqrt{2E\gamma/\pi c}$ where c is the crack depth and α is a constant close to unity. The cleavage fracture surface energy, 2γ, for silicon is 2.5 J/m^2 at -196°C [19,20] and is presumably slightly less at room temperature as the elastic moduli decrease. The measured fracture stress is thus consistent with crack-like defects ~1-2 μm deep. Such defects are possibly conceivable for our case where the diameter of the largest (and weakest) particles is ~20 μm [11]. Nevertheless, it has to be admitted that no direct observations have been made of such defects [5] and it may be that the particles are more perfect than suggested here. However, locally high stresses may exist, affecting small volumes within the particles and which are, therefore, undetected by the neutrons. It is, for example, known that the stress in a particle is highly non-uniform when the particles are in closely spaced clusters such as in the inter-dendritic spaces [21]. A second possibility is the existence of stress concentrations ahead of dislocation pile-ups. Such stress concentrations mean that the critical defect size is reduced from the values given above. These questions are important because if the low strength could be explained it would open the possibility for having stronger particles and, therefore, more ductile castings – a point previously made by Campbell [4].

Conclusions

The following conclusions can be drawn from this work:
(1) Stress transfer to silicon particles at low stresses is consistent with models for elastic composites.
(2) The analysis for the plastic misfit stress is complicated by stress relaxation associated with particle fracture and with relief of the thermal misfit stress.
(3) The fracture strength of silicon particles appears to depend on the yield strength of the casting.
(4) The fracture strength of the weakest silicon particles is $200 - 400$ MPa. This may indicate small defects in the particles or it may reflect local stress concentrations.

Acknowledgements

We are grateful to: (i) the CCLRC for the provision of beamtime, (ii) General Motors R&D for making the cast plates, (iii) TRF, JRG and DV acknowledge support from the Access to Major Research Facilities Programme, a component of the International Science Linkages Programme established under the Australian Government's innovation statement, Backing Australia's Ability.

References

1. A. Gangulee and J. Gurland, "On the fracture of silicon particles in Al-Si alloys", *Trans. Metall. Soc. AIME*, 239 (1967), 269-272.
2. D.S. Saunders, B.A. Parker and J.R. Griffiths, "The fracture toughness of an aluminium casting alloy", *J. Aust. Inst. Metals*, 20 (1975), 33-38.
3. C.H. Cáceres, C.J. Davidson, J.R. Griffiths and Q.G.Wang, "The effect of Mg on the microstructure and mechanical behavior of Al-Si-Mg casting alloys", *Metall. Mater. Trans.*, 30A (1999), 2611-2618.
4. J. Campbell, "Entrainment defects", *Mater. Sci. Technology*, 22 (2006), 127-143.
5. J.R. Griffiths, "Role of bifilms in fracture", *Mater. Sci. Technology*, 22 (2006), 1001-1003.
6. A.A. Benzerga, S.S. Hong, K.S. Kim, A. Needleman and E. Van der Giessen, "Smaller is softer: an inverse size effect in a cast aluminum alloy", *Acta Mater.*, 49 (2001), 3071-3083.
7. C.H. Cáceres and J.R. Griffiths, "Damage by cracking of silicon particles in an Al-7Si-0.4Mg casting alloy", *Acta Mater.*, 44 (1996), 25-33.
8. R.W. Coade, J.R. Griffiths, B.A. Parker and P.J. Stevens, "Inclusion stresses in a two-phase alloy deformed to a plastic strain of 1%", *Phil. Mag.*, 44A (1981), 357-372.
9. S.J. Harris, A. O'Neill, J. Boileau, W. Donlon and X. Su, "Application of the Raman technique to measure stress states in individual Si particles in a cast Al-Si alloy", *Acta Mater.*, (2006), in press.
10. A. Pyzalla, A. Jacques, J. Feiereisen, T. Buslaps, T. D'Almeida and K.-D. Liss, "*In situ* analysis of the microstrains during tensile deformation of an AlSi-MMC at room temperature and elevated temperature", *J. Neutron Research*, 9 (2001), 435-442.
11. Q.G. Wang, C.H. Cáceres and J.R. Griffiths, "Damage by eutectic particle cracking in aluminum casting alloys A356/357", *Metall. Mater. Trans.*, 34A (2003), 2901-2912.
12. V.A. Lubarda, "On the effective lattice parameters of binary alloys", *Mechanics of Materials*, 35 (2003), 53-68.
13. J.R. Santisteban, M.R. Daymond, J.A.J. James and L. Edwards, "ENGIN-X: a third generation neutron strain scanner", *J. Appl. Cryst.*, (2006), in press.
14. A. Larson and R.B. Von Dreele, "GSAS – General Structure Analysis System", *Technical Report LA-UR-86-748*, Los Alamos National Laboratory, USA, (1984).
15. T.W. Clyne and P.J. Withers, *An introduction to metal matrix composites*, Cambridge University Press, (1993). Also www.msm.cam.ac.uk/mmc/publications/soft.html.
16. E. Orowan, "Fracture and strength of solids" *Reps. Progress Physics*, 12 (1948-49), 185-232.
17. A. Kelly, *Strong Solids*, Clarendon Press, 1966.
18. K. Gall, M.F. Horstemeyer, M. Van Schilfgaarde and M.I. Baskes, *J. Mech. Phys. Solids*, 48 (2000), 2183-2212.
19. J.J. Gilman, "Direct measurements of the surface energies of crystals", *J. Appl. Physics*, 31 (1960), 2208-2218.
20. R.J. Jaccodine, "Surface energy of germanium and silicon", *J. Electrochem Soc.*, 110 (1963), 524-527.
21. J. Segurado and J. LLorca, "Computational micromechanics of composites: the effect of particle spatial distribution", *Mechanics of Materials*, 38 (2006), 873-883.

THE EFFECT OF SI CONTENT ON THE SIZE AND MORPHOLOGY OF FE-RICH AND CU-RICH INTERMETALLICS IN AL-SI-CU-MG ALLOYS

C.H. Cáceres[a], B. Johannesson[b], J.A. Taylor[a], A. Canales-Nuñez[c], M. Cardoso[c], J. Talamantes[c]

[a]CAST CRC, School of Engineering, The University of Queensland, Brisbane QLD 4072
Australia
[b] Technological Institute of Iceland (IceTec), Keldnaholt, 112 Reykjavik, Iceland

[c]Corporativo Nemak, Garza Garcia, NL 66221, Mexico

Keywords: Al-Si-Cu-Mg casting alloys; alloy A319; tensile ductility; Iron intermetallics, Cu intermetallics.

Abstract

Recent studies have shown that the β-Al$_5$FeSi and θ-Al$_2$Cu phase intermetallic particles are refined and dispersed in the presence of high silicon, hence improving the ductility of Al-Si-Cu-Mg alloys. In alloys with low Si content, the Fe- and Cu-rich particles form long and closely intertwined clusters which lead to low ductility. At high Si contents, the intermetallic phases appear more dispersed and the clusters of particles are small and isolated from each other, hence the increased ductility. In this work experiments involving ternary Al-Si-0.8Fe alloys and commercial A319 alloy with different contents of Si are presented. It is shown that, contrary to the conclusions of the earlier work, Si by itself is not responsible for the refining of the Fe-rich intermetallics. This seems to suggest that a synergistic interaction with other elements, possibly Sr and Mn, may be necessary for the refining of the intermetallics to occur. Alternatively, it is speculated that the presence of a low melting point eutectic, in this case the Al-Si-Al$_2$Cu eutectic, is the reason for the refining and dispersion of intermetallics in the higher Si alloys.

Introduction

In a recent publications [1,2] it has been shown that high levels of Si increase the ductility of Al-Si-Cu-Mg casting alloys (in particular alloy family 319) that contain high levels of Fe. The flow curves shown in Figure 1, and the quality index chart of Figure 2 illustrate the effect. The compositions of the alloys presented in Figures 1 and 2 are listed in Table 1. Alloy #1, with low Si and Fe content is quite ductile, while alloy #2, with its increased Fe content, exhibits very limited ductility. Alloy #14, on the other hand, with an increased Si content, is almost as ductile as alloy #1 despite its high Fe level. A similarly striking effect is observed in the sequence of alloys #1 → #5 → #17 in which an increased level of Si (#17) improves the ductility of the alloy with high Cu content (#5). In the alloy sequence #1 → #6 → #18, an even more dramatic effect is seen in the presence of high levels of both Cu and Fe, as the increased ductility occurs with a large increase in strength.

Detailed metallography [2] of the alloys of Table 1 showed that increased Si content in Al-Si-Cu-Mg alloys indeed causes a size refining effect on the β-Al$_5$FeSi platelets, ascribed to the reduced tendency in the high Si alloys to form large pre-eutectic β-Al₅FeSi and α-Al₁₅(Mn,Fe)₃Si₂ particles during solidification due to a reduction in the available growth period. The size refining effect of a high Si content is also evident in other intermetallics that form from the eutectic liquid in Al-Si-Cu-Mg alloys, such as α-Al₁₅(Fe,Mn)₃Si₂ and θ-Al₂Cu. That is, the evidence suggests that increased silicon content tends to not only refine the size of intermetallic particles but also to redistribute them into a more uniform dispersion within the interdendritic and intergranular regions compared with lower silicon content which promotes long clusters of intertwined particles along the grain boundaries. The growth and propagation of microcracks nucleated by the cracking of the intermetallics is therefore more difficult and involves more local plasticity when the particles are more dispersed, increasing the tensile ductility of the alloys for high Si contents.

It is known that the Si content may radically change the primary aluminium grain structure, from a globular morphology at Si contents below about 6% to an orthogonal dendritic structure at higher Si levels [3]. It has been suggested [1] that these Si-induced morphological changes in the solidification structure are responsible for the observed refining effect on the Cu- and Fe-rich intermetallic phases during solidification, leading to the observed increase in tensile ductility.

The experiments of Figures 1 and 2 and the metallographic study detailed above were performed in the multicomponent alloys of Table 1, so is was deemed necessary to verify the main hypotheses and conclusions using less complex alloys. In addition, the metallographic evidence of intermetallic refining was collected in specimens heat treated to a T6 condition, and a more detailed study of the refining of intermetallics in the as-cast condition was also considered important in order to fully understand the phenomenon. Thus, experiments were organised to cast and examine, on the one hand, commercial alloy A319 in the as-cast condition, and on the other, ternary Al-Fe-Si alloys, in both cases for different Si contents. The experiments involving alloy A319 were carried out at Nemak, Monterrey, involving the ternary alloys were performed at IceTech, Reykjavik.

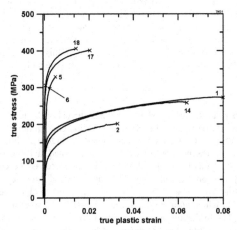

Figure 1. Flow curves of the various alloys studied (see Table 1 for composition), illustrating the effect of the different compositions on the strength-ductility behaviour [1,2].

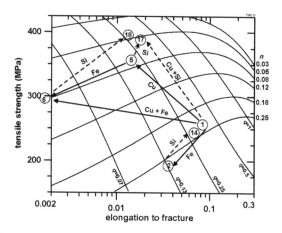

Figure 2. Quality index chart for alloy A319 of various compositions (see Table 1). The dashed arrows indicate the effect of increased Si or Si and Cu on alloys #1, #2, #5 and #6. The solid arrows indicate the effect of increased Fe, Cu, or both, on alloys #1 and #5 [1,2].

Table 1. Nominal chemical compositions of the alloys of Figures 1 and 2, together with the Fe- and Cu-rich intermetallic phases identified in the T6 heat-treated alloys [1,2].

Key	Nominal composition (mass %)	Intermetallics
1	4.5Si-1Cu-0.1Mg-0.2Fe-0.0Mn	β-Al$_5$FeSi
2	4.5Si-1Cu-0.1Mg-0.5Fe-0.25Mn	β-Al$_5$FeSi
14	9Si-1Cu-0.1Mg-0.5Fe-0.25Mn	α-Al$_{15}$(Fe,Mn)$_3$Si$_2$
5	4.5Si-4Cu-0.1Mg-0.2Fe-0.0Mn	β-Al$_5$FeSi
17	9Si-4Cu-0.1Mg-0.2Fe-0.0Mn	θ-Al$_2$Cu
6	4.5Si-4Cu-0.1Mg-0.5Fe-0.25Mn	β-Al$_5$FeSi
18	9Si-4Cu-0.1Mg-0.5Fe-0.25Mn	α-Al$_{15}$(Fe,Mn)$_3$Si$_2$ θ-Al$_2$Cu

Materials and Experimental procedure

Alloy 319: Industrial secondary alloys were used for the study, Sr modified, with two Si levels. The compositions are given in Table 2. These alloys were cast in a silica sand wedge-type mould with a metallic chill at the bottom. This provides the casting with a quasi-directional solidification pattern, resulting in DAS values between 16 and 90 µm. To maintain consistency with the ternary alloys, specimens with SDAS ≈ 45 µm were used for this part of the study.

Ternary alloys: The compositions (mass %) of the ternary alloys of the present study are listed in Table 3. The alloys were prepared in an electric furnace by dissolving the required amounts of

137

Si and Fe, the latter in the form of ALTAB Fe tablets[1], in molten aluminium kept at 740°C. The melt was gently stirred after adding the Si and Fe, allowed to settle for 10 minutes, and stirred again prior to pouring. Castings were made in a steel mould preheated to about 350°C. The mould cavity was about 150 mm deep, and 60 mm in diameter, with a slight taper. The pouring temperature (listed in Table 3) was predetermined so as to have a constant superheat of approximately 50°C for each composition. Specimens for metallographic observation were sectioned 25 mm from the bottom of the cast cylinders.

Table 2. The chemical composition (mass %) of alloy 319.

Si %	Cu %	Mg %	Fe %	Mn %	Sr ppm
10.74	3.46	0.28	0.78	0.44	120
5.07	3.26	0.29	0.80	0.47	114

Table 3. The chemical composition (mass %) of the Al-Si-Fe alloys, and the respective pouring temperatures.

Si %	Fe %	Pouring Temperature (°C)
4.54	0.86	685
8.12	0.87	665
11.31	0.86	630

The cast plates of alloy 319 and the cast cylinders of the ternary alloys were sectioned and metallographically polished. The secondary dendrite arm spacing (SDAS) was measured by a linear intercept method on an optical microscope. The specimens were subsequently photographed in an SEM using a back scattered electrons (BSE) detector to image the Fe-rich intermetallics with high contrast. In the case of alloy 319, in order to preserve the resolution while covering a large surface area, a montage of mosaic images was composed for each area observed. The β-Al_5FeSi interparticle spacing of the ternary alloys was measured by a linear intercept method on printed SEM images.

Results

Alloy 319: Figures 3 and 4 are images of alloy 319, showing the effect of the high Si level on the intermetallics, which in the higher Si alloy appear finely divided and more uniformly dispersed. Copper particles were also comminuted by the higher Si content, showing as small bright dots in the dendritic channels all over the image. Thus, as reported in the previous works, increased Si in the industrial alloy does refine the Fe-Mn phase, and disperses the Cu-rich intermetallics.

---

[1] *London and Scandinavian Metallurgical Co. Ltd., UK.*

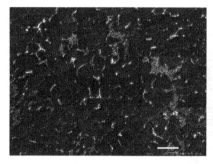

Figure 3. BSE image of alloy 319 with 4.5%Si. The intermetallics, which appear as bright spots and particles, tend to be lumped together forming long agglomerates. Scale mark: 200 μm

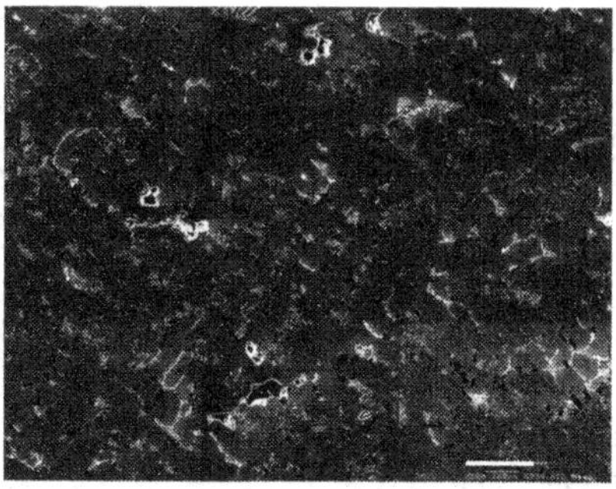

Figure 4. BSE image of alloy 319 with 11%Si. The intermetallics appear finely divided and much more uniformly dispersed in comparison with Figure 3. Scale mark: 200 μm.

Al-Si-Fe ternary alloys. Figures 5 through 7 show the microstructures of the ternary Al-Si-Fe alloys. In this case, contrary to the expectations, an actual increase in the size of the β-Al₅FeSi platelets was observed. Consistently, Table 4 shows that as the Si content increased from 4.5 to 11.3%, the inter-particle spacing more than doubled, at constant SDAS.

Table 4. The SDAS and β-Al₅FeSi inter-particle spacing for the ternary alloys.

Alloy	SDAS (μm)	Inter-particle spacing (μm)
4.5 %Si	49	40.3
8.1 %Si	50	67.5
11.3 %Si	--	96.6

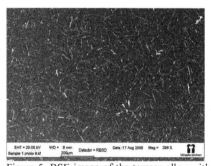

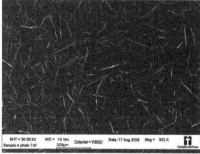

Figure 5. BSE image of the ternary alloy with 4.5% Si. The β-Al₅FeSi plates appear as white needles.

Figure 6. BSE image of the ternary alloy with 8.2%Si. The β-Al₅FeSi platelets are much longer than in Figure 5, and are further apart.

139

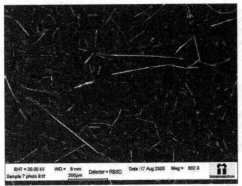

Figure 7. BSE image of the ternary alloy with 11% Si.
The β-Al₅FeSi platelets are much longer than in
Figures 5 and 6.

Proposed Mechanism of Intermetallic Particle Refinement

The present results are somewhat puzzling, as it is clear that, since the concentrations of all other elements in the full 319 alloy except Si remain constant, it is the increased Si that causes the refining of Cu-rich and Fe-rich intermetallics. However, and contrary to the earlier suggestions that an increased amount of Al-Si eutectic liquid should increase the number of liquid pools in which β-phase platelets can nucleate and hence become refined, increasing the Si content in the ternary alloys had the opposite effect on the size and distribution of the Fe-rich intermetallics.

This suggests that there may be a synergistic interaction between Si and some of the other alloy components in 319, such as Sr or Mn, that is necessary to cause the intermetallic refining effect at high concentrations of Si. This suggestion remains unconfirmed at present. Alternatively, it is possible that it is the presence of copper that is the key factor in the refining influence of Si. This is examined below.

In the absence of Cu (as in the ternary alloys), Fe, which is barely soluble in Al, partitions strongly into the liquid phase and then forms classic β-Al₅FeSi platelets before and/or during the formation of Al-Si eutectic. In this case, the β phases increase in size as the Si content increases.

In the presence of Cu (as in the secondary 319 alloy) it may be that Fe is less able to form the β phase in association with the Al-Si eutectic reaction, and instead continues to partition strongly into the remaining Cu-rich liquid phase until the point at which the Al-Si-Al₂Cu eutectic forms at ~524°C [4]. If this is the case, then the Fe-rich intermetallics (whether α, β or π) would be precipitated out under conditions controlled by the dispersion and scale of these final Al-Si Al₂Cu pools. In other words, copper content, and the attendant Al-Si-Al₂Cu eutectic, encourage intermetallic particle comminution and dispersion.

For given Cu-content in a low Si alloy, due to the high fraction solid of α-Al phase, the final Cu-rich liquid forms as networks around the grain boundaries rather than extensively in the interdendritic spaces. The Al-Si-Al₂Cu eutectic (and the accompanying Fe-rich intermetallics) therefore form around the grains in the characteristic observed stringers of relatively long particles, leading to low ductility as reported earlier [1,2].

In the higher Si alloys, the α-Al fraction solid is lower and the Cu-rich liquid pools are dispersed within the intricate network of interdendritic channels and fine intergranular spaces common to in the highly dendritic grains. The attendant Fe phases are therefore likewise dispersed and fine.

The role of Si is therefore complex. It serves to change grain and dendritic morphology but this of itself does not cause intermetallic particle refinement. It appears to require the additional presence of copper in order to delay the precipitation of Fe, and hence produce the refining and dispersion of the intermetallic phases with the accompanying improvement in ductility.

Conclusions

- Metallographic studies show that increased Si content in Al-Si-Cu-Mg alloys causes a size refining effect on both θ-Al$_2$Cu and the β-Al$_5$FeSi platelets.
- The size refining effect of a high Si content is also evident in other intermetallics that form from the eutectic liquid in these alloys, such as α-Al$_{15}$(Fe,Mn)$_3$Si$_2$.
- Increased silicon content tends to not only refine the size of intermetallic particles but also to redistribute them into a more uniform dispersion within the interdendritic and intergranular regions compared with lower silicon content.
- In Al-Si-Fe ternary alloys, the size of the β-platelets increases with the Si content.
- It is speculated that the presence of a low melting point eutectic, in this case the Al-Si-Al$_2$Cu eutectic, is the reason for the refining and dispersion of intermetallics. At high contents of Si the larger amount of eutectic liquid causes the final liquid pools of Cu-rich liquid to be finely divided and dispersed, Fe remains dissolved in these liquid pools, and so the Fe-rich particle refining is co-dependent on that of the Cu intermetallics. Al low Si content, the larger fraction of primary Al phase causes the final Cu and Fe enriched liquid to be distributed around the grains in larger, and fewer, liquid pools, and hence the intermetallic particles precipitate in these regions as long interconnected particle stringers.

References

1   Cáceres, C. H.; Svensson, I. L.; Taylor, J. A. *Int. J. Cast Metals Res.*, 2003; Vol. 15; pp 531-543.
2   Cáceres, C. H.; Taylor, J. A. In: *Shape Casting: The John Campbell Symposium*; Tiryakioglu, M., Crepeau, P. N., Eds.; TMS, Warrendale: San Francisco, 2005; pp 245-254.
3   Hutt, J. E. C.; Easton, M.; Hogan, L. M.; StJohn, D. In *4th Int. Conf. on Solidification Processing (SP97)*: Sheffield, UK, 1997; pp 268-272.
4   Edwards, G. A.; Sigworth, G. K.; Cáceres, C. H.; StJohn, D.; Barresi, J. *AFS Trans.*, 1997; Vol. 105; pp 809-818.

Shape Casting: The 2nd International Symposium *Edited by Paul N. Crepeau, Murat Tiryakioğlu and John Campbell*
TMS (The Minerals, Metals & Materials Society), 2007

ON THE USE OF GENERAL EXTREME VALUE (GEV) DISTRIBUTION IN FRACTURE OF AL-SI ALLOYS

Murat Tiryakioğlu

Robert Morris University, Dept. of Engineering, Moon Township, PA 15108, USA

Keywords: Extreme values, Gumbel distribution, porosity, defects

Abstract

The use of the General Extreme Value distribution in the fracture of Al-Si alloys was demonstrated. Analysis of data from the literature showed that the size and aspect ratio of damaged Si particles in Al-Si alloys follow the Gumbel distribution. Thirteen of the fourteen datasets from the literature for size of defects on fracture surfaces of cast Al-Si alloys were also found to follow the Gumbel distribution. The use of the correct statistical distribution should improve the accuracy of fracture-mechanics based models to predict fatigue life of cast Al alloys.

Background

When major structural defects such as porosity and oxide inclusions are present in the structure, they control fracture in Al-Si alloys. In such cases, fracture takes place prematurely and there are almost no Si particles on the fracture surface [1]. In the absence of major structural defects, the alloy continues to deform into the late stages of work hardening and fracture is initiated by the damage to (fracture of) Si particles. Consequently, Si particles are abundant on the fracture surface. Regardless of whether the fracture is initiated by structural defects such as porosity and oxide inclusions, or microstructural features such as eutectic or intermetallic phases, properties can be expected to be controlled by the defect or feature which leads to the largest stress concentration. This paper illustrates the use of Extreme Value statistics to model the damage to Si particles with strain and largest defect areas on fracture surfaces in Al-Si alloys.

For largest values from any distribution, Gnedenko [2] defined three types of limiting extreme value distributions: the Gumbel distribution (type 1), the Fréchet distribution (type 2) and the Weibull distribution (type 3). The Gumbel distribution function is written as;

$$P(X \leq x) = \exp\left(-\exp\left(\frac{x-\mu}{\sigma}\right)\right) \qquad (1)$$

The Fréchet distribution (type 2) function is,

$$P(X \leq x) = \begin{cases} 0 & , x < \mu \\ \exp\left(-\left(\frac{x-\mu}{\sigma}\right)^{-m}\right) & , x \geq \mu \end{cases} \qquad (2)$$

and the Weibull distribution (type 3) function is

$$P(X \leq x) = \begin{cases} \exp\left(-\left(\frac{\mu-x}{\sigma}\right)^{m}\right) & , x \leq \mu \\ 0 & , x > \mu \end{cases} \qquad (3)$$

where μ, σ and m are parameters. Note that Equations 1-3 are for largest values (upper tail of the distribution) and that there exist similar equations for smallest values (lower tail). Gnedenko also showed that the distribution of largest (or smallest) values is determined by the distribution from which the sample comes. For distributions decreasing exponentially at upper tails, such as in exponential, normal and lognormal distrbutions, the distribution of largest values will be Gumbel. If the distribution decreases following the power law following x^m or x^{-m}, then the distribution of the largest values is Fréchet and Weibull, respectively.

Jenkinson [3] introduced a General Extreme Value (GEV) distribution which groups the three types distinguished by Gnedenko:

$$P(X \le x) = \exp\left(-\left(1 + \xi\left(\frac{x-\mu}{\sigma}\right)\right)^{\frac{-1}{\xi}}\right) \qquad (4)$$

where ξ is the tail index. The tail index is related to the shape parameter, m, by $\xi = -1/m$, and determines the type of distribution: $\xi > 0$ indicates a Fréchet distribution, $\xi < 0$ corresponds to a Weibull distribution, and $\xi \to 0$ indicates a Gumbel distribution. The advantage of using the GEV distribution is that, the data themselves determine the most appropriate type of tail behavior through inference on ξ, and there is no need to make subjective assumptions on which extreme value distribution should be adopted [4]. For small values of ξ, the type 2 and 3 distributions are very close to the type 1 distribution [5], and confidence intervals need to be calculated to determine whether the point 0 is included in the interval.

Application of GEV Distribution to Fracture in Al Alloys

The GEV distribution was applied to (i) damage to Si particles and (ii) area of defects leading to fracture using data from the literature. Data were analyzed using *extRemes* toolkit [6] which employs the maximum likelihood method to estimate ξ, μ and σ, and profile likelihood method to estimate the confidence interval on ξ.

Damage to Si Particles by Deformation in Al-Si Alloys

Probability of damage to Si particles in Al-Si alloys has been modeled by the Weibull distribution for smallest values [7,8,9,10,11]. The critical stress for particle cracking is related to aspect ratio, α [9], and to $d^{-\frac{1}{2}}$ (for spherical particles) [12] where d is particle diameter. Hence damage can be expected to be more probable with increasing size and/or aspect ratio, which was confirmed by experimental observations [7].

The size distribution of Si particles can be represented by the lognormal distribution for both metal matrix composites (MMCs) [10] and cast alloys [13,14,15]. The statistical distribution of particle aspect ratio in extruded MMCs was addressed by Llorca [16] who introduced a distribution function that decreases exponentially. For cast alloys, the statistical distribution of aspect ratio was not addressed to the author's knowledge. However, major and minor axes of a particle growing in a metallic matrix are expected to follow a lognormal distribution [17]. Consequently the distribution of aspect ratio of Si particles can be expected to also follow a lognormal distribution with a threshold (since minimum value of α is 1). It was shown that [18,19] the largest values from a lognormal distribution follow a Gumbel distribution.

Poole and Charras [20] studied the effect of deformation on the damage to Si particles in an MMC with 20%Si and 0.5%Mg in W and T6 tempers. They also distinguished the damage as

144

single and multiple cracks in Si particles. The results of the analysis of size (in terms of equivalent diameter, d_{eq}) and aspect ratio data in the T6 condition summarized in Table 1. For both d_{eq} and α, the point of 0 is within the confidence interval, indicating a Gumbel distribution.

Table 1. ξ values of GEV distribution fits to T6 size (deq) and aspect ratio (α) data of Ref. 20.

	ξ	95% confidence interval		Inferred
		Lower	Upper	Distribution
d_{eq}	0.1152	-0.0592	0.3203	Gumbel
α	0.0316	-0.1361	0.2412	Gumbel

The d_{eq} and α data of Poole and Charras are plotted in Gumbel probability plots in Figure 1. Note that damaged particles follow the same distribution regardless of whether the particles are cracked once or multiple times.

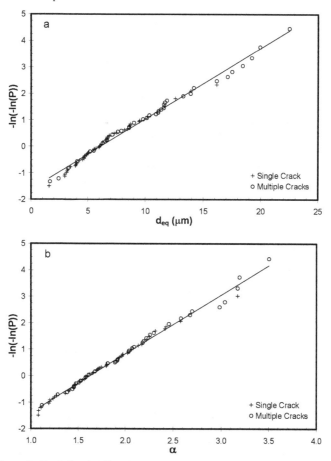

Figure 1. Gumbel probability plots for (a) d_{eq} and (b) α using the data of Ref. 20.

145

Wang *et al.* [11] investigated the damage to Si particles in A356 and A357 alloys at different strains. The histograms for sizes of all Si particles as well as the damaged particles in two specimens deformed to two different strains are presented in Figure 2.a. The lognormal fits to all Si particles for the two specimens are identical, as presented as a cumulative probability plot in Figure 2.b. The size of damaged particles follows the Gumbel distribution, Figure 2.b.

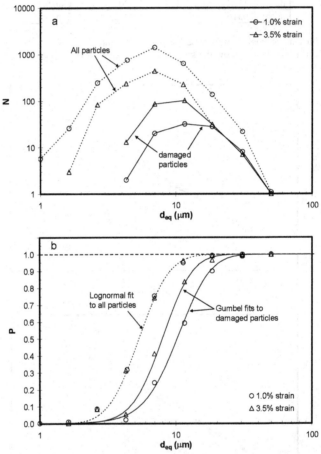

Figure 2. Analysis of size data from Ref. 11. (a) Histograms of all and damaged particles at 1.0% and 3.5% strain, and (b) cumulative probability plots showing lognormal fit to all particles for both cases and Gumbel fits to damaged particles at 1.0% and 3.5% strain.

Wang *et al.* also investigated the aspect ratios of all and damaged Si particles in two specimens deformed to two different levels of strain. The histograms of all and damaged Si particles are shown in Figure 3.a. Figure 3.b. shows the cumulative probability plots for the 3-parameter lognormal fit to aspect ratio data for all Si particles. For damaged particles, the data follow the Gumbel distribution for both strain levels.

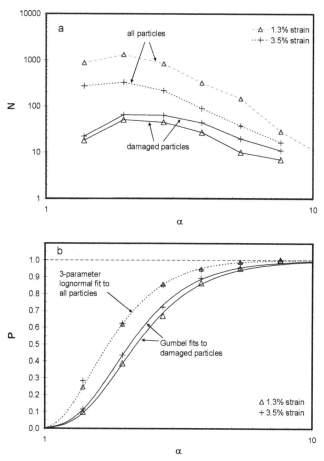

Figure 3. Analysis of aspect ratio data from Ref. 11. (a) Histograms of all and damaged particles at 1.3% and 3.5% strain, and (b) cumulative probability plots showing 3-parameter lognormal fit to all particles for both cases and Gumbel fits to damaged particles at 1.3% and 3.5% strain.

The analysis presented above shows that the damage to Si particles follows a Gumbel distribution for Al-Si alloys, regardless of whether they are MMCs or cast alloys.

Failure Initiating Defects on Fracture Surfaces

Dvorak and Shwegler [21] showed theoretically that defect sizes in castings should follow a lognormal distribution. The validity of this statement is confirmed when the pore size histograms of Casellas *et al.* [22] for Sr-modified and unmodified Al-9.8Si-0.27%Mg-0.18%Mn alloy castings are reanalyzed. Lognormal fits to their pore size data are presented in Figure 4.

147

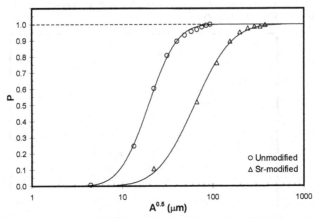

Figure 4. Cumulative probability and lognormal fits to the pore size distributions of Ref. 22.

The sizes of defects initiating fatigue failure in cast Al-Si alloys have been measured by many researchers [23,24,25,26,27], who have used lognormal [23,24,25], Weibull for minima [26,27] and power-law [22] distributions to model the distribution of defect area on the fracture surface.

Fourteen datasets from the literature for four different Al-Si alloys were analyzed in this study. The estimated ξ for each dataset, 95% confidence intervals and inferred distributions are presented in Table 2. For all datasets except for one, the inferred distribution is Gumbel.

Table 2. Analysis of GEV fits to the datasets from the literature.

Alloy	Notes	Ref.	ξ	95% Confidence Intervals		Inferred Distribution
				Lower	Upper	
A356	Low-H	28	0.0179	-0.2410	0.3297	Gumbel
	High-H		0.0198	-0.1548	0.2365	Gumbel
	Filtered	26	-0.2117	-0.4113	0.0573	Gumbel
	Unfiltered		-0.0902	-0.2731	0.1361	Gumbel
	Unfiltered+HIP		0.0947	-0.1599	0.5821	Gumbel
	SDAS=20-25µm	29	-0.3184	-0.6496	0.0693	Gumbel
	SDAS=70-75µm		0.0076	-0.3595	0.4648	Gumbel
A357		30	-0.2501	-0.6932	0.0578	Gumbel
319	Filtered	31	0.0941	-0.2126	0.5537	Gumbel
	Unfiltered		0.4936	0.1046	1.0506	Fréchet
		32	-0.0282	-0.2559	0.3012	Gumbel
Al-9%Si-3%Cu	Large pores*	33	-0.0670	-0.3484	0.2727	Gumbel
	Small pores*		0.0151	-0.1356	0.2105	Gumbel
	Die cast	25	0.0355	-0.1923	0.3253	Gumbel

*: determined from tensile specimens

Discussion

The analysis of both size and aspect ratio of damaged Si particle data from an Al-20%Si-0.5%Mg MMC and A356-A357 cast alloy showed that the Gumbel distribution should be used to model the damage. However to the author's knowledge, the Gumbel distribution has never been used to model the damage to Si particles. Instead, researchers [7,8,9,10,11] used the Weibull distribution, not for largest values, but smallest values. The form that they used is indeed what was originally proposed by Weibull [34]. Although the Weibull distribution for smallest values has been used to model largest values [35], the Weibull distribution serves only as an approximation to the Gumbel distribution [36]. For small values of ξ, all three types of extreme value distributions are almost identical [5]. However, it is possible to estimate directly the parameters of only the Gumbel distribution for largest values from a lognormal distribution [19]. Hence it may be possible to predict the damage to Si particles from their distribution. It should be noted that the statistical analysis of damaged particle data did not consider the joint (bivariate) distribution of size and aspect ratio. More research is needed in this area.

The size of inclusion defects measured on fracture surfaces of steel castings has been shown to follow the Gumbel distribution by Murakami and his coworkers [37,38,39]. In the present study, strong evidence was presented for the use of the Gumbel distribution to model the most critical defects in the structure of cast Al-Si alloys. The use of the correct statistical distribution should improve the accuracy of fracture-mechanics based models to predict fatigue life of cast Al alloys.

Conclusions

- The size distribution of Si particles is lognormal in cast Al-Si alloys.
- For aspect ratio, 3-parameter lognormal distribution should be used.
- Pore size distribution in Al-Si alloys is lognormal.
- Both the size and aspect ratio of damaged Si particles follow the Gumbel distribution.
- The size of defects found on fracture surfaces of cast Al-Si alloys follows the Gumbel distribution.
- The use of the correct statistical distribution should improve the accuracy of fracture-mechanics based models to predict fatigue life of cast Al alloys.

Acknowledgements

The author would like to thank Profs. N.R. Green (University of Birmingham), P.D. Lee (Imperial College), H. Mayer (Universität für Bodenkultur Wien) and W.J. Poole (University of British Columbia) for sharing their data; Drs. Carlos H. Caceres (The University of Queensland), John Campbell (The University of Birmingham), Ralph T. Shuey (Alcoa), James T. Staley, Sr. (retired) for their comments.

References

1. M. Tiryakioğlu, J. Campbell, J.T. Staley: *Scrip. Mater.*, v. 49, pp. 873-878, 2003.
2. B.V. Gnedenko: *Annals of Math.*, vol. 44, pp. 423-453, 1943.
3. A. F. Jenkinson: *Q. J. Royal Meteorology Soc.*, vol. 87, pp. 145-158, 1955.
4. S. Coles: An Introduction to Statistical Modeling of Extreme Values, Springer, 2001.
5. F. M. Longin: J. Business, vol. 69, pp. 383-408, 1996
6. http://www.assessment.ucar.edu/toolkit/
7. J.-S. Chou, C.W. Meyers: AFS Trans., pp. 165-173, 1991.

8. M.T. Kiser, F.W. Zok, D.S. Wilkinson: Acta Mater., vol. 44, pp. 3465-3476, 1996.
9. C.H. Caceres, J.R. Griffiths: Acta Mater., v. 44, pp. 25-33, 1996.
10. E. Maire, D.S. Wilkinson, J.D. Embury, R. Fougeres: Acta Mater., v. 45, pp. 5261-5274, 1997.
11. Q.G. Wang, C.H. Caceres, J.R. Griffiths: Metal. Mater. Trans. A, v. 34A, pp. 2901-2912, 2003.
12. J. Gurland, J. Plateau: Trans. ASM, v. 56, pp. 442-454, 1963.
13. B.A. Parker, D.S. Saunders, J.R. Griffiths: Metals Forum, v. 5, pp. 48-53, 1982.
14. F.N. Rhines, M Aballe: Metal. Trans. A, v. 17A, pp. 2139-2152, 1986.
15. S. Shivkumar, S. Ricci, Jr., B. Steenhoff, D. Apelian, G. Sigworth: AFS Trans., pp. 791-810, 1989.
16. J. Llorca: Acta Metall. et Mater., v. 43, pp. 181-192, 1995.
17. R.T. Shuey, personal communication, Alcoa, 2006.
18. E.J. Gumbel: Statistics of Extremes, Columbia University Press, New York, 1958.
19. N.D. Singpurwalla: Technometrics, v. 14, pp. 703-711, 1972.
20. W.J. Poole, N. Charras: Mater. Sci. Eng. A, v. A406, pp. 300–308, 2005.
21. H.R. Dvorak, E.C. Schwegler: Intl. J. Fracture Mech.. Vol. 8, pp. 110-111. Mar. 1972
22. D. Casellas, R. Pérez, J.M. Prado: Mater. Sci. Eng. A, v. A398, pp. 171-179, 2005.
23. J.Z.Yi, Y.X. Gao, P.D. Lee, H.M. Flower, T.C. Lindley: Metall. Mater.Trans. A. v. 34A, pp. 1879-91, 2003.
24. H. Mayer, M. Papakyriacou, B. Zettl, S.E. Stanzl-Tschegg: Intl. J. Fatigue, v. 25, pp. 245–256, 2003.
25. H. Mayer, M. Papakyriacou, B. Zettl, S. Vacic: Intl. J. Fatigue, v. 27, pp. 1076–1088, 2005.
26. C. Nyahumwa, N.R. Green, J. Campbell: Metall. Mater. Trans. A, v. 32A, pp. 349-358, 2001.
27. Q.G. Wang, P. Jones, M. Osborne: In Advances in Aluminum Casting Technology II, M. Tiryakioğlu, J. Campbell, eds., ASM International, pp.75-84, 2002.
28. J.Z. Yi: Ph.D. Thesis, Imperial College, UK, 2004.
29. Q.G. Wang, D. Apelian, D.A. Lados: Journal of Light Metals, v. 1, pp. 73-84, 2001.
30. C.J. Davidson, J.R. Griffiths and A.S. Machin: Fatigue and Fracture of Engineering Materials and Structures, v. 25, pp. 223-230, 2002.
31. G.E. Byczynski: The Strength and Fatigue Performance of 319 Aluminium Alloy Castings, Ph.D. Thesis, University of Birmingham, 2002.
32. J.C. Ting, F.V. Lawrence, Jr.: Fatigue and Fracture of Engineering Materials and Structures, v. 16, pp. 631-647, 1993.
33. N. Roy, P.R. Louchez, F.H. Samuel: Journal of Materials Science, v. 31, pp. 4725-4740, 1996.
34. W. Weibull: Journal of Applied Mechanics, v.18, pp 293-297, 1951.
35. N. Devictor, M. Marquès, P. Van Gelder: In Lifetime Management of Structures, pp. 83-97, A. Lannoy, ed., Det Norske Veritas, 2003.
36. K. Trustrum, A. De S. Jayatilaka: J. Mater. Sci., v. 18, pp. 2765-2770, 1983.
37. Y. Murakami, T. Toriyama, E.M. Coudert: J. Testing Eval., v. 22, pp. 318-326, 1994.
38. S. Beretta, Y. Murakami: Fatig. Fract. Eng. Mater. Struc., vol. 21, pp. 1049-1065,
39. Y. Murakami, S. Beretta: Extremes, v. 2, pp. 123-147, 1999.

EFFECT OF PROCESSING ON THE STRUCTURE AND PROPERTIES OF SQUEEZE CAST Al-7Si-0.3Mg ALLOY

Kumar Sadayappan[1] and Werner Fragner [2]

1 – CANMET – Materials Technology Laboratory, Ottawa, Canada
2 – ARC – Lightmetals Competence Centre, Ranshofen, Austria

Keywords: Al-Si alloy, squeeze casting, tensile properties

Abstract

Squeeze casting of Al-Si alloy was carried out using a UBE 350 HVSC machine. Effects of process variables such as melt modification, melt and die temperatures and casting pressure on the mechanical properties were evaluated using a step plate casting. The operating temperatures of the melt and tools were found to have significant influence on properties of the casting than the casting pressure. The segregation of the eutectic liquid which can be effectively controlled by process temperatures is one of the reasons cited for the scatter in properties. The results from this investigation are presented and discussed.

Introduction

Squeeze casting is one of the emerging manufacturing processes capable of producing near-net shape components with minimum casting defects such as shrinkage and gas porosity. Squeeze casting process can be classified as direct or indirect depending on the way the liquid metal is introduced into the die. The metal is poured into the die cavity and pressurized by an upper die in the direct squeeze casting process. On the other hand, liquid metal is injected into the die cavity from a horizontal or vertical shot sleeve in indirect squeeze casting process. Both processes are reported to have resulted in enhanced mechanical properties due to fine, defect free microstructure. The differences in the processes along with the advantages and disadvantages are described elsewhere [1, 2].

The pressure applied during the solidification of the metal ensures that casting is virtually shrinkage free. The pressure also increases the heat transfer between the mold and solidifying metal by preventing air-gap formation between the two surfaces. The resulting defect free fine microstructure is the reason for enhanced mechanical properties. There are reports that squeeze cast samples exhibit 30% or higher tensile properties compared to gravity permanent mold castings [3].

However, as in other manufacturing processes, squeeze cast components suffer from various defects such as inclusions, segregation and shrinkage due to inadequate feeding. The defects and their effects on the properties are not very clear even though they are the subject of few publications [4-6].

In this work the effects of melt processing and other casting variables on the properties of an indirect squeeze cast Al-Si-Mg alloy were evaluated.

Experimental Work

250 kg of Al-7Si-0.3Mg pre-alloyed ingots were melted in a resistance furnace. The molten metal was degassed using rotary degasser and the melt quality of the melt was evaluated using reduced pressure test and brightimeter. Al-10% Sr master alloy was used for modification and Al-5Ti-1B was used as the grain refiner. The composition of the metal was analyzed by Optical Emission Spectroscope. The compositions of the alloy used in this investigation are presented in Table I.

Table I. Composition of Alloys prepared

Alloy	Si	Mg	Fe	Mn	Ti	Sr
A356	7.47	0.3958	0.1576	0.0322	0.1310	0.0003
A356 + Sr	7.39	0.3969	0.1638	0.0300	0.1350	0.0153

An UBE 350 ton HVDC indirect squeeze casting machine was used for the casting trials. The step plate casting, shown in Figure 1 was used for the casting trials. The casting is 110 mm wide and has 6 steps, each 40 mm long. The thicknesses of the steps were 14, 10, 6, 5, 4 and 3 mm. At the top end, a small overflow is placed. The casting was attached to the shot sleeve through a fan gate.

Figure 1. Step plate casting

The tool steel mould was heated to 225°C before the casting trials. However due to the variation in the metal flow volume, the final temperature distribution was different. The temperature was higher near the gate section compared to the thinner section size at the top. The distribution of the temperature was measured frequently and kept within a close range.

The casting cycle begins with coating of the shot sleeve and tool steel die with lubricant. After this measured amount of metal was poured into the shot sleeve. Immediately the shot sleeve was tilted to align with the die and the plunger is activated. The time delay between pouring and plunger activation was kept at minimum, less than 2 sec. All the castings were produced using the same plunger speed.

The casting variables, tested in this investigation were melt modification, grain refinement, melt temperatures and casting pressure. Test castings were subjected to T6 heat treatment. The process variables and other parameters used in this investigation are presented in Table II.

Tensile test samples were obtained from 14, 10 and 6mm sections of the castings. Three 5 mm specimens were machined from 6mm plate, three 6mm specimens were machined from the 10mm plate, and two 8mm specimens were obtained from 14mm section. For each process variable specimens were obtained from two castings. The tensile strength, yield strength at 0.2% extension and elongation at fracture were measured. The extension was measured using extensometer. Metallographic samples were obtained from various positions of the step plate casting. The fracture surfaces of the tensile test samples were examined.

Table II. Casting Variables

Variable	Values
Melt treatment	Unmodified, modified, grain refined
Melt temperature, °C	700, 730
Casting pressure, bar	250, 500, 800
Delay after pour, s	0, 2, 5
Casting speed, m/s	Approach: 0.2 Casting: 0.11
Mold temperature, °C	Top: 200 – 225 Bottom: 225 – 275
Casting weight, g	1500
Cycle time, s	120

Results

The results from the tensile testing are presented in Table III, IV and V. These tables represent three variables: melt processing, applied pressure and section size. Each result in these tables is an average of four or six samples.

Table III presents the results of the unmodified castings while the results from modified castings are presented in Table IV. The properties of grain refined castings are shown in Table V. All the properties are for heat treated alloys in T6 condition. The following points can be mentioned from these tables:

- The ultimate tensile strength of the alloys range from 270 to 310 MPa in unmodified condition; there is no significant improvement in this after modification where UTS remains between 290 and 315MPa. However the strength increases significantly after grain refinement. The UTS of grain refined alloy varies from 320 – 345 MPa.
- The yield strength illustrates a trend similar to that observed for UTS.
- The elongation has shown significant amount of scatter.
- The tensile properties of the squeeze cast alloys are significantly higher than standard values reported for castings produced in permanent molds. The minimum specifications for the alloy A356 are UTS 260 MPa, YS 180 MPa with an Elongation of 5%.
- Melt processing has significant impact on the properties of squeeze castings. Modification improved the tensile elongation compared to unmodified alloys. On the other hand

153

modification combined with grain refinement improved strengths compared to unrefined alloys.
- The range of casting pressure studied in this investigation has little influence on properties.
- The scatter in the data is significant. In Table IV the elongation values range from 2.0% to 9.3%. Similarly the strength values show inconsistency. Some increase in the strength is observed when the section size was reduced from 14 mm to 10 mm but this is not always true for 6 mm sections. The scatter is present irrespective of the melt processing condition.

Table III. Properties of unmodified castings

Pressure, bar	Section size, mm	UTS, MPa	YS, MPa	Elongation, %
250	6	311	263	6.4
	10	304	260	3.3
	14	309	264	2.9
500	6	284	256	2.4
	10	303	263	2.8
	14	268	262	0.8
800	6	313	263	9.6
	10	307	263	4.7
	14	301	264	2.3

Table IV. Properties of modified castings

Pressure, bar	Section size, mm	UTS, MPa	YS, MPa	Elongation, %
250	6	316	268	8.5
	10	309	263	6.1
	14	293	257	3.0
500	6	294	264	2.4
	10	300	262	5.1
	14	289	257	2.0
800	6	310	268	8.0
	10	313	265	9.3
	14	303	258	5.1

Table V. Properties of modified and grain refined samples

Pressure, bar	Section size, mm	UTS, MPa	YS, MPa	Elongation, %
250	6	335	282	7.6
	10	342	292	6.8
	14	326	282	4.3
500	6	336	281	9.2
	10	342	290	8.7
	14	328	279	5.9
800	6	332	281	7.0
	10	340	292	6.6
	14	322	282	3.4

The microstructure of the castings was evaluated. The low magnification micrograph in Figure 2a illustrates the dense and fine structure. The finer distribution of phases is shown in Figure 2b. This defect free structure is the reason for the improved properties of the alloy. However, the structure is not always uniform. In places severe segregation was noticed. One such area is shown in Figure 3a. The cluster of primary dendrites and islands of eutectic can be clearly observed. An island of eutectic is shown at a higher magnification in Figure 3b.

In the corners, particularly between two steps, channels of eutectic phases can be observed. Due to the restriction caused by the change in section size hot tears can form in these corners and these cracks will be filled by the eutectic liquid during pressurization. One such observation is shown in Figure 4.

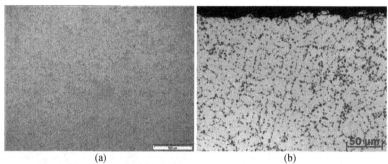

(a)   (b)
Figure 2. Typical microstructure of squeeze cast Al-Si alloy

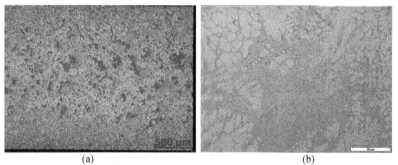

(a)   (b)
Figure 3. Microphotographs showing the segregation of phases

The segregation can reduce the properties of the castings. However, the fracture surfaces of samples with low elongation values have revealed that casting defects cause severe damage rather than the segregated microstructure. Most often inclusions, oxide films, shrinkage and large metallic particles which are not well bonded with the rest of the matrix were observed in the fracture surfaces. Two such features are shown in Figure 5. Figure 5a shows a large particle surrounded by a regular metallic matrix. The boundary between the particle and the matrix contain some inclusion and oxide layer. An oxide layer is shown in Figure 5b.

155

Figure 4. Photograph showing eutectic channels in corners

The source for the large primary particles observed as shown in Figure 5a was not clear. Inspection of the metal biscuit at the bottom of the step casting revealed that a thick shell had formed on the walls of the shot sleeve and this shell is broken up in pieces during the injection process. The process is shown in detail in Figure 6.

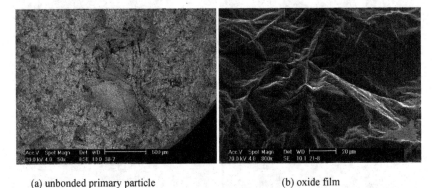

(a) unbonded primary particle   (b) oxide film

Figure 5 Fractographs showing casting defects

These particles were reported to reduce the properties of squeeze casting significantly [6]. This problem can be reduced by controlling the process parameters such as shot sleeve temperature, pouring temperature and time delay but will always be present although in reduced scale. The proper design of gating system may prevent these solidified chunks entering the casting cavity.

A source for inclusions is the pre-solidified layer formed on top of the melt surface in the shot sleeve. The oxide films observed on fracture surfaces might also have evolved in the same pre-solidified layer. The lubrication applied on the shot sleeve and die surface is another major source of the inclusions.

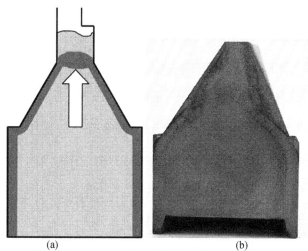

(a)	(b)

Figure 6. Formation of a solidified shell in the shot sleeve

Conclusions

The indirect squeeze casting produced premium quality castings with high tensile properties. However, various casting defects were found to reduce the properties considerably. The reproducibility of properties depends on successful control of these defects. Optimization of process parameters and effective casting design can help to obtain consistent properties in squeeze casting so that the promised potential of this process can be achieved.

Acknowledgements

The authors wish to thank Dr. Jennifer Jackman and Dr. Helmut Kaufmann, directors of CANMET – Materials Technology Laboratory (CANMET – MTL) and ARC – Lightmetal Competence Centre (ARC – LKR) respectively for the approval of this international collaborative project. The authors are also grateful to Dr. Mahi Sahoo and Dr. Franka Pravdic, the managers of the casting programs in the respective laboratories for their unstinted support. The contribution of following staff during this project is greatly appreciated; Mr. Andrea Schiessl, Ms.Ulrike Galovsky, Ms. Tina Ventura and Mr. Thomas Stadler of ARC – LKR; and Ms. Renata Zavadil of CANMET – MTL. The authors are grateful to UBE industries for the financial support for part of the work.

References

1. S. Okada et al., "Development of a Fully Automatic Squeeze Casting Machine," *AFS Transactions*, 90, (1982), 135-146.
2. M. Gallerneault, G. Durrant, and B. Cantor, "The Squeeze Casting of Hypoeutectic Binary Al-Cu," *Metallurgical and Materials Transactions A*, 27A, (1996), 4121 - 4132.
3. D.Y. Maeng et al., "The Effects of Processing Parameters on the Microstructure and Mechanical Properties of Modified B390 Alloy in Direct Squeeze Casting," *Journal of Materials Processing Technology,* 105 (2000) 196-203.

4. S. Rajagopal and W.H. Altergott, "Quality Control in Squeeze Casting of Aluminum," *AFS Transactions*, 93 (1985), 145-154.
5. A.M. Gokhale and G.R. Patel, "Quantitative Fractographic Analysis of Variability in Tensile Ductility of a Squeeze Cast Al-Si-Mg Base Alloy," *Materials Characterization*, 54 (2000), 13-20.
6. R. Kimura et al., "Influence of Abnormal Structure on the Reliability of Squeeze Castings," *Journal of Materials Processing Technology*, 130-131, (2202), 299-303.

EFFECT OF VARIOUS HIP CONDITIONS ON BIFILMS
AND MECHANICAL PROPERTIES IN ALUMINUM CASTINGS

JT Staley, Jr[1], M Tiryakioğlu[2], J Campbell[3]

[1]Bodycote Materials Testing; 7530 Frontage Rd.; Skokie, IL 60077, USA
[2]Robert Morris University; 6001 University Blvd.; Moon Township, PA 15108, USA
[3]The University of Birmingham; Edgbaston, Birmingham B15 2TT, UK

Keywords: Aluminum, Castings, HIP, Tensile, Fatigue, Bifilms

Abstract

This research investigates the effects of various HIP conditions on bifilms and mechanical properties of aluminum castings. A206 castings were non-HIPed or HIPed at (i) temperature typical for HIP, (ii) at solution heat-treat and (iii) at eutectic melting temperatures. Average and threshold tensile and fatigue properties improved as a result of HIP although elongation and fatigue life variability did not decrease. We propose that this is due to HIP's ability to heal porosity and thin bifilms more effectively than thicker bifilms.

Introduction

The automotive and aerospace industries desire to replace wrought components with aluminum castings to reduce manufacturing and operating costs. However, mechanical properties of aluminum castings are impaired to a great extent by the presence of oxide bifilms incorporated into their structure as a result of poor process design and/or melt quality [1]. These oxides are essentially cracks in the structure and not only decrease average properties but also increase variability [1,2,3,4,5]. This is why castings have been used only sparingly in critical applications [6]. Research has shown that other quality issues such as hot tear and porosity are a result of the presence of bifilms. [7,8]. Bifilms lead to porosity because of solidification shrinkage and/or diffusion of hydrogen gas, which causes the bifilms to expand [1,7]. Both mechanisms increase the maximum defect size in castings and consequently decrease mechanical properties, such as tensile strength, elongation and fatigue life [1,4,5,9,10].

Hot isostatic pressing (HIP), uses high temperature and inert gas pressure over a period of time to close porosity [11,12]. The typical parameters for HIP of aluminum castings are 2 to 6 hours at 510 to 521°C with an applied pressure of 103 MPa [13]. This high hydrostatic pressure causes pores that are not connected to the surface to be "forged" shut through plastic flow of metal. Diffusion across the closed pore is necessary to eliminate internal surfaces of pores. It has been shown that HIP can close porosity and improve the mechanical properties of castings [9,14,15]. Nevertheless, in many cases, too few specimens were tested for the statistical characterization of the HIPed castings. Researchers have discussed the possibility of using HIP to mitigate the effects of oxides in aluminum castings by a healing mechanism [1,14,16,17]. In some cases, typical HIP conditions have been shown to improve average properties but also increase variability in aluminum castings mainly due to typical HIP's inability so far to completely mitigate oxide effects. [18,19]. Hence, there is a need to investigate non-typical HIP conditions as a means of healing bifilms. This research investigates the effects of different HIP conditions on oxides and the mechanical properties of aluminum castings by using Weibull statistics [20].

Experimental

Chill cast pig type ingots of A206 were procured for testing since it was expected that ingots produced by conventional processing techniques would exhibit many oxides and much porosity. The chemical composition (wt%) is: 4.62 Cu, 0.33 Mn, 0.31 Mg, 0.20 Ti, 0.02 Si, 0.07 Fe. Four different conditions were tested: (i) non-HIPed, (ii) typical HIPed (516°C, 103MPa, 4hrs), (iii) solution heat-treatment HIPed (529°C, 103MPa, 4hrs) and (iv) eutectic melting HIPed (543°C, 103MPa, 2hrs). Ingots were cut into 25x25x125 mm bars that were then solution treated at 530°C ± 5 for 8 hours, quenched in water and artificially aged at 185°C ± 5 for 5 hours ± 0.25 to T71 condition. Tensile specimens of 12.8 mm diameter and 50.8 mm gage length and fatigue specimens of 8 mm diameter and 108 mm length were machined from the heat-treated bars. Tensile tests were conducted at room temperature following ASTM E8 on castings that were non-HIPed (35 tests), typically HIPed (20 tests), heat-treat temperature HIPed (12 tests) and eutectic temperature HIPed (21 tests). Specimens were strained at a rate of approximately 0.001/s with an extensometer clipped to the gage length until failure. Ten fatigue tests for each of the four conditions were conducted at room temperature following ASTM E466. Specimens were tested at a maximum stress of 170 MPa, R=0.1 under load control using a sine waveform at a frequency of 60Hz. Fracture surfaces of specimens were examined using a stereomicroscope, scanning electron microscopy (SEM) and energy dispersive X-Ray spectroscopy (EDS). In addition, metallography was performed on oxides in the various HIP conditions.

Results

Figure 1A shows the microstructure of as-cast ingots in the polished condition where many oxides and much porosity can be seen. For the most part, the oxides are associated with the porosity and are located at dendritic solidification fronts in eutectic areas or the last areas to solidify. The crack-like nature of bifilm oxides is easily seen with a variety of space existing between the two thin oxide layers that make up the bifilm. Figure 1B compares polished cross-sections of ingots in various HIPed conditions. The HIPed samples have essentially no measurable porosity. In addition, and especially for the eutectic HIPed casting, the HIPed oxides appeared more broken up and intermixed with copper rich eutectic.

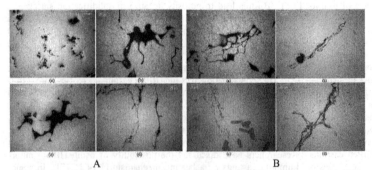

A   B

Figure 1. A) Microstructure of non-HIPed, as-polished ingots; (a) overall area, (b) bifilms and associated porosity, (c) mostly porosity and (d) mostly bifilms. B) As-polished cross-sections of bifilms in castings; (a) non-HIP, (b) typical HIP, (c) heat-treat HIP and (d) eutectic HIP.

160

Figure 2 shows tensile strength and elongation results for all specimens tested. There appear to be at least two groups (populations) within each HIP condition. For HIPed samples tensile strength, minimum and maximum values are higher and range smaller than for non-HIPed samples. However, for elongation, minimum and maximum values are higher but the range (variability) is larger.

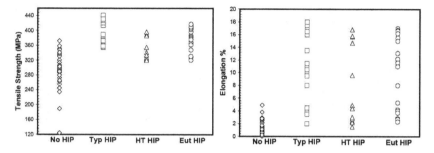

Figure 2. Tensile strength and elongation results for all specimens tested.

After HIP, average elongation improved by as much as a factor of 8 and elongations of over 18 percent were achieved. Minimum tensile strength increased by a factor of 2.5 and minimum elongations for HIPed samples are near the average for the non-HIPed condition. Typically HIPed specimens show the highest average tensile strength and elongation followed by eutectic HIPed specimens. Average yield strength is similar for all HIPed samples. Figure 3 shows fatigue results for all specimens tested. Again, there appear to be at least two groups within each condition. The minimum fatigue lives for HIPed specimens are higher than the maximum fatigue lives for non-HIPed specimens although the range is smaller for non-HIPed samples. Minimum fatigue life increased by a factor of 30, but average life improved by as much as a factor of 136 and lives of over 10,000,000 cycles were achieved. Among the three HIP conditions, eutectic HIPed specimens show the highest average and maximum fatigue life followed by typically HIPed specimens. Minimum fatigue life is similar for all HIPed samples.

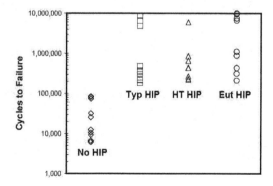

Figure 3. Fatigue results for all specimens tested.

Figure 4a shows digital camera fractographs of non-HIPed tensile tested specimens with various percent areas (dark) of $MgAl_2O_4$ spinel bifilm oxide (corroborated by EDS). For the approximately 20 samples examined, spinel oxide area appeared to range from 1 to 55%. Note that the spinel bifilms were all in areas of porosity and that other areas of porosity contained Al_2O_3 alumina bifilms. The amount of spinel on fracture surfaces seems to reduce the elongation of non-HIPed specimens, as shown in Figure 4b. The scatter in Figure 4b indicates that factors other than the amount of spinel (such as alumina bifilms) affect the elongation.

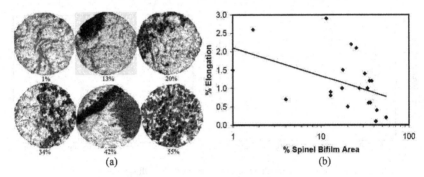

(a)   (b)

Figure 4. (a) Non-HIPed tensile tested specimens with various areas of spinel bifilm oxide (dark), and (b) elongation plotted versus % spinel bifilm area.

Figure 5 shows SEM fractographs of a non-HIPed and typically HIPed tensile tested specimen with low (0.1% and 2.0%) elongation, respectively. For the non-HIPed sample, much porosity and many bifilms and chunky spinel oxides covered dendrites on the fracture surface. EDS revealed a relatively high amount of magnesium and oxygen associated with the oxides. All fracture surfaces examined for HIPed samples with low elongation showed what appeared to be partially healed spinel oxides with reduced amounts of magnesium and oxygen. Limited areas of unhealed bifilms were observed.

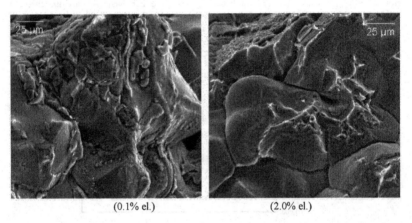

(0.1% el.)   (2.0% el.)

Figure 5. SEM fractographs of a non-HIPed (left) and typically HIPed tensile tested specimen with low elongation. Note the spinel oxides on the dendrites and partial healing due to HIP.

162

Analysis and Discussion

Much of the fracture surface for non-HIPed specimens and nearly the entire fracture surface for HIPed specimens with high elongation exhibited ductile fracture features free of porosity and oxides. For any tensile specimen, the tensile strength and elongation are controlled by the largest effective defect. It is apparent that average tensile properties increased due to HIP. Hence, on average, HIP reduced the largest effective defect. It appears possible that the two groups of strength and elongation data are due to the presence of two different defects; thin alumina bifilms that are mostly healed by HIP and thick spinel bifilms that are only partially healed by HIP. Due to the bi-distribution of properties, 3-parameter bi-Weibull statistics were conducted [21]:

$$P = p\left(1 - \exp\left(-\left(\frac{\sigma - \sigma_{T1}}{\sigma_{01}}\right)^{m1}\right)\right) + (1-p)\left(1 - \exp\left(-\left(\frac{\sigma - \sigma_{T2}}{\sigma_{02}}\right)^{m2}\right)\right) \qquad (1)$$

where P is probability of failure, p is fraction of the Weibull distribution no.1, σ is the property and σ_T, σ_0 and m are the threshold value, scale parameter and shape parameter (Weibull modulus), respectively. Figure 6 shows 3-parameter bi-Weibull plots of elongation for non-HIPed and eutectic HIPed castings.

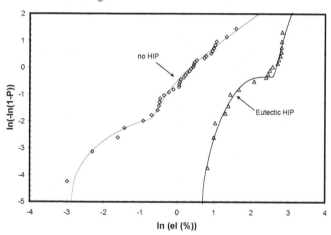

Figure 6. 3-parameter bi-Weibull plots for %el of non-HIPed (left) and eutectic HIPed castings.

The threshold values for elongation are presented in Table 1. Note that threshold increased due to HIP and that the lower group thresholds for HIP are equal or higher than the higher group threshold for non HIP, suggesting that HIPing made the castings safer. The tensile strength and elongation thresholds are similar for all HIP conditions, which suggests that additional healing of defects did not occur due to increased HIP temperature.

Table 1. Threshold values for percent elongation for the four conditions.

Group	No HIP	Typ. HIP	HT HIP	Eut. HIP
Lower	0.1	0.5	1.5	1.7
Higher	0.5	9.3	14.3	13.0

As with the tensile specimens, non-HIPed fracture surfaces of fatigue specimens revealed various amounts of oxides. For the non-HIPed samples, fatigue initiated from multiple sites of porosity with various amounts and types of bifilms. For HIPed specimens, defects varied from surface connected, small $MgAl_2O_4$ chunky particles to interior, larger areas of thick Al_2O_3 bifilm coated dendrites. Analogously to tensile testing, for any fatigue specimen, the fatigue life is controlled by the largest effective defect (bearing in mind that the effectiveness of the defect is strongly dependent on its location in the specimen since stress intensity increases the closer the defect is to the surface). It was shown that average fatigue life increased after HIP. Hence, on the average and in common with the tensile specimens, HIP reduced the largest effective defect. Figure 7 shows SEM fractographs of fatigue initiating defects for typically HIPed specimens with low and high fatigue lives. Both defects are spinel bifilms. The specimen with lower life also had interior bifilm oxides of $MgAl_2O_4$ and Al_2O_3.

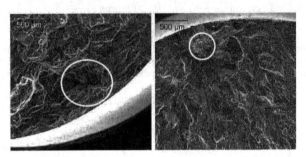

Figure 7. SEM fractographs showing fatigue initiating defects (circled) for typically HIPed specimens with low (183,949 cycles) (left) and high (6,537,164 cycles) fatigue lives.

Weibull statistics were used to determine whether the safety of the castings increased due to HIP, as measured by threshold fatigue life. Although there is evidence for the presence of two populations, single three-parameter Weibull distribution was fitted to data due to small sample size. Figure 8 shows 3-parameter Weibull plots for fatigue life of the castings.

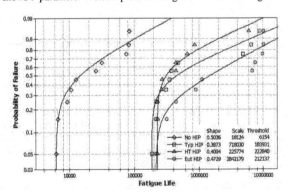

Figure 8. 3-parameter Weibull plots for fatigue life of castings in various HIPed conditions.

The significant increase in average and threshold fatigue life as a result of HIP again shows that HIPing made the castings safer.

Conclusions

1. Commercial ingots of A206 contained many bifilms and much porosity suggesting that there was significant potential for improvement to be made in the chill cast pig type ingot production process.

2. Microstructural examination showed that HIP carried out at typical, heat-treat and eutectic temperatures all substantially closed porosity. In addition, and especially for the eutectic HIPed casting, HIPed oxides and bifilms appeared more fractured and intermixed with copper rich eutectic.

3. Due to HIP, average elongation improved by as much as a factor of 8, elongations of over 18 percent were achieved and minimum tensile strength increased by a factor of 2.5. However, minimum elongations for HIPed samples are near the average for non-HIPed samples and there is more spread in the elongation data for HIPed samples.

4. HIP improved average fatigue life by as much as a factor of 136 and lives of over 10,000,000 cycles (runouts) were achieved. Minimum fatigue lives increased by a factor of 30 for HIPed specimens and are higher than the maximum fatigue lives for non-HIPed specimens.

5. There were at least two groups of property values within each HIP condition. This is attributed to different amounts and types of bifilms.

6. Spinel bifilm area on the fracture surface of non-HIPed specimens ranged from 0 to 55%. Spinel bifilms were all in areas of porosity and other areas of porosity contained various amounts of Al_2O_3 bifilms.

7. There is evidence suggesting that bifilms and properties were not significantly affected by increased HIP temperature.

8. Nearly all fracture surfaces examined for HIPed samples with low elongation and fatigue life showed spinel bifilm defects. Limited samples with areas of unhealed alumina bifilms (relatively thick) were observed. Thus, it appears that HIP was able to heal porosity with thin bifilms but not with thicker bifilms.

9. The mechanism for healing bifilms by HIP appeared to be the breaking up of bifilms and the intermixing (especially for eutectic HIP) of bifilms with copper rich eutectic.

10. HIP reduced the largest effective defect and specimens with higher tensile and fatigue properties had fewer and smaller defects.

11. The significant increased average and threshold for tensile strength, elongation and fatigue life as a result of HIP suggests that HIPing made the castings safer.

Acknowledgements

The authors acknowledge the helpful comments of Dr. JT Staley, Sr. (retired), tensile testing by Bobby Archibald of Bodycote and HIP by Dr. Steve Mashl of Bodycote.

165

References

1   Campbell J, *Castings*, 2nd Ed, Butterworth-Heinemann Ltd, ISBN 0 7506 4790 6.

2   Mi J, Harding R and Campbell J, *Effects of the Entrained Surface Film on the Reliability of Castings*, Met and Mat Trans A, Volume 35A, September 2004, p2893.

3   Green N.R. and Campbell J, *Influence of Oxide Film Filling Defects on the Strength of Al-7Si-Mg Alloy Castings*, AFS Transactions, 1994, V114, pp 341-347.

4   Nyahumwa C, Green NR and Campbell J, *Effect of Mold-Filling Turbulence on Fatigue Properties of Cast Aluminum Alloys*, AFS Transactions, 1998, V58, pp 215.

5   Wang Q, Crepeau P, Griffiths J and Davidson C, *A Review of the Effects of Oxide Films and Porosity on Fatigue of Cast Aluminum Alloys*, The John Campbell Symposium, Edited by M. Tiryakioglu and P. Crepeau, TMS, 2005.

6   Tiryakioğlu M, Campbell J and Green NR, *Review Of Reliable Processes For Aluminum Aerospace Castings*, Transactions of the American Foundrymen's Society, V96-158, 1996, pp 1069-1078.

7   Campbell J, *The Origin Of Porosity In Castings*, Proceedings of the Fourth Asian Foundry Congress, 1996, pp 33-50.

8   Fox S and Campbell J, *Visualisation Of Oxide Film Defects During Solidification of Aluminium Alloys*, Scripta Materiala, V43, June 2000, pp 881-886.

9   Griffin J, Church J and Weiss D, *Effect of Micro-porosity on the Fatigue Life and NDE Response of A206-T4 Aluminum*, Copyright 2004 American Foundry Society, AFS Library copy 20040502A, 17 pages.

10   Yi J, Gao Y, Lee H, Flower H and Lindley T, *Scatter in Fatigue Life due to Effects of Porosity in Cast A356-T6 Aluminum-Silicon Alloys*, Met and Mat Trans A, Volume 34A, Sept. 2003, pp 1879-1890.

11   Hebeisen J, *HIP Casting Densification*, ASME-PVP Conference Proceeding - Boston, MA, August 1-5, 1999, Publisher; ASME.

12   Atkinson H and Davies S, *Fundamental Aspects of Hot Isostatic Pressing: An Overview*, Met Trans A, Volume 31A, December 2000, pp 2981-3000.

13   Hunt W, *HIP Process Makes Aluminum Castings Structurally More Viable*, Aluminum International Today, March 2002.

14   Staley JT Jr., Tiryakioğlu M and Campbell J, *The Effects of HIP on Bifilms in Aluminum Castings*, 67th World Foundry Congress Proceedings, ICME, June 5-6, 2006.

15   LaGoy J, Weihmuller L and Cox B, *The Influence of HIP on Fatigue Life of Aluminum Castings: Two Case Studies*, Proceedings of AFS International Conference on High Integrity Light Metal Castings, ed. S.P. Thomas, 2005, pp 167-177.

16   Campbell J, *Entrainment Defects*, Materials Science and Technology, V22 No2, February 2006, pp 127-145.

17   Nyahumwa C, Green NR and Campbell J, *Influence of Casting Technique and Hot Isostatic Pressing on the Fatigue of an Al-7Si-Mg Alloy*, Met and Mat Trans A, Volume 32A, February 2001, pp 349-358.

18   Wakefield G and Sharp R, *Effect of Casting Technique on Fatigue Properties of Hot Isostatically Pressed Al-10Mg Castings*, MS&T, V12 , June 1996, pp 518-522.

19   Mashl S and Diem M, *The Response of Two Al-Si-Mg Castings to an Integrated HIP and Heat-treat Process*, SAE Technical Paper Series, Vol. 2004-01-1019, 2003.

20   Weibull W, *A Statistical Distribution Function of Wide Applicability*, Journal of Applied Mechanics, V18, 1951, pp 293-297.

21   Jiang S, Keçecioğlu D, *Graphical Representation of Two Mixed-Weibull Distributions*, IEEE Trans on Reliability, v.41, 1992, pp. 241-247.

Shape Casting: The 2nd International Symposium *Edited by Paul N. Crepeau, Murat Tiryakioğlu and John Campbell*
TMS (The Minerals, Metals & Materials Society), 2007

THE INFLUENCE OF OXIDE INCLUSIONS
ON THE POST-HIP FATIGUE LIFE OF TWO AL-SI-MG CASTINGS

Stephen J. Mashl

Bodycote HIP, 155 River Street, Andover MA 01810, USA

Keywords: Aluminum Casting, Hot Isostatic Pressing, HIP, Porosity, Oxides, Fatigue Life

Abstract

Hot isostatic pressing (HIP) has a long been used in aerospace applications where, through the elimination of microporosity, the fatigue life of a casting is improved. In recent years, process and equipment improvements have resulted in reduced processing costs, especially as applied to aluminum castings, thus making HIP a viable option in price critical, non-traditional markets. Past trials that evaluated the effect of HIP on aluminum castings have produced mixed results. Post-HIP improvement in fatigue resistance could be large, small or non-existent. Investigation revealed that oxides can have a strong, negative, effect on post-HIP fatigue properties. This study reviews existing theory on the influence of porosity and inclusions on fatigue life and, using experimental data from a program aimed at integrating HIP and heat treat into a single process, shows how variations in oxide concentration can decrease or eliminate the improvement in fatigue life normally associated with hot isostatic pressing.

Background

The Hot Isostatic Pressing (HIP) process was developed in the mid 1950's at Battelle Memorial Laboratories, Columbus, Ohio. While initial studies focused on the cladding of nuclear reactor fuel rods, the advantages of using the HIP process to eliminate the macro- and micro-porosity in castings soon became apparent. The HIP process quickly became embedded in the process specifications for fatigue critical aerospace castings such as the gas turbine compressor components shown in Figure 1. Over the years, continual process and equipment development efforts have resulted in significant decreases in the cost of HIP processing so that now HIP is seeing use in price critical applications, e.g., high performance automotive components such as the aluminum heads for the V-12 gasoline engine shown in Figure 2.

The most common reason for employing HIP to eliminate porosity from castings is to improve mechanical properties. Given a good quality casting, historical results have shown that the use of HIP typically yields a 3-5 fold increase in tensile ductility and a 7-10 fold increase in high-cycle fatigue life. Greater commercial interest in low-cost aluminum HIP processes has resulted in an increase in the number and types of aluminum foundries evaluating HIP as a means of providing a premium product. With this trend, the HIP industry is now handling a diversity of parts produced by myriad casting techniques and concurrently, experiencing increased variability in the quality of the castings. As a result, commercial HIP service providers have encountered growing numbers of castings that do not achieve the expected post-HIP property improvement.

This paper reviews portions of research programs [1-2] conducted by Bodycote HIP that, while not specifically performed to evaluate the relative influence of oxides upon the mechanical properties, obtained information that provided insight into the influence of oxides on the post-HIP fatigue properties of aluminum castings. These results can also be compared to recent, non-HIP related, research on the effect of inclusions and porosity on the fatigue behavior of aluminum castings. The data presented in this paper suggests an alternative interpretation of the role of porosity and oxide inclusions as property limiting variables in fatigue critical applications. Specifically, the data presented here indicates that, in some cases, oxides can play a greater role in limiting the fatigue properties of aluminum castings than that of residual porosity.

Figure 1. Cast titanium-6Al-4V compressor section components for a Pratt & Whitney – Volvo gas turbine engine.

Figure 2. Automotive aluminum cylinder heads for the V-12 engine used in the BMW 760i/760Li automobiles being loaded on tooling prior to HIP processing.

Introduction

The detrimental effect of porosity and non-metallic inclusions on mechanical properties have long been known and continue to be studied [3-7]. While many experimental projects have focused on either porosity or inclusions, one excellent research project conducted by Wang, Apelian and Lados [7] quantitatively evaluated the influence of both porosity and oxide inclusions on fatigue life. In this work the researchers assume that linear elastic fracture mechanics apply and employ the Paris-Erdogon equation for the calculation of the number of cycles required to propagate a crack to failure, i.e.,

$$\frac{da}{dN} = C\left(\Delta K_{eff}\right)^m \tag{1}$$

Where: a = crack length
N = number of cycles
C = maximum defect size
ΔK_{eff} = alternating effective stress intensity factor range
m = Paris-Erdogon crack growth law exponent.

Integrating Equation 1 and making some simplifying assumptions, Wang, et al., then develop an expression that allows the calculation of the number of cycles to propagate a crack to failure (N_P) as a function of the fatigue stresses and the initial defect size (a_i) That is,

$$a_i N_P = B \cdot \left(\Delta \sigma\right)^{-m} \tag{2}$$

Where: B = is approximated as a constant
$\Delta\sigma$ = the difference between the maximum and minimum fatigue stresses.

Wang et al. then use Weibull statistical analysis to develop an equation that calculates the statistical probability of specimen failure at or less than a given number of cycles in a fatigue testing sequence. This is:

$$F_W\left(N_p\right) = 1 - \exp\left[\left(\frac{N_P}{N_c}\right)^b\right] \tag{3}$$

Where: $F_W(N_P)$ = the statistical probability of specimen failure at a given cycle or lower
N_c = the characteristic fatigue life at which 63% of specimens have failed
b = the Weibull modulus.

These relationships allow for the possibility of cyclic loading prior to fatigue testing, thus N_f, the total number of cycles to failure would be equal to N_i, the initial number of cycles to failure prior to testing, plus N_p, the number of cycles to failure during testing. In typical laboratory fatigue testing $N_P = N_f$. Taking the natural log of Equation 3 twice, transforms that expression into one that, when the quantity ln ln [1/(1-F_W)] is plotted versus N_f, should display linear behavior with the slope of the line being equal to the Weibull modulus.

In the experimental portion of the Wang, Apelian, Lados work, a series of castings were poured from the same melt which, through manipulation of the casting process and the selective use of HIP, yielded specimens containing either: porosity and oxides; oxides without porosity; or specimens that were essentially oxide and pore free (or at least possessed defects of such a minimal size and distribution that they did not play a role in fatigue failure). Following fatigue testing, Wang, et al. analyzed the fracture initiation site on all specimens and classified each specimen according to the characteristics of the crack initiation site. This information was then matched with the fatigue life for a given specimen and plotted on a graph of the probability of failure (ln ln [1/(1-F_W)]) versus the natural log of the number of cycles to failure, using different symbols to identify the feature responsible for crack initiation. This plot is reproduced in Figure 3. The data is very well behaved. The steep slope of the lines indicates that there was minimal scatter in their data and the presence of three distinct linear bands of data indicates that the mode of failure was constant and consistent for a given group of data.

While Wang, et al. recognize that both porosity and oxide inclusions have a negative effect on fatigue life, they state that porosity is more detrimental to fatigue performance than any other microstructural defect, including oxide films. This interpretation seems justified because, comparing all samples at given probability of failure, specimens in which fracture initiated at pores always failed at the lowest number of cycles. Graphically, this is demonstrated by drawing a horizontal line across Figure 3 at any point on the y-axis. Moving along this line from left to right, i.e. in the direction of increasing number of cycles, the line representing failure initiated by porosity is always encountered before the line of data that represents crack initiation at oxides or slip bands.

Experimental Description, Results and Discussion

This paper presents fatigue data from the evaluation of two different, commercially produced, castings having similar but not identical compositions. The first group of castings evaluated in this project were anti-lock brake system (ABS) housings, an example of which is shown in Figure 4.

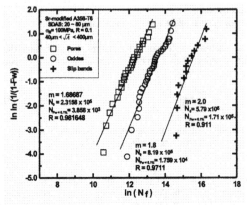

Figure 3. A two parameter Weibull plot of fatigue life data for strontium modified A356 castings containing two types of defects; pores and oxide particles. When no defects are present, cracks initiate due to dislocation pile-up along slip bands. The pore free material was HIPed or Densal treated by Bodycote. After Wang et al [7].

These permanent mold castings were originally believed to possess a standard A356 composition; it was learned during the course of the project that additional magnesium had been added to the melt to improve the machinability of the part. The second group of castings were the wedge shaped, chilled sand castings produced by a commercial foundry, using their facility's best practice. In this case, best practice included the placement of porous ceramic filters in the mold runners to entrap any oxides that passed into the mold during pouring.

Figure 4. An ABS housing permanent mold casting having a modified A356 composition. (~0.9% Mg vs. a typical ~0.4%Mg)

Figure 5. A chilled wedge, sand casting of standard A356 composition produced by a commercial foundry, using "best practice" techniques.

From an initial sampling of 32 castings (16 ABS housings and 16 wedges), 16 castings (8 of each type) were kept in the as-cast condition and subjected to the standard T6 heat treatment described in Table I. The remaining castings were put through Bodycote's Densal® process and then given an identical T6 treatment. Densal is an aluminum specific HIP process, the parameters of which are proprietary. Densal processing typically yields an improvement in fatigue life equivalent to that obtained using published commercial HIP conditions for aluminum. Following heat treat, specimens from both groups of castings were sectioned and tensile and fatigue specimens were machined from the pieces.

Tensile testing of specimens machined from these castings yielded the data shown in Tables II and III and is compared to that for published property values . The higher than expected strength levels and lower than anticipated ductility for the ABS casting samples brought the chemical composition of the alloy into question.

Table I. Processing conditions

Description	HIP Process	Solution Heat Treat	Age
Baseline Process: As-Cast+T6	None	10 Hours at 538°C – Water Quench	4 hours at 155°C
Current HIP Practice: Densal, followed by T6	Densal®	10 Hours at 538°C – Water Quench	4 hours at 155°C

Note: Densal® is Bodycote's aluminum specific HIP process. Densal parameters are proprietary.

Additional investigation found that these parts had a higher than normal magnesium content than is typical in an A356 alloy (0.9 wt.% Mg vs. a typical 0.4 wt.%), explaining, in part, the elevated strength and minimal ductility levels observed in these castings. The tensile properties of the sand cast wedges are excellent, with outstanding ductility. HIP generally results in an increase in tensile ductility but in these two cases, the material appeared to have either such a low inherent ductility that pore removal did not affect a change or, such a high level that HIP did not provide much improvement. It is worth noting that it is easy to minimize shrinkage porosity in the wedge shape and evaluation of the microstructure using optical microscopy bore this out.

Table II – Average tensile results of permanent mold, ABS castings

Process Route	0.2% Yield Strength, MPa (ksi)	Ultimate Strength, MPa (ksi)	Percent Elongation	% Reduction in Area
Cast - T6	292.7 (42.5)	325.4 (47.2)	0.7	2.1
Cast – Densal - T6	282.5 (41.0)	320.3 (46.5)	0.7	2.0
Typical Properties A356.0	186 (27)	262 (38)	5.0	

Table III – Average tensile results of sand-cast Sr-modified A356 samples.

Process Route	0.2% Yield Strength, MPa (ksi)	Ultimate Strength, MPa (ksi)	Percent Elongation	% Reduction in Area
Cast - T6	192.4 (27.9)	287.5 (41.7)	17.2	24.0
Cast – Densal - T6	187.2 (27.2)	287.5 (41.7)	17.2	23.9
Typical Properties A356.0	165 (24)	228 (33)	3.5	

Fatigue testing of the specimens machined from ABS housings was performed at 130MPa maximum load, with an R value of 0.1 at a cyclic loading rate of 50 Hertz. As shown in Figure 6, these specimens showed no improvement in fatigue life with HIP processing. The fatigue specimens machined from the wedge castings were tested at a maximum applied load of 193 MPa, R=0.1, 50Hz. In this case and as displayed in Figure 7, the HIPed specimens showed, on average, an order of magnitude increase in the number of cycles to failure. The behavior of the latter group of castings were typical of aerospace material, while the fatigue data from the ABS housings did not fit the expected pattern, i.e. improvement with HIP. In an effort to gain an explanation for the mixed results, the fracture surfaces of the fatigue specimens were analyzed.

For the fracture analysis study, samples were given new identification labels in order to provide impartial, double-blind conditions. In almost all cases, the fracture initiation site was easily identified using a combination of optical microscopy and either secondary electron imaging or backscattered electron imaging, and, as shown in Figures 8 and 9, either the remains of an open pore or a layer of oxide was found at the fracture initiation site.

171

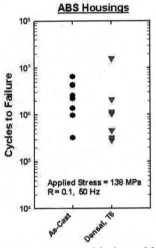

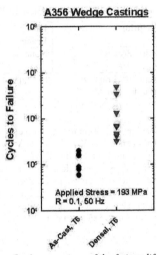

Figure 6. A comparison of the fatigue life data obtained from the testing of the ABS housings shown in Figure 4. The data show the lack of improvement in fatigue life following HIP, which is atypical for aluminum castings.

Figure 7. A comparison of the fatigue life data obtained from the testing of the wedge castings shown in Figure 5. The data show an average, order-of-magnitude improvement in the fatigue life following HIP.

Correlation of the crack initiation defect with the specimen and its fatigue life is shown in Figures 10 and 11. Reviewing the fatigue and fractography data from the ABS housing specimens in Figure 10, the heavy influence of oxides on the fatigue failure is obvious. What is most interesting, is that in the as cast, un-HIPed specimens, 3 of 7 samples (one of the original eight specimens failed in the testing grips) failed at oxides rather than at the pores which must have been present. This indicates that, in these specimens, the size and distribution of the oxides were such that oxides, not pores, actedas the property limiting defect. This behavior contradicts the conclusion drawn by Wang, et al., that porosity is the most detrimental defect in an aluminum casting, overriding the influence of oxide films. Analysis of the correlated wedge casting data in Figure 11 shows different behavior. These specimens exhibited a nominal 10 fold increase in fatigue life following HIP. This level of post-HIP property improvement is typical of that observed in aerospace alloys. Further, all fracture initiation in as-cast samples began at pores, an indication that that the ceramic filters placed in the gating system were, at least partially effective in reducing oxides in these castings. It should be noted that in the HIPed data, oxides are the predominant crack initiating defect, indicating the potential for further improvement. If the size and distribution of oxides were reduced to a point where they were no longer a factor, crack initiation at slip planes or eutectic particles would be expected.

The presence of pores at the crack initiation sites of two, HIPed A356 wedge samples is interesting (indicated as P* in Figure 11). HIP is generally perceived as eliminating all porosity. In both of these instances the pores were significantly different from the typical shrinkage pore shown in Figure 9. These two pores were much smaller than the other porosity observed (measured in 10's of microns rather than 100's of microns) and the pore interior seemed to be lined with plate-like precipitates rather than rounded secondary dendrite arms.

Figure 8 – BEI image of oxide at fracture initiation site at edge of sample.

Figure 9 – SEI image of pore fracture initiation site, dendrites visible.

It seems possible that a cluster of plate like inclusions could locally "reinforce" a small volume of material inhibiting pore collapse via plastic flow. Additionally, the fact that Densal is designed to be as cost-effective as possible would imply that pressures lower than the conventional 105MPa are used. Perhaps a combination of clustered precipitates and processing conditions resulted in the observed behavior.

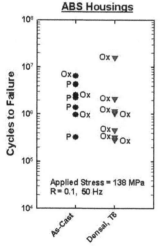

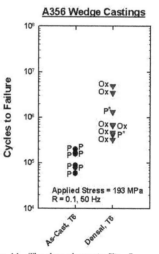

Figure 10. The data shown in Fig. 6, annotated to indicate the type of defect found at the crack initiation site. "P", indicates initiation at a pore, "Ox", indicates that the fatigue crack began at an oxide.

Figure 11. The data shown in Fig. 7, annotated to indicate the type of defect found at the crack initiation site. "P", indicates initiation at a pore, "Ox", indicates that the fatigue crack began at an oxide.

Considering these results it is clear that porosity is not always the casting defect most detrimental to fatigue life. Obviously, control of both porosity and inclusions, specifically oxides, is necessary to produce a cast aluminum component having premium properties. The data presented here also suggest a different interpretation of the original Wang, Apelian, Lados data. Figure 3 can be re-evaluated from a "vertical" perspective. By separating the data with vertical

173

lines between the data points having the highest and lowest probability of failure for a given group of data, as is done in Figure 12, four distinct regimes are created. As explained in the caption of Figure 12, in regions I, III, and IV, a single type of microstructural feature dominates fatigue crack initiation. In region II however, either oxides or porosity can be the property limiting defect. The author proposes that this is the regime in which the specimens machined from ABS housings in this study reside.

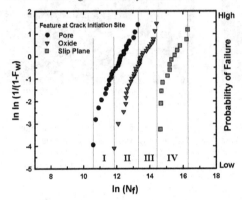

Figure 12. The data from Figure 3, divided vertically into four regions over which one or more microstructural feature is the fatigue life limiting defect. In region I, fatigue life is always controlled by porosity. In region III, oxide inclusions are the property limiting defect. In region IV, microstructural features such as preferential slip planes or eutectic particles control life but in region II either pores or oxides can be the predominant defect in fatigue crack initiation.

Conclusions

From an analysis of the data in this study, it can be concluded that either oxides or porosity can act as the property limiting defect in high cycle fatigue failure. If a given batch of aluminum castings contain oxide inclusions of a size and distribution sufficient to dominate the fatigue crack initiation event, removal of porosity via post-cast HIP processing will do nothing to improve fatigue properties.

References

1. S.J. Mashl and M. M. Diem, *The Response of Two Al-Si-Mg Castings to an Integrated HIP and Heat Treat Process*. SAE Tech. Pub. #2004-01-1019. SAE Inc., Warrendale, PA. (2004)

2. M.M. Diem, S. J. Mashl, R.D. Sisson, *Evaluating a Combined HIP and Solution Heat Treat Process for Aluminum Castings*, Heat Treating Progress, Vol.3, No.4, pp. 52-55 (2003).

3. D.A. Gerard and D.A. Koss, *The Dependence of Crack Initiation on Porosity During Low Cycle Fatigue*, Mat. Sci. & Eng. A Vol 129, 1990, pp.77-85

4. J. Campbell, *"The Origin of Porosity in Castings"* Proceedings of the Fourth Asian Foundry Congress, Australian Foundry Institute, 1996, pp.33-50

5. J.F. Major, *"Porosity Control and Fatigue Behavior in A356T61 Aluminum Alloy"*, AFS Transactions, 105, (1997) 901-906

6. J. Campbell, *"The New Metallurgy of Cast Metals"*, Proceedings from the 2nd International Aluminum Casting Technology Symposium, 7-9 October 2002, Columbus, Ohio, ASM International, 2002.

7. Q.G. Wang, D. Apelian, D.A. Lados, *"Fatigue Behavior of A356-T6 Aluminum Cast Alloys. Part I. Effect of Casting Defects"*, Journal of Light Metals 1, 2001, pp. 73-84

8. S.J. Mashl, J.C. Hebeisen, D. Apelian, Q.G. Wang, *"Hot Isostatic Pressing of A356 and 380/383 Aluminum Alloys: An Evaluation of Porosity, Fatigue Properties and Processing Costs"*, SAE Tech. Pub. #2000-01-0062. SAE Inc., Warrendale, PA. (2000)

Shape Casting: The 2nd International Symposium *Edited by Paul N. Crepeau, Murat Tiryakioğlu and John Campbell*
TMS (The Minerals, Metals & Materials Society), 2007

STRUCTURE, PROPERTY, PROCESSING RELATIONS IN CAST, SOLUTIONIZED, AND AGED AL-SI-CU-MG AND AL-SI-MG ALLOYS

Jian Fang, Yong Ma, Harold D. Brody

Institute of Materials Science, University of Connecticut, Storrs, CT 06268-3136

Keywords: Aluminum Alloys, Casting, Age Hardening, Tensile strength, Process Design

Abstract

An empirical approach is used to find design rules to relate tensile strength of cast and age hardened aluminum alloys to microstructural features (dendrite arm spacing and volume fraction porosity) and to process parameters (local solidification time, melt hydrogen concentration, and solution treatment time). End-chilled plate castings are used to provide a wide range of as-cast microstructures in Al-Si-Cu-Mg (based on 319) and Al-Si-Mg (based on 356) alloys. Coupons cut from the plates are given brief to extended solution treatments, quenched and given a uniform aging treatment. Tensile properties of the cast and heat treated coupons are measured and fit to an empirical relation. The empirical relation is validated by casting, heat treating, and testing additional sets of plate castings and the empirical relations are used to make microstructure-property maps that can be used in computer assisted process design.

Introduction and Background

The core of materials science and engineering is the development and application of the relations among the four stems of the discipline: processing, structure, properties, and performance. The work in the 1950's by Passmore, Flemings, and Taylor [1-3] on premium quality aluminum casting exemplifies the discipline. They used an empirical approach to relate casting and heat treating process parameters, to microstructure, to strength and ductility properties, and to the performance for aerospace applications of premium quality cast and age hardened 195 and 356 alloys. Passmore et al used an end chilled horizontal plate casting to obtain a range of casting parameters and as-cast microstructures in a each plate; namely, high cooling rates and steep temperature gradients near the chill and longer solidification times and lower temperature gradients away from the chill. Concomitantly, dendrite arm spacing and dendrite cell size, constituent particle size, micropore size; i.e., all features of the as-cast microstructure were finer in locations near the chill. In premium quality casting practice extremely low impurity content, especially iron, effective degassing, and streamlined gating. The primary heat treatment process parameter to be varied was solution treatment time. Coupons cut from regions of the plates close to the end chill and given extended solution treatment times exhibited tensile and yield strengths about double those achieved in typical commercial practice at the time. Measured elongations to failure were seven to eight times the ductility achieved in typical practice. Premium quality aluminum castings were being adopted for aerospace components previously reserved for forgings where high specific strength and high reliance were required. Distance from a chill and solution treatment time are the primary process parameters Passmore et al employed in process-property maps prepared to guide casting and heat treatment process. The use in design of parameters such as cooling rate, local solidification time, and/or temperature gradient, were not practical with the computing technology available at that time.

As part of a overall program to develop computer aided design and analysis software [4-7] we followed the lead of Passmore et al to develop process-microstructure and process-property maps, firstly, for Al-Si-Cu-Mg alloys based on 319 and, recently, for Al-Si-Mg alloys based on

356. The scope of the collaborative university-industry-government program is to develop, verify and market an integrated system of software, databases, and design rules to enable quantitative prediction and optimization of the heat treatment of aluminum alloy castings. The system is built upon a quantitative understanding of the kinetics of microstructure evolution in complex multicomponent multiphase alloys, on a quantitative understanding of the interdependence of microstructure and properties, on validated kinetic and thermodynamic databases, and validated quantitative models.

The objective of the work discussed herein is to determine and validate process-property and process-microstructure maps that that can be used as design rules to predict the tensile strength of cast and precipitation hardened Al-Si-Cu-Mg and Al-Si-Mg alloys. The design rules are based on the minimum set of casting and heat treatment process parameters needed to adequately predict tensile strength. The procedure used to determine the design rule for this family of aluminum alloys is general and can be followed to develop design rules for other families of cast aluminum alloys and for other cast alloy systems. The work is guided by the processing, property, and performance requirements of the automotive industry. Validated process-property maps need to be tailored to the alloys and practices of individual foundries.

Approach

End-chilled plate castings are used to prepare test coupons that experience a wide range of solidification conditions and exhibit a wide variety of as-cast microstructures and mechanical properties. [4-7] Coupons cut from the cast plates are solution treated for times ranging from 0 to 32 hours, quenched, and given the same aging treatment. Microstructural parameters, including dendrite arm spacing, dendrite cell size, and volume percent porosity, are measured. Tensile properties are measured for the full range of solidification conditions and solutionizing times. A multi-dimensional least square analysis is used to develop empirical relation, i.e. a design rule relating tensile properties and a minimum set of microstructure and processing parameters. The empirical relation is validated by casting, heat treating, and testing additional sets of test plates and comparing the second set of measured properties to the properties predicted by the relation determined from the first set of measurements.

We sectioned and measured several microstructural parameters of several automotive aluminum structural castings made in permanent molds, sand molds with and without copper chill blocks, and by the lost foam process. [7] Dendrite spacing in the commercial automotive castings ranged from 20 mm in regions adjacent to a large copper chill to 100 microns in thick wall sand castings. Volume percent porosity ranged from less than 0.1% to 3%. Generally, in regions of castings where high strength and toughness are specified the casting process is designed to produce fine dendrites and low porosity. An end-chilled horizontal plate casting was designed to simulate a broad range of casting practice and to provide six standard round (0.505" dia) tensile samples from each plate. A vertical plate casting was designed to simulate the solidification conditions in those regions where performance requirements are stringent. Two tensile specimens can be cut from each vertical plate. In order to simulate the as-cast microstructure adjacent to a copper chill block or in a thin walled permanent mold casting and to have enough thickness to provide standard 0.505" tensile samples, three chills were placed around the end of the vertical plate casting. Volume percent porosity in well degassed melts and the size of the pores decrease in parallel with the dendrite spacing, as local solidification time is reduced. Dendrite spacing and volume % porosity porosity do not vary independently in these castings.

Experimental

For the Al-Si-Cu-Mg alloys horizontal plate (180 x 230 x 25 mm) molds were prepared, one plate per mold, with a copper block placed in the mold against one end (180 x 25 mm) of the pattern and with a tapered cylindrical riser placed over the other end of the plate. For the Al-Si-

Mg alloys vertical plate (150 x 100 x 25 mm) molds were prepared, four plates per mold, with one or three copper blocks placed vertically in the mold against one end (150 x 25 mm) of the plate. A central, tapered cylindrical riser is used to feed the four plates.

An unpressurized gating system is used to bring the molten alloy into the sodium silicate bonded silica sand (AFS 80) molds. The ceramic filter, runner and ingate are in the drag. The plate, riser, tapered rectangular sprue, and copper chill(s) are in the cope. For the horizontal plates the flaskless molds are set on bottom boards with a ten-degree tilt and the alloy runs uphill from the riser end toward the chilled end.

All charge materials are preheated to at least 200°C before they are added to the crucible. Melts are degassed with an argon purge from a preheated graphite lance inserted to the bottom of the melt. When used, Al-10%Sr modifier is added after degassing. The Al-Si-Cu-Mg alloys poured into horizontal plate molds were melted in a 3000 Hz induction furnace in an alumina crucible, and the molten alloy was tilt poured (730°C, approximately) into the mold through a preheated alumina funnel placed directly over the pouring basin in the sprue. Generally, seven end-chilled plate molds were poured from a single 60 kg melt so that each plate in the series would have the same composition, including the same dissolved hydrogen content. The Al-Si-Mg alloys poured into vertical plate molds were melted in a gas fired furnace in a clay graphite crucible. Generally, two molds containing a total of eight plates were hand poured from a single 18 kg melt.

Plate castings are allowed to cool overnight before they are removed from the mold. Plates are cut into two sets of coupons: one set is used for tensile testing and the second set for microstructural analyses. Six tensile coupons are cut from each horizontal plate and two tensile coupons from each vertical pl;ate.

The nominal alloy composition selected for the 319 study is Al-7%Si-3.5%Cu-0.33%Mg with 0.5%Fe. The nominal alloy composition selected for the 356 study is Al-7%Si-0.33%Mg with 0.1%Fe. Microstructures and properties for alloys richer and leaner in Si, Cu, Mg and Fe also have been examined. The nominal alloy is not modified and not grain-refined. Either modified, grain-refined, or both modified and grain-refined alloys also have been examined.

Most Al-Si-Cu-Mg coupons are solution treated at 505°C and most Al-Si-Mg alloys are solution treated at 540°C for solutionizing times ranging from ¼ hr to 32 hrs prior to a warm water quench. Solutionizing and aging are carried out in a forced air resistance heated furnace. Coupons are tied in groups by wire with a separation of at least 25 mm between coupons. At the end of the solutionizing period a bundle of samples are pulled from the furnace and quenched within a few seconds in warm water. Some samples are aged in the as-cast condition (0 hrs. solution treatment time). All Al-Si-Cu-Mg coupons are given the same aging heat treatment, held at 230 °C for 3 ½ hrs; and all Al-Si-Mg coupons are aged at 165°C for 8 hrs. Typically, alloys are held at room temperature for 24 hours between quenching and aging.

Tensile coupons are machined after heat treatment into standard cylindrical test specimens (0.505" or 12.8 mm dia with 2" or 50.8 mm gauge length) and pulled with an extensometer attached to the gauge length at Westmoreland Testing. If the reported elastic modulus is not within 10% of the handbook value, the tensile test data for the coupon is not used. The extensometer is used to measure total strain to fracture. Elastic strain is subtracted from total strain to compute plastic strain.

Dendrite arm spacing and dendrite cell spacing in the as-cast alloys are measured on the coupons prepared for microstructural analysis. Dendrite cell size is measured by inscribing random lines over the microstructure and measuring the average distance between intersections of the lines with interdendritic regions. Dendrite arm spacing is measured by identifying primary arms with several perpendicular secondary arms in the plane of view and measuring the average perpendicular distance between secondary arms. Volume percent porosity is measured by comparing the density of sections cut from the metallography coupons of the end-chilled cast plate to the density of a control sample of the same composition (barring macrosegregation along the length of the plates). The control sample is made by remelting and then chill casting a small piece of the cast plate.

Results

For the Al-Si-Cu-Mg alloys cast in end chilled horizontal plates the measured dendrite arm spacing increases from < 30 μm at 13 mm from the end chill to > 60 μm at 200 mm from the end chill. The volume percent of porosity measured over the same range increased from <0.1% near the chill to 1.2% away from the chill.

For the Al-Si-Mg alloys cast in end chilled vertical plates the measured dendrite arm spacing increases from < 25 μm at 13 mm from the end chill to > 50 μm 100 mm from a single end chill. For plates cast with three chills dendrite arm spacing increases from < 20mm near the chill The volume percent of porosity measured over the same range increased from <0.1% near the chill to < 0.4% away from the chill.

The tensile properties of the nominal Al-Si-Cu-Mg alloy (not grain-refined and not modified) are plotted simultaneously against dendrite arm spacing and solution treatment time at 505°C in Figure 1.

The contour plots illustrate a two-fold (100%) increase in tensile strength from coarse dendrites and two-hour solution treatment to fine dendrites and 32-hour solution treatment. Sharply lower values are found for samples that were aged in the as-cast (slow cooled) condition. Considering samples that were aged without prior solution treatment, the overall increase in tensile strength is three-fold (200%). Qualitatively similar results are found for other Al-Si-Cu-Mg compositions and conditions studied.

The yield strength increases sharply for the first two-hour increment of solution treatment and then reaches a plateau. The increase of yield strength from two-hour solution treatment and coarse dendrites to fine dendrites and 32-hour solution treatment is less than 20%..

Elongation properties peak at fine dendrite arm spacing and long solution treatment times. Increasing solution treatment time at 505°C has a significant effect on ductility (% plastic elongation at fracture) for locations with fine dendrite arm spacing. Samples from locations with coarse microstructures and long local solidification times have very low ductility. Ductility is not improved significantly by extended solution heat treatment.

The tensile properties of the nominal Al-Si-Mg alloy (grain-refined and not modified) are plotted versus solution treatment time at 540°C in Figure 2. The open symbols represent measured tensile strengths and the closed symbols represent measured (0.2% offset) yield strengths. The squares, diamonds, triangles, and circles represent samples with dendrite arm spacing of 17, 35, 44, and 61 μm, respectively. No samples were aged without solution treatment. The minimum solution treatment time, ¼ hour, was used to reverse the effect of slow cooling in sand molds.

Tensile strengths for coupons given extended solution treatments are about 20% greater than those given brief, ¼ hour solution treatments.

Yield strength increases for the first couple of hours of solution treatment time and reaches a plateau.

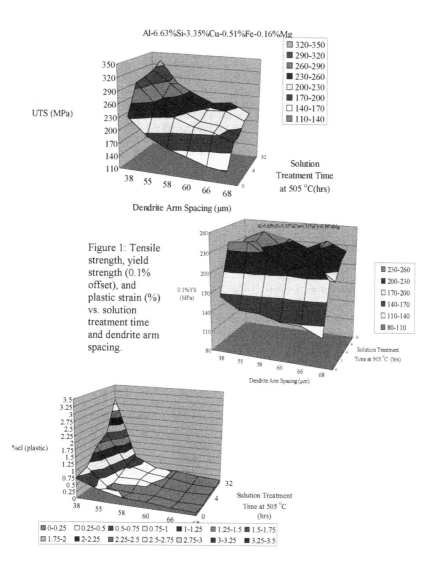

Figure 1: Tensile strength, yield strength (0.1% offset), and plastic strain (%) vs. solution treatment time and dendrite arm spacing.

Elongations increase substantially even for samples with relatively coarse (> 60 μm) dendrite arm spacing.

Hypothesis

Our starting hypothesis is that for a fixed alloy composition, pouring practice, quenching practice and aging treatment, the primary factors that influence the local tensile properties of the cast and heat treated alloy are the local thermal parameters during solidification (local solidification time and temperature gradient), the amount of dissolved hydrogen in the melt as the alloy enters the mold, and the solution treatment thermal cycle. The local solidification time determines the scale of the as-cast microstructure: dendrite arm spacing, silicon-phase particle size, iron-rich intermetallic particle size, volume percent porosity, and size of porosity. Solidification simulation software can be used to predict the local solidification time and dendrite arm spacing; or, as in the discussion below, measured dendrite arm spacing can be used to represent local solidification time and the size scale of the as-cast microstructure.

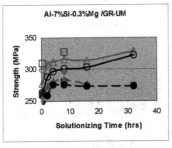 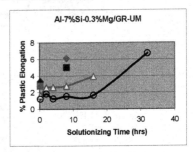

Figure 2: Tensile strength, yield strength (0.2% offset-dashed line), and plastic strain (%) vs. solution treatment time (hrs) for dendrite arm spacings: 17 μm (squares), 35 μm (diamonds), 44 μm (triangles), 61 μm (circles).

Volume percent porosity in the as-cast microstructure is a direct result of dissolved hydrogen, local solidification time, and temperature gradient and is a predominant factor in limiting alloy ductility. Solidification simulation routines are getting better at predicting size and volume of microporosity. Volume percent porosity in the as-cast microstructure can be determined by quantitative metallography or, as in the discussion below, by density measurement.

The solution treatment thermal cycle influences the effectiveness of the dissolution of nonequilibrium phases, the formation of equilibrium phases, the distribution of strengthening elements through the terminal solid solution matrix (α-aluminum dendrite cores) and the rounding of constituent particles. Time at the solution treatment temperature is the single parameter used to represent the solution treatment process.

Analysis

A multidimensional least square procedure is used to express the tensile strength as a function of dendrite arm spacing (DAS), volume percent porosity (%P), and solution treatment time (θ_S):

$$UTS = a \, (DAS) + b \, (\ln[1+\%P]) + c \, (\ln[\theta_S]) + d$$

where **a**, **b**, **c**, and **d** are the parameters to be fit by least square analysis of the measured tensile strength data. Figure 3 compares the values of tensile strength computed with the above equation to the measured values for the nominal Al-Si-Cu-Mg alloy (not grain-refined and not modified) cast in horizontal end chill plate molds. The comparison produces a 45 degree line (slope =1) and an R^2 value above 0.97. Figure 4 compares the values of the tensile strength computed with the above equation to the measured values for the nominal Al-Si-Mg alloys (grain-refined and not modified) cast in vertical end chill plate molds. The comparison, again, produces a 45 degree line but the R^2 value is only 0.9. Slightly different equations and similar fits are obtained for variations about the nominal composition and for combinations of grain refinement and modification. The fits obtained for Al-Si-Cu-Mg alloys cast in the horizontal plate molds is consistently superior to the fits for the Al-Si-Cu alloys cast in the vertical plate molds.

The comparisons in Figures 3 and 4 are between the empirical equation and the data used to obtain the equation. To test the validity of the empirical equation as a predictive tool another set of end-chill horizontal plate castings was cast, sectioned, heat treated, and tested in tension. Figure 5 shows the comparison of the tensile properties predicted by the empirical equation for the nominal Al-Si-Cu-Mg alloy and the tensile strength measurements from the second set of plate castings. The fit is represented by a line of slope 0.7 and an R^2 value of 0.95. Similar comparisons of the empirical equations determined for the Al-Si-Mg alloys cast in vertical plates with new sets of castings awaits determination of the reason for the poorer fit with the non-copper containing alloys. In addition to the change in alloy family (from low ductility to moderate ductility), the mold configuration was changed from horizontal to vertical, data was collected over a smaller range of process variables, fewer data points were collected per condition, and the aging temperature was decreased from relatively high value selected for the copper containing alloy. Experiments where we change one variable at a time may clarify the reason(s) for the poorer predictability of the tensile properties for the non-copper containing alloys.

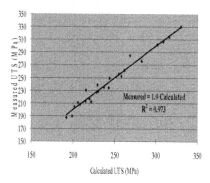

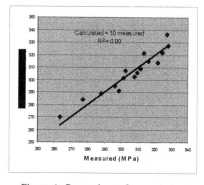

Figure 3: Comparison of computed and measured tensile strengths for nominal Al-Si-Cu-Mg alloy.

Figure 4: Comparison of computed and measured tensile strengths for nominal Al-Si-Mg alloy.

The empirical relations can be used to form a microstructure-properties map. Using the empirical relation for the nominal Al-Si-Cu-Mg alloy with a fixed solutionizing time constant strength lines can be drawn on axes of dendrite arm spacing versus log of one plus volume percent porosity. The iso-strength lines are straight with a negative slope. In Figure 5 the iso-strength lines are drawn at 50 MPa intervals. Also plotted in Figure 5 are measured values of tensile strength for a series of horizontal plate castings in which the hydrogen dissolved in the melt was varied from low to moderate to high. Volume fraction porosity increased in proportion and tensile strength decreased with increased hydrogen content. The measured microstructural parameters are plotted as points on the DAS vs ln(1+%P) map and the measured strengths for each point are labeled. Over a very large range of porosity and dendrite arm spacing the empirical relation and the derived microstructure-property map are very good predictors of tensile strength in this alloy.

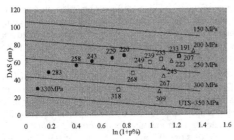

Figure 5: Comparison of measured tensile strength and iso-strength lines computed with the empirical relation for as-cast dendrite arm spacing and volume percent porosity, nominal Al-Si-Cu-Mg alloy.

Summary

Tensile strength in cast and age hardened Al-Si-Cu-Mg alloys is influenced strongly by as-cast microstructure and solutionizing time. Yield strength is much less sensitive to these parameters. Elongation to failure can be improved significantly by extended solution treatment for regions with fine dendrite arm spacing and is little changed for regions with coarse dendrite arm spacing. For Al-Si-Mg alloys tensile strength increases only moderately for extended solution treatment and fine dendrite arm spacing. Elongation to failure can be increased significantly by extended heat treatment even for regions with coarse dendrite arm spacing. Empirical relations that predict, quantitatively, tensile strength as functions of dendrite arm spacing, volume fraction porosity, and solutionizing time for Al-Si-Cu-Mg using horizontal end chill plate castings are reliable. Relations determined for Al-Si-Mg alloys for smaller end chill vertical plate castings are less reliable predictors of tensile strength.

References

1. M.C. Flemings et al, "Rigging Design of a Typical High Strength, High Ductility Aluminum Casting," *Trans. AFS*, 65 (1957) 550-555.
2. E.M. Passmore et al, "Fundamental Studies on Effect of Solution Treatment, Iron Content, and Chilling of Sand Cast Aluminum-Copper," *Trans. AFS*, 66 (1958) 96-104.
3. M.C. Flemings, *Solidification Processing* (New York, NY: McGraw-Hill, 1974).
4. Z. Yao et al, "Integrated Numerical Simulation and Process Optimization for Aluminum Alloy Solutionizing," *ASM Heat Treating Society Conference Proceedings* (2005).
5. Jian Fang et al, "Empirical Model for Tensile Property Prediction in Cast and Heat Treated Al-Si-Cu-Mg Alloys," Proceedings World Foundry Conference (2006).
6. Feng Yi et al, "Simulation of Solute Redistribution during Casting and Solutionizing of Multi-phase, Multi-component Aluminum Alloys," Proceedings World Foundry Conference (2006).
7. J.E. Morral et al, "Effect of Solution Treatment on Microstructure and Properties of Aluminum Casting Alloys," (Final Report, Univ. of Connecticut to CHTE, 2004).

Acknowledgements

The authors are grateful to our sponsors: the Department of Energy (under contract DOE DE-FC36-01ID14197) and the Center for Heat Treating Excellence; to the members of the industry focus groups for these programs, chaired by Dr. Scott MacKenzie, Houghton, for the DOE project and Dr. Paul Crepeau, GM Powertrain, for the CHTE program; and to the program teams at WPI, UConn, and OSU. Important contributions to this project have been made by former UConn team members: Prof. John Morral, Dr. Dingfei Zhang and Sudhir Adibhatla.

182

DEVELOPMENT OF NOVEL Al-Si-Mg ALLOYS WITH 8 TO 17 Wt% Si AND 2 TO 4 Wt% Mg

Xiaochun Zeng[1], Makhlouf M.Makhlouf[2], Sumanth Shankar[1]

[1]Light Metal Casting Research Centre (LMCRC), McMaster University, Hamilton, ON, Canada
[2]Advanced Casting research Center, Worcester Polytechnic Institute, Worcester, MA, USA

Keywords: Al-Si-Mg alloys, Alloy Development, New Al alloys

Abstract

A new family of Al alloys is being developed to obtain superior castability and properties of the cast component as compared to conventional Al-Si cast alloys. Alloys with compositional ranges varying from 8 wt% to 17 wt% Si coupled with 2 wt% to 4 wt% Mg are being investigated. These alloys were developed based on our understanding of the effect of various iron containing phases and Sr additions on the nucleation of the eutectic phases in Al-Si hypoeutectic alloys. Higher than conventional levels of Mg was added for two reasons. Firstly, to prevent the formation of β(Al,Fe,Si) intermetallic phase by promoting the formation of π(Al,Fe,Si,Mg) phase during solidification, and secondly, to alter the rheological properties of the alloy leading to better fluidity and mould filling characteristics of these alloys. In this paper, the mechanical properties and casting behavior of a few compositions in the above-mentioned ranges will be presented.

Introduction

Al-Si alloys have been in the forefront of automotive and aerospace casting research for several years due to their high strength to weight ratio and easy of castability. In recent years, there has been increasing research and development activities to adopt Al alloys to produce high integrity cast components for automotive applications resulting in rapid advances in technologies to develop novel Al alloy compositions and casting processes. The most popular among the Al alloys used for automotive cast components is the Al-Si hypoeutectic alloys. Most Al-Si cast alloys available today were developed three to four decades ago and have not gone though any major revisions in chemistry. In order to develop better Al alloys it is imperative to fully understand the evolution (nucleation and growth) of the various phases formed during solidification. Recently, a detailed understanding of the mechanisms of nucleation of eutectic phases in Al-Si hypoeutectic alloys and its morphological modification by Sr additions [1,2] has created new avenues to develop novel Al-Si alloys with higher than usual additions of Mg. It was observed that all commercial Al-Si alloys are in effect ternary Al-Si-Fe alloys. It was found that as low as 0.0038 wt% of Fe in Al-Si binary alloys can result in a ternary eutectic structure consisting of α-Al + Si + β (Al,Fe,Si) phases [1]. Moreover, the critical step in the nucleation mechanism of unmodified Al-Si hypoeutectic alloys is the mechanism of nucleation of eutectic Si on β (Al,Fe,Si) intermetallic phase (not on the α-Al phase) [1]. Moreover, in alloys with less than 0.0030 wt% Fe, the morphology of the eutectic phases was well refined (modified) [3]. In the Sr modified Al-Si alloy, the morphological change of the eutectic phases is mainly due to alteration in the nucleation mechanism. It was shown that Sr additions to Al-Si melts not only changes the eutectic Si morphology from a plate like to fibrous morphology, but also refines the eutectic Al grains by over an order of magnitude [2]. It was proposed [2] that the critical step in this mechanism is the change in rheological properties, apparent viscosity and interface surface energies of the inter-dendritic liquid that initiated an alternate nucleation mechanism resulting in the modification of the morphologies of the eutectic phases.

Based on this understanding, it is apparent that the objective is to avoid the formation of β(Al,Fe,Si) phase during solidification of Al-Si alloys and to add a potent surfactant to amply alter the rheological properties of the Al-Si melt in order to obtain a refined morphology of the eutectic phases. It was found that addition of Mg to Al-Si alloys at levels between 2 to 3.5 wt % help achieve both these objectives. In this publication we present the results of the development of new Al – Si – Mg alloy family with 8 to 17 wt% Si and 2 to 4 wt% Mg. Alloy phase diagrams, thermal analyses during solidification, typical microstructures and results of castability studies for these alloys are presented in this publication.

Materials and Procedure

Alloys

All the alloys used in this study were prepared from commercial purity Al, Al – 50 wt% Si master alloys and Al – 50 wt% Mg master alloy. The iron content in the resultant alloys had nominal values of 0.2 wt%. All alloy compositions are nominal compositions. The actual measured compositions varied by ±3% for Si, ±7% for Mg and ±12% for Fe levels. All alloys were prepared in freshly coated ceramic crucibles. Unless otherwise mentioned, the microstructures shown in this publications were obtained from melts that were allowed to solidify in a 200 g crucible at about 50 °C/min.

Thermal Analysis

Thermal analysis during solidification was performed by inserting an exposed K-type thermocouple at the center of a 200-gram ceramic crucible with melt solidifying at about 50 °C/min. Data acquisition was done by means of a SCXI – 1100 system from National Instruments© fitted with a SCXI – 1303 module. Labview™ 8 was used as data acquisition software with a DAQ-mx interface.

Tensile Properties Measurement

Standard ATSM B 557 tensile test bar specimen were cast in a quiescent bottom filled cast iron mould cavity. Figure 1 shows a schematic of the the test bar. The tensile test was performed in an Instron™ 5500 universal tester with a 50 kN load cell.

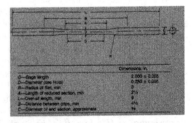

Microscopy

A Nikon Epiphot™ metallograph was used for optical microscopy. Samples from the thermal analysis and the tensile test (at gage) were metallographically analyzed after standard mounting, grinding and polishing procedures.

Figure 1: Schematic of the tensile test specimen.

Alloy Phase Diagrams

All the alloy phase diagrams in this publication were simulated in Pandat® with Pan-Al® database. Figure 2 shows a plot of temperature (T) versus wt % Mg isopleths of a typical Al-Si-Mg-Fe phase diagram. The iron content in all the phase diagrams was 0.2 wt %. Figures 2 (a), (b) and (c) shows T vs wt %Mg for 10, 13.5 and 15 wt % Si, respectively. Figure 3 shows the T versus wt%Si isopleths of the Al-Si-Mg-Fe phase diagram. The iron content is at 0.2 wt% and Figures 3 (a) and (b) show the two isopleths at 2.5 wt% and 3.5 wt% Mg in the alloy, respectively. Figure 4 shows a T versus wt%Fe isopleths of the Al-Si-Mg-Fe phase diagram. The alloy shown in Figure 4 has 15 wt%Si and 3.5 wt%Mg.

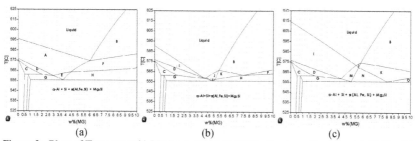

(a)   (b)   (c)

Figure 2: Plots of T versus wt% Mg isopleths of Al-Si-Mg-Fe phase diagram. (a) Al-10wt% Si-0.2wt% Fe with increasing levels of wt% Mg, (b) Al-13.5wt%Si-0.2wt%Fe with increasing levels of wt% Mg, and (c) Al-15wt%Si-0.2wt%Fe with increasing levels of wt% Mg.

Table 1 presents the nomenclature of phases in all the phase diagrams in this publication.

Table 1: Nomenclature for phases in all the phase diagram in this publication.

Notation	Phases
A	Liquid + Al
B	Liquid + Mg₂Si
C	Liquid + Al+ Si+ β(Al,Fe,Si)
D	Liquid + Al + Si
E	Liquid + Al + π(Al,Fe,Si,Mg)
F	Liquid + Al + Mg₂Si
G	Liquid + Al + Si + π(Al,Fe,Si,Mg)
H	Liquid + Mg₂Si + Al + π(Al,Fe,Si,Mg)
I	Liquid + Si
J	Liquid + π(Al,Fe,Si,Mg)
K	Liquid + Mg₂Si + π(Al,Fe,Si,Mg)
M	Liquid + Si + π(Al,Fe,Si,Mg)
N	Liquid + Mg₂Si + Si + π(Al,Fe,Si,Mg)
O	Liquid + Mg₂Si + Al + π(Al,Fe,Si,Mg)
P	Liqud + Al + β(Al,Fe,Si)
Q	Liqud + Al + β(Al,Fe,Si) + Mg₂Si
R	Liquid + Al + Si + Mg₂Si

It can be seen from Figure 2 that between 0 to 2 wt% Mg the β(Al,Fe,Si) intermetallic phase forms as a solidifying phase and when the Mg content increases beyond 2 wt%, the β(Al,Fe,Si) phase changes to the π(Al,Fe,Si,Mg) phase. In Al-10 wt%Si-0.2wt%Fe alloys (Figure 2 (a)), between 2 and 3 wt% Mg the first solidifying phase is Al followed by Si.

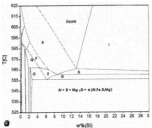 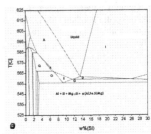

Figure 3: Plots of T versus wt% Si isopleths of Al-Si-Mg-Fe phase diagram. (a) Al-2.5wt% Mg-0.2wt% Fe with increasing levels of wt% Si, (b) Al-3.5wt%Mg-0.2wt%Fe with increasing levels of wt% Si.

In alloys with greater than 3 wt%Mg, the second solidifying phase is π(Al,Fe,Si,Mg) and subsequently Mg₂Si. Hence, in Al-10 wt% Si-0.2wt%Fe alloys the composition of Mg was varied between 2 wt% and 3 wt%. As the wt%Si increases in the alloy the level of Mg addition can increase as shown in Figure 2 (b). In Al-15 wt%Si-0.2wt%Fe alloys (Figure 2 (c)), the optimum level of wt%Mg addition is between 2wt% and 4wt%. At these levels, the first solidifying phase is Si and the second is Al.

Figure 3 shows that the β(Al, Fe, Si) phase forms until about 8 wt% Si. Figure 3 also shows that between 8 wt% and about 13.5 wt% Si the first and second solidifying phases are Al and Si, respectively and in alloys with more than 13.5wt% Si, the first and second solidifying phases are Si and Al, respectively.

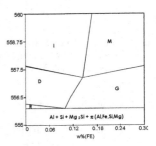

Figure 4: Plots of T versus wt% Fe isopleths of Al-Si-Mg-Fe phase diagram. The alloy here is Al-15 wt%Si-3.5wt% Mg with increasing levels of wt% Fe.

Microstructure

Figure 5 shows typical microstructures of Al-10wt%Si-0.2wt%Fe alloy with increasing levels of Mg from 0 to 4 wt%. As predicted in Figure 2 (a), Figure 5 shows that the Al-10wt%Si alloys show a modified morphology of the eutectic phases at 2.8wt%Mg (which lie between 2 and 3 wt% Mg). Alloys outside this range of Mg content do not show any morphological modification of the eutectic phases. In alloys with less than 2 wt% Mg the eutectic Si phase may have been nucleated by the β(Al,Fe,Si) phase and in alloys with greater than 3 wt% Mg the π(Al,Fe,Si,Mg) and the Mg₂Si phases may precipitate after Al during solidification and aid further nucleation of the other eutectic phases. A similar result was observed in Al-15wt%Si-0.2wt%Fe samples, where the morphology of the eutectic phases were modified when the Mg content was between 2 to 4.2 wt% as predicted by Figure 2 (c).

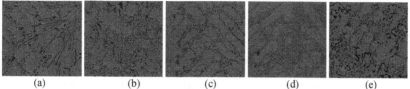

(a) (b) (c) (d) (e)

Figure 5: Typical microstructures of Al-10wt%Si samples cooled in a crucible at 50 °C/min. (a) to (e) represents alloys with 0wt% Mg, 1wt%Mg, 1.9wt%Mg, 2.8wt%Mg and 4wt%Mg, respectively.

Figures 6 (a) and (b) show a microstructure comparison of Al-10.5wt%Si-2.5 wt%Mg alloys with 0.06 wt%Fe and 0.2 wt%Fe, respectively. As predicted in Figure 4, Figure 6 shows that the alloy with 0.06 wt% Fe has the Mg$_2$Si phase solidified before the eutectic phases (Mg$_2$Si appears as the black phases (Chinese script) in Figure 6 (a)) and this phase is absent in alloys with 0.2 wt%Fe. Hence, most castability and property studies were performed with 0.2 wt%Fe to avoid the detrimental effects of Mg$_2$Si as a solidifying phase.

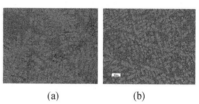

(a) (b)

Figure 6: Typical optical images of microstructures of Al-10.5wt%Si-2.5 wt%Mg with (a) 0.06 wt%Fe and (b) 0.2 wt%Fe.

Figure 7 shows a typical microstructure of Al-14.5wt%Si-2.6wt%Mg-0.2wt%Fe at three different magnifications. Although, the phase diagram in Figure 2(c) predicts that Si is the first solidifying phase, the Si phase was not observed until the Si levels reached about 15.4 wt% as shown in Figure 8, where the Si phase appear as small spots of primary Si.

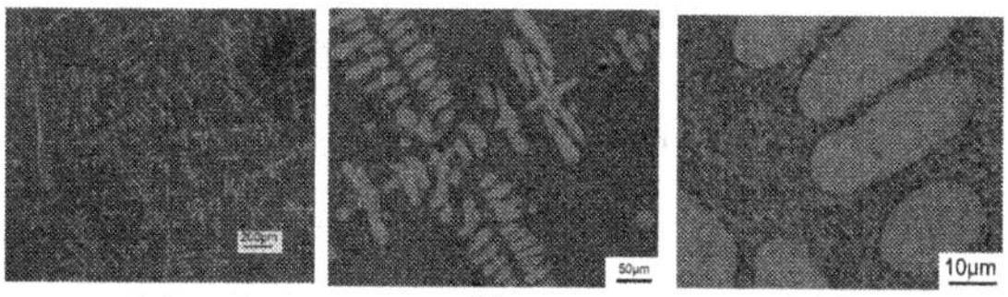

(a) (b) (c)

Figure 7: Typical microstructure of Al-14.5 wt%Si-2.6 wt%Mg-0.2 wt%Fe. (a) to (c) shows increasing magnification of microstructure taken in an optical microscope.

Pandat® software predicts that the volume fraction of Si in Al-15wt%Si-3wt%Mg-0.2wt%Fe before the second solidifying phase (Al) evolves is less than 0.2 %. This may be the reason for the absence of Si phase in the alloys with 13.5 to 15wt% Si.

Thermal Analysis

Figures 9 (a) and (b) show typical temperature versus time thermal curves obtained during solidification of Al-10wt%Si-2.3wt%Mg-0.2wt%Fe and Al-15wt%Si-3.2wt%Mg-0.2wt%Fe

Figure 8: Typical microstructure of Al-15.4wt%Si-3.2wt%Mg-0.2wt%Fe. The tiny dark spots in the microstructure are the primary

alloys, respectively. The plots in Figure 9 also feature the derivative curve for the temperature, which helps in identifying the nucleation events of the solidifying phases. In Figure 9 (a), the nucleation event of the first solidifying phase, Al (588 °C) was observed as predicted in Figure 2 (a). However, the nucleation event of Si (around 568 °C) and that of π(Al,Fe,Si,Mg) phase (around 558 °C) (as predicted by Figure 2 (a)) is not observed. Since Si does not nucleate on primary Al [1] and the absence of β(Al,Fe,Si) phase resulted in a suppression of the evolution of the Si phase and resulted in evolution of highly refined and morphologically modified eutectic phases. In Figure 9 (b), the nucleation of the Al phase (558 °C) and the eutectic phases (555 °C) are at temperatures predicted by the phase diagram in Figure 2 (c). However, the nucleation event of the primary Si phase was not present in any of the thermal curves for this alloy. The explanation for this phenomenon is still unknown and currently under investigation. Typical microstructure of the sample with thermal analysis shown in Figure 9 (a) is shown in Figure 5 (d) and that of the sample with thermal analysis shown in Figure 9 (b) is shown in Figure 7.

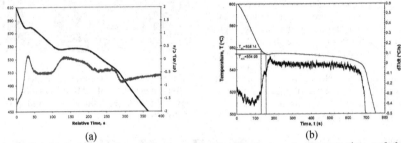

(a)   (b)

Figure 9: Typical thermal analysis curves showing the temperature versus time and the derivative of the temperature versus time. (a) Al-10wt%Si-2.3wt%Mg-0.2wt%Fe, and (b) Al-15wt%Si-3.2wt%Mg-0.2wt%Fe.

Castability

The castability tests consisted of the following studies:

> ➢ Dross Formation Assessment
> ➢ Fluidity Measurements
> ➢ Mechanical Properties Measurements
> ➢ Hardness Measurements (Natural Ageing)

<u>Dross Formation Assessment</u>

Al-10wt%Si-2.6wt%Mg-0.2wt%Fe, Al-15wt%Si-3.2wt%Mg-0.2wt%Fe and A356 (Al-7.5wt%Si-0.3wt%Mg-0.06wt%Fe-0.02wt%Sr) were separately melted in an electric furnace and held for about 10 hours each at 730 °C. The dross formed on the surface of the melt was removed and collected by skimming every one hour and the total dross was weighed. The initial amount of each melt was maintained at 40 lb (18.18 kg). Table 2 shows results of these experiments.

Table 2 shows that the amount of dross formed during holding the new alloys is comparable to that obtained from standard A356 alloy.

188

Table 2: Dross Formation study on 40 lb alloys held at 730 °C for 10 hours

Alloy	Dross (lb) after 10 hours
Al-10wt%Si-2.6wt%Mg-0.2wt%Fe,	1.7
Al-15wt%Si-3.2wt%Mg-0.2wt%Fe	1.8
A356	1.8

Fluidity Measurements

The fluidity tests were performed in standard N-Tec® fluidity moulds. Figure 10 shows the results of the fluidity tests for Al-15wt%Si-3.2wt%Mg-0.2wt%Fe and A356 alloys. In Figure 10, the more fingers the metal fills the better is the fluidity. Figure 10 shows that the pouring temperature of the Al-15wt%Si-3.2wt%Mg-0.2wt%Fe alloy was lower than that of A356 by over a 100 °F and yet the fluidity was greater than that ofA356.

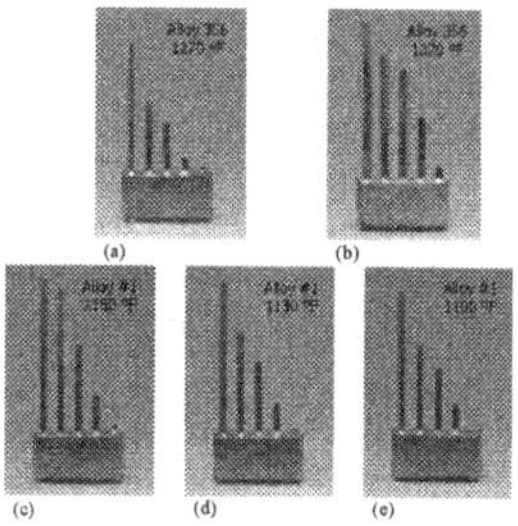

Mechanical Properties Measurements

The alloys cast were Al-10wt%Si-2.6wt%Mg-0.2wt%Fe, Al-15wt%Si-3.2wt%Mg-0.2wt%Fe and A356. The pouring temperature for Al-10wt%Si-2.6wt%Mg-0.2wt%Fe was 700 °C, Al-15wt%Si-3.2wt%Mg-0.2wt%Fe was 630 °C and A356 was 735 °C. The alloys were degassed using a rotary impeller at 400 rpm and Ar gas at 4 L/min for 30 min. The tensile tests were performed 24 hours after casting the test bars. 10 bars of each alloy were used for the tests. Table 3 shows the average results of the tensile tests. Table 3 shows that the new alloys show superior as-cast mechanical properties when compared to as-cast A356. Heat treatment procedures and mechanical properties of heat treated samples are currently being investigated. Figure 11 shows typical microstructure of the three alloys shown in Table 3 for samples taken from the center of the gage in the tensile bar.

Figure 10: Fluidity tests results. (a) and (b) A356. (c) to (e) Al-15wt%Si-3.2wt%Mg-0.2wt%Fe alloy. The pouring temperatures are shown in each figure.

Table 3: As-cast mechanical properties of the alloys. Ten bars were broken for each alloy.

Alloy	Ultimate Tensile Strength	Yield Strength	Elongation
Al-10wt%Si-2.6wt%Mg-0.2wt%Fe	33 ksi (228 MPa)	22 ksi (152 MPa)	8%
Al-15wt%Si-3.2wt%Mg-0.2wt%Fe	33 ksi (228 MPa)	11 ksi (76 MPa)	13%
A356	28 ksi (193 MPa)	15 ksi (103 MPa)	5%

Hardness Measurements

Due to the extraordinarily high amounts of Mg added to the alloy, the natural ageing tendencies of the alloy (hardness measurements) were evaluated on the tensile test bars. Hardness samples were taken

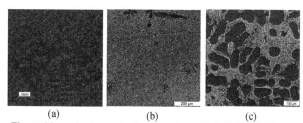

(a)   (b)   (c)

Figure 11: typical as cast microstructure of (a) Al-10wt%Si-2.6wt%Mg-0.2wt%Fe, (b) Al-15wt%Si-3.2wt%Mg-0.2wt%Fe and (c) A356.

from the gage of the tensile test bar and the thickness of the cylindrical sample was about 12 mm. Figure 12 shows a plot of the hardness data for the Al-10wt%Si-2.6wt%Mg-0.2wt%Fe and Al-15wt%Si-3.2wt%Mg-0.2wt%Fe samples taken over a long period of time. Six hardness measurements were performed every 24 hours for both alloys. Figure 12 shows that the alloys did not show any appreciable change in hardness, which may be due to the lack of any natural ageing phenomenon.

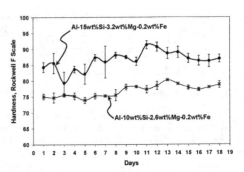

Figure 12: Hardness measurements.

Conclusions

- The proposed theories on nucleation of eutectic Si on β(Al,Fe,Si) and the effect of Sr additions to the inter dendritic liquid in Al-Si hypoeutectic alloys [1-3] have been verified.
- The addition of 2 to 4 wt% Mg to Al-Si alloys changes the β(Al,Fe,Si) to π(Al,Fe,Si,Mg) and also changes the rheological properties of the melt favorable to result in a highly modified morphology of the eutectic phases.
- Based on phase diagram simulation, thermal analysis, microstructure analysis and mechanical property tests, new casting alloys with Al, 8 to 17 wt% Si, 2 to 4wt% Mg and 0.2 wt% Fe have been developed with superior castability and as-cast mechanical properties

Summary

In this publication, we have presented preliminary efforts to design new Al alloys Currently, work is going on to better define the compositional ranges of Si, Mg and Fe in these alloys along with investigations on design of both primary and secondary alloys. Further, trials to cast components using these alloys via various processes such as high pressure die casting, low pressure casting, gravity casting and sand casting are underway. A few phenomena such as the lack of evolution of primary Si in the alloy with 15wt% Si, and the extraordinarily fine microstructure of the eutectic phases in these alloys are being investigated.

Acknowledgements

Xiaochun Zeng and Sumanth Shankar will like to express their gratitude to Orlick Industries, Hamilton, ON, Canada, Burlington Technologies, Inc., Burlington, ON, Canada and Ontario Centre of Excellence, Mississauga, ON, Canada for their continued support in this project. Makhlouf M. Makhlouf expresses his gratitude to the member companies of the Advanced Casting Research Center, WPI, Worcester, MA, USA for their continued support in this project.

References

1. Shankar S, Riddle Y and Makhlouf M. M, *Acta Materialia*, 2004, v52, n15, p4447.
2. Makhlouf M. M, Shankar S and Riddle Y, AFS Transactions, Paper 05-088(02), April 2005.
3. Shankar S, Riddle Y and Makhlouf M. M, Metall. Mater. Trans. A, June 2005, v36A, p1613.

SHAPE CASTING:
2nd International Symposium

Modeling

Session Chairs:
Mark R. Jolly
Jacob Zindel

Shape Casting: The 2nd International Symposium *Edited by Paul N. Crepeau, Murat Tiryakioğlu and John Campbell*
TMS (The Minerals, Metals & Materials Society), 2007

RATIONALIZATION OF MATERIAL PROPERTIES FOR
STRUCTURAL AND DURABILITY MODELING OF CASTINGS

Paul N. Crepeau

General Motors Corporation
895 Joslyn Road, MC/483-710-270
Pontiac, MI 48340 USA

Keywords: Material properties, modeling, aluminum, cast iron

Abstract

Finite element analysis models crucially depend on mechanical properties in constituitive equations for stress-strain fields and fatigue damage. Choosing appropriate values of these properties is most difficult as knowledge of properties must be inferred from prior testing, which may not represent the future. Part geometry and availability often limit sampling. Billets and coupons are more conveniently sized, but measured properties must correlate back to parts. Properties vary due to process- and alloy-sensitive gradients in microstructure. Moreover, properties are random variables exhibiting mean and standard deviation estimated with significant uncertainty. This paper will discuss these and other issues in determining mechanical properties and in selection of values of mechanical properties for structural and durability analysis of cast automotive engine components.

Introduction

Finite element analysis (FEA) modeling[1] necessarily requires many simplifying assumptions the first of which is that the mesh reproduces the geometry with enough refinement to provide accurate estimations of behavior and the last of which is a short list of properties to be fed into constituitive equations describing material behavior. Some assumptions have low impact on final results or can be tested. For example, mesh geometry can be checked against the original part model; mesh fineness can be tested against results for convergence.

Boundary conditions such as imposed loads or constraints are often the result of measurements and other FEA modeling. In modeling cylinder blocks in automotive engines a multitude of sources of boundary conditions exist including:
- Cast iron cylinder liners (in aluminum blocks) that exert stress through differential thermal contraction
- Head gasket(s) with raised embossments
- Bolts connecting the cylinder head to the block
- Bolts connecting bearing caps or bedplates to the crankcase
- Bearing caps
- Crankshaft and piston assemblies that exert time-dependent loads due to firing and inertia

Usually residual stress imparted from the manufacturing process is assumed to be zero. This important subject will not be discussed here.

---

[1] Durability analysis, a post-processing operation, will be considered here as integral to FEA.

A major set of assumptions surrounds the set of material properties to use. Unlike many of the inputs required in FEA, the mechanical properties of the part being modeled cannot be directly measured except in forensic and other post-mortem studies (which are outside the scope of this paper). Although properties may be audited on many parts and in many regions in the parts, what to use in an FEA model requires the assumption that what has been measured in the past represents what will exist in the future. Properties are very predictable if parts are simple and if they are made under made under rigorously controlled and stable processes. Unfortunately this is usually not the case in castings.

A common assumption is that a single set of properties applies to the entire part. This is clearly untrue in castings. In fact, process rules to provide a sound casting virtually guarantee that properties will vary throughout the part: In order to avoid gross shrinkage a significant temperature gradient must be established in castings so that the last metal to solidify, usually with unfed gross shrinkage, is in a non-critical location, preferably a riser [1]. Even in alloys such as gray iron that exhibit little if any solidification contraction, a thermal gradient must exist to permit feeding of liquid metal to the part as it cools from pouring temperature to solidification temperature. The temperature gradient is inseparable from local solidification time and rate which, in turn, affect most microstructural characteristics that control properties, features such as porosity, dendrite arm spacing, and both refinement and distribution of constituent phases. Further, post casting heat treatment may add gradients of hardening phases: local solution treatment response depends on distribution of soluble as-cast phases, which in turn affects local quench sensitivity; all stages of heat treatment depend on local heating and cooling, where centers of parts lag behind the surfaces. The well-known effect of microstructural characteristics on properties remains the subject of extensive study [2].

One approach is to model the properties in castings *a priori* [3,4]. The necessary tools currently exist to some degree of maturity: solidification modeling [5], rate-dependent phase formation during solidification [6], porosity formation during solidification [7,8], quench thermal transfer [9], heat treatment response [10] to name a few. Each of these separate analyses requires its own set of assumptions on various boundary conditions such as melt gas content, alloy chemistry, heat transfer coefficients, mold geometry, and the like. From the standpoint of the modeler these foundry variables are frequently unknown, uncontrolled, variable, or proprietary. Further, the current state of the art in the governing constitutive models may not be sufficient to provide the level of predictability needed to justify the expenditure of modeling resources. While this is arguably the case for cast aluminum, some commercial solidification software codes currently provide estimates of tensile properties in cast iron [11].

Thus, structural and durability FEA must rely on predetermined properties. While the choice of what properties to use is frequently difficult, it is a decision that must be made – the modeling will be conducted with whatever data are available if only to compare the relative effects of one geometry with another.

Material Property Data Issues

Extensive databases of mechanical properties exist in various forms and formats [12,13]. The task of managing mechanical property data has spawned specialized software codes to facilitate entry and retrieval of data [14,15]. However, the value of much of these data and the conventional methods of material identification and data retrieval, particularly in castings, is questionable.

It is useful to consider the relationship between a population that has been sampled, samples that have been tested, and the future state of the population, schematically depicted in Figure 1.

194

Most metallurgical testing is destructive, meaning that populations of parts can only be sampled, and that samples once tested are permanently removed from the population. Sampling populations is a subject upon itself [16]. The test results are intimately associated with the samples. (However, test error and variation add to apparent property behavior). The purpose, of course, is prediction of the properties exhibited by the future parts. This is only possible if nothing changes in the process, which is rarely true in the foundry environment. Raw materials change. Personnel change. "Continuous improvement" is an ongoing fact of life in foundries. And improvement is measured against only 2 measures: cost, which is self-explanatory, and quality (here customer service and delivery are included in quality). Quality is measured against requirements and specifications.

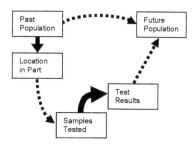

1. Samples are taken from a past population, subject to statistical error.
2. Test results are intimately related to the samples, subject to test error.
3. Both past population and test results are casually related to the future population.

Figure 1. Schematic relationship between test results and samples, and past and future populations.

Castings, however, tend to be underspecified. Fit and form, weight and dimensions, surface imperfections, and visual and radiographic porosity are readily measured, but mechanical properties are easily misunderstood and often misrepresented. Mechanical properties should be specified at a particular location; the properties measured can only apply to that location and the condition of the part when it is sampled [17]. Properties that are not specified do not figure into decisions on continuous improvement. There is no reason to believe that unspecified properties or properties in unspecified locations will not drift.

Further, the foundry result for each casting design is unique. During casting development processes are optimized. Gating is revised. Geometry is modified. All of this happens with an eye on the specifications, some of which are surpassed on the first iteration and some that require painful course changes to be just barely met. Sometimes arbitrarily high specifications are relaxed to reflect a foundry's best efforts. Thus similar parts experiencing different developmental journeys in different foundries will exhibit different properties.

In light of the above the question remains: How should mechanical property values be determined for structural and durability FEA?

Database Development

A number of questions should be answered before a material property database is populated:

- To what degree are users of the database expected to interpret the data?
- What engineering functions will be using the database?
- What are the goals of the ensuing FEA projects?
- What level of accuracy is needed in the FEA model?
- What properties will be used in the FEA model?
- What statistical representations of properties are desired?

A database that is a collection of test results permits critical interpretation of data. While this may be useful for research, each time the data are interpreted for FEA modeling a different conclusion will be found. A rigid set of rules will give consistent interpretation, but defining these rules is difficult and subject to endless exception. A far better strategy is for expert metallurgists to interpret available data using consistent guidelines and conclude on scalar or vector (i.e., UTS vs. temperature) data entries for each material property in each material condition. This permits FEA specialists to concentrate on the myriad other inputs and boundary conditions.

It is important to understand the goals of the modeling and the mission of users of the results. For example, structural analysis is commonly performed with expected average properties. It is currently unfeasible to calculate all the permutations of maxima and minima in elastic modulus, for example, to determine the distribution of stresses at each point at the finite element mesh. Hence, only averages need be provided for elastic modulus, Poisson's ratio, coefficient of thermal expansion[1], and, for elastic-plastic analysis, either stress-strain curves or strength coefficient plus strain hardening exponent. Durability modeling on the other hand requires more esoteric properties that are at the same time more difficult to measure and sensitive to processing, e.g., fatigue strength, strongly affected by porosity. Structural and durability analysis meant to determine whether geometry revision increases or decrease stresses in a part may only need broad generalizations of properties. However, if the goal is to reduce the scale of physical validation of parts or component assemblies it is important to use property values that realistically represent the future.

The inputs required by FEA software must be understood by the database developer. For example the definition Abaqus uses for yield strength is actually the proportional limit, not the more common 0.2% offset [18]. Databases commonly list UTS and YS in engineering stress while FEA codes expect true stress (although is not a large source of error). Further, the full capability of FEA codes is rarely exploited; companies will generally standardize a subset of available methods most applicable to their parts. Hence, rather than waste resources populating and maintaining unused data, it is preferable to maximize reliability and the number of material conditions in a database.

Finally, the statistical representation of the properties must be determined. As mentioned above averages are usually desired for basic linear properties. In durability analysis, the traditional approach has been to model average behavior and to seek a conservative safety factor. A more desirable approach may be to use minimum properties, but how should these defined? The UTS may be represented by a 3-parameter Weibull distribution with a minimum, but faced with the way defects distribute in castings, there is no true minimum. Thus the standard for a minimum could be the "−3sigma"[2] property. Minimum properties cannot be defined without such an arbitrary standard. There is no definition for minimum in the ISO tensile standard [20]. The definition of minimum in ASTM tensile specifications is not very helpful for establishing database minima: where a test result fails to surpass the specified minimum, retests are permitted, and if the next 2 pass, the entire lot passes[20 -21]. The result is that a minimum specified as such differs markedly from a −3sigma property limit, Figure 2.

Statistical treatment of properties is seductive, but difficult to apply in the real world. The problem, in addition to the previous discussion of predicting future states, is uncertainty in the current state – the results of test data. For example, a common sample size n=15 with standard deviation S=15 has an upper 95% confidence interval (95%CI) on S of 6.9 so S could be as high as 21.9, Figure 3. The −3sigma estimation would have more than 3 times this uncertainty given additional uncertainty in the mean. If the sample size is small, say n=5, the upper 95%

---

[1] Elastic modulus, Poisson's ratio, coefficient of thermal expansion, plus density are basic linear properties that are fortunately insensitive to process and their values are most easily populated into databases.

[2] In this text "−3sigma" will signify the result of [mean − 3(standard deviation)].

confidence interval on S could inflate a test result of 15 by 20.6 to 35.6; hence, useful variation data requires large sample sizes.

In the domain of aircraft design, however, resources are made available for extensive characterization and statistical analysis of each material used[1] [23]. This is rarely possible in other industries.

Determination of Database Entries

After the discussion above one wonders how a database is ever populated. The simple answer is that the values must be chosen arbitrarily. The choice to follow a rigorous statistical treatment is in itself arbitrary, since statistics cannot truly predict the future, and because there are rarely, if ever, sufficient test data to overcome inherent uncertainty.

The author follows the general guideline that databases should be populated with property values that are "appropriate" to each end user. This implies that even for the same material different data may be provided to different end users.

A case in point is using part specifications as a basis for database entries. A certain part may have a print specification for 10^7-cycle fatigue strength at a designated location of 70 MPa at -3sigma per GMN7152-Class 2 [17]. It is expected that the foundry will surpass the spec by a comfortable margin. The implication of the specification is, however, that if fatigue strength drifts downward the parts are usable so long as the specification is met. Thus durability analysis with the standard

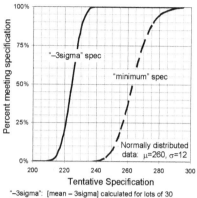

Figure 2. Derived probability of meeting 2 tentative specifications. (Previously unpublished.)

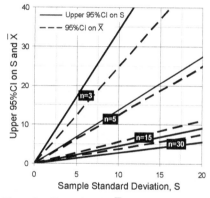

Figure 3. Uncertainty in $\bar{X}$ and S (quantified by 95% confidence interval) versus S at 4 sample sizes n. (Previously unpublished derivation from normal statistics [21].)

of "-3sigma" as the minimum must consider the fatigue strength that parts could exhibit in the future, not necessarily what they exhibit in the present. (If the parts need to have higher fatigue strength to achieve target safety factors then the specified properties should be raised.) This philosophy is taken from the MMPDS Handbook where the upper A-Basis design limit is fixed by the material property specification [23].

The single data point for 10^7-cycle fatigue strength described above is useful for more than just a finite-life fatigue analysis. Testing continues both to certify compliance with the specification

---

[1] The Metallic Material Properties Development and Standardization Handbook provides Herculean rules for statistically defining design properties for aircraft structures that understandably require large data sets [23].

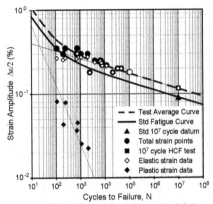

Figure 4. Strain-life test data and standard fatigue curve drawn through standard 10^7-cycle fatigue strength (as strain amplitude). (Previously unpublished.)

and to benchmark other properties in each part. A recent study on 319-F provided an average strain-life curve from which a parallel curve was drawn through the database standard 10^7-cycle fatigue strength, Figure 4. In this example σ_f' and ε_f' in the nominal strain-life curve were transformed (slopes b and c of the elastic and plastic lines were kept constant) to generate a standard property curve running through the standard 10^7-cycle fatigue strength.

However, data derived from a part specification only apply to the designated control location in the part. While some parts may have large areas specified to some material property minimum, the reality is that producers and users of castings agree where the properties are actually measured and the consequence is that the results do not necessarily apply elsewhere. Properties on some parts, iron castings in particular, may be controlled with separately cast test bars and keel blocks. While these kinds of samples may be sensitive to changes in the base metal, the actual part properties usually differ. Further, few specifications are statistical so a "minimum" is not the same as "−3sigma". In summary, quality control data are not design data.

The author has observed that when a durability analysis project is assigned the FEA specialist will complete the project with whatever data are available. The role of the metallurgist then is to provide the appropriate material property data. A painful exercise is to prepare a database in advance populated with the properties that are needed for any possible project. The author has done this in a proprietary database for cylinder heads and cylinder blocks engines made by GM Powertrain in North America. Many of the entries were based on large amounts of data; many on very little. Few parts were fully characterized with test data for fatigue strength at multiple temperatures. Properties had to be inferred from similar material conditions wherever data was lacking. Once the database was fully populated the entries having the poorest supporting data became the target of ensuing mechanical property benchmarking programs.

Property Mapping

An exciting development in FEA modeling is the capability of using different mechanical properties in each element. Solidification software codes are available that estimate average mechanical properties in cast iron as a function of parameters such as solidification rate and shake-out time [11]. As discussed previously, however, modeling may never be able to comprehend all of the foundry variables that affect properties, particularly fatigue strength, which is porosity-sensitive. Whether a sufficient subset of variables can be identified remains to be seen.

On the other hand, much as the shape of the strain-life curve from Figure 4 can be used to propagate given fatigue strength to other lifetimes, the gradients in the parameter result fields from solidification analysis can be applied to propagate given properties throughout a part. One parameter, solidification rate (or its counterpart solidification time) is fundamentally related to

198

dendrite arm spacing (DAS), a measurable microstructural feature [24] widely reported to correlate with mechanical properties in cast aluminum [25]. Micro-structural fineness and notably porosity also correlate with solidification rate [26] making DAS an interesting candidate for this exercise. While porosity, a predominant factor in fatigue [27], can vary independently of DAS, the field of porosity concentration may correlate loosely with predicted DAS, Figure 5. Hence, a field of predicted DAS can also serve as the field of mechanical properties; what is missing is derivation of appropriate scaling factors. The approach promulgated here is to associate 2 DAS values with 2 property levels at extremities of the part. This permits self-consistent interpolation of standard properties, for example, a chilled location and a slowly cooled location throughout the part, and facilitates usage of standard properties

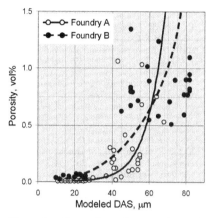

Figure 5. Comparison of measured porosity versus predicted DAS in a cylinder block made at 2 foundries. The crankcase was chilled and the head bolt bosses, risered. (Previously unpublished.)

regardless of their origin [28]. This approach to property mapping is currently under investigation in a durability software code [29].

Summary and Conclusions

In this paper issues surrounding the developing material property databases for structural and durability analysis were discussed and some solutions were provided.

Castings exhibit gradients of properties due to inherent design rules needed to avoid shrinkage porosity. Castings are also plagued by myriad conditions that effect properties, circumstances that cloud the predictability of benchmark testing to establish design properties.

Building a database that contains the appropriate material properties for structural and durability analysis requires establishment of entries based, at times, on little or non-existent data. Even background data existing in abundance may offer limited predictability of future conditions. However, material properties must be made available by one means or another so that modeling can proceed. Once tabulated the properties listed in a database can then be verified though subsequent testing.

Prediction of properties *a priori* holds some promise for the future but must overcome enormous stumbling blocks involving basic modeling inputs. A simpler and currently feasible approach of property mapping was presented whereby simplified process modeling is used to propagate given properties from given locations to other regions in a part.

Acknowledgments

The author would like to thank Richard Osborne and Mark Osborne (no relation) for helpful comments in the preparation of this manuscript.

References

1. J. Campbell, *Castings, 2ed*, Elsevier Butterworth-Heinemann, Oxford, UK, 2004.
2. J.Z. Yi, Y.X Gao, PD. Lee, T.C. Lindley, "Microstructure-Based Fatigue Life Prediction for Cast A356-T6 Aluminum-Silicon Alloys," *Met. Trans. B*, **37B** (Apr. 2006) 301-311.
3. J. Allison, "Linking Design, Manufacturing and Materials," September 22, 2005, TMS/SAE Webcast.
4. D.M. Maijer, Y.X. Gao, P.D. Lee, T.C. Lindley, T. Fukui, "A Through-Process Model of an A356 Brake Caliper for Fatigue Life Prediction," *Met. Trans. A*, **35A** (2004)3275-3288.
5. Jolly, M, "Casting simulation: How well do reality and virtual casting match. State of the art review," *Int. J. Cast Metals Res.*, **14**(2002)303-313.
6. X.M. Pan, C. Lin, J. E. Moral, H.D. Brody, "An Assessment of Thermodynamic Data for the Liquid Phase in the Al-Rich Corner of the Al-Cu-Si System and Its Application to the Solidification of a 319 Alloy," *J. Phase Equil. and Diff.*, **26**(2005)225-233.
7. L. Lu, K Nogita, S.D. McDonald, A.K. Dahle, "Eutectic Solidification and its Role in Casting Porosity Formation," *JOM*, **56**(2004)52-58.
8. J.D. Zhu, S.L. Cockcroft, D.M. Maijer, "Modeling of Microporosity Formation in A356 Aluminum Alloy Casting," *Met. Trans. A.*, **37A** (2006) 1075-1085.
9. O. Macchion, S. Zahrai, J.W. Bouwman, "Heat Transfer From Typical Loads Within Gas Quenching Furnace," *J. Matl. Proc. Tech.*, **172**(2006)356-362.
10. S.C. Weakley-Bollin, W. Donlon, C. Wolverton, J.W. Jones, and J.E. Allison, "Modeling the Age-Hardening Behavior of Al-Si-Cu Alloys", *Met. Trans. A*, **35A**(2004)2407-2417.
11. C. Heisser, L. Carmack, "Improvement of Mechanical Properties of Ductile Iron Castings Using New Casting Process Optimization Tool," *AFS Trans.*, **113**(2005)581-586.
12. D. Cebon, M.F. Ashby, "Information Systems For Material And Process Selection," *Adv. Matl. Proc.*, **157**(2000)44-48.
13. ASM International, *Material Property Data Online*.
14. "MSC/MVISION Material Database," The MacNeal-Schwendler Corp., Los Angeles, CA, 2002.
15. J.G. Kaufman, E.F. Begley, "A Data Interchange Markup Language," *Adv. Mat. Proc.*, Nov (2003)35-36.
16. W.G. Cochran, *Sampling Techniques, 3ed.*, Wiley, New York, NY, USA, 1977.
17. "GMN7152: Specification and Verification of Tensile and Fatigue Properties in Cast Components," *General Motors Engineering Standards*, 2001.
18. *Analyists User's Manual, Vol III: Materials*, Abaqus, Inc., Providence, USA, 2004, p. 11.1.1-7.
19. "ASTM E8-04: Test Methods of Tension Testing of Metallic Materials," *Annual Book of ASTM Standards*, ASTM International, West Conshohocken, PA, USA, 2004.
20. "ASTM B557-06: Tension Testing Wrought and Cast Aluminum- and Magnesium-Alloy Products," *Annual Book of ASTM Standards*, ASTM International, West Conshohocken, PA, USA, 2006.
21. W.H. Hines, D.C. Montgomery, *Probability and Statistics in Engineering and Management Science*, *2ed.*, Wiley, New York, NY, USA, 1980.
22. "ISO 6892: Metallic Materials – Tensile Testing at Ambient Temperature," International Organization for Standardization, Geneva, Switzerland, 1998.
23. R.C. Rice, J.L Jackson, J.G. Bakuckas, S.R. Thompson, *Metallic Material Properties Development and Standardization Handbook*, U.S .Department of Transportation, Federal Aviation Administration, report no. DOT/FAA/AR-MMPDS-01, Jan, 2003.
24. R.E. Spear, G.R. Gardner, "Dendrite Cell Size", *AFS Trans.*, **71**(1963)209-215.
25. Y.X. Gao, J.Z. Yi, P.D. Lee, T.C. Lindley, "A Micro-Cell Model of the Effect of Microstructure and Defects on Fatigue Resistance in Cast Aluminum Alloys," *Acta Mater.*, **52**(2004)5435-5449.
26. Q.T. Fang, D.A. Granger, "Porosity Formation in Modified and Unmodified A356 Alloy Castings" *AFS Trans.*, **97**(1990)989-1000.
27. D.L. McDowell, K. Gall, M.F. Horstemeyer, J. Fan, "Microstructure-Based Fatigue Modeling of Cast A356-T6 Alloy," *Eng. Frac. Mech.*, **70**(2003)49-80.
28. P.N. Crepeau, B.J. McClory, "Method for Estimation of Mechanical Properties," *Research Disclosure*, publ. no. 488044, Oct. 2005.
29. P.N. Crepeau, "Proposal for Property Mapping," presentation at Fe-Safe User Group Meeting, Ottawa Lake, MI, USA, Sept. 19, 2006.

Shape Casting: The 2nd International Symposium *Edited by Paul N. Crepeau, Murat Tiryakioğlu and John Campbell*
TMS (The Minerals, Metals & Materials Society), 2007

MODELING THE ONSET AND EVOLUTION OF HYDROGEN PORES DURING SOLIDIFICATION

Sergio Felicelli, Claudio Pita, Enrique Escobar de Obaldia

Department of Mechanical Engineering and Center for Advanced Vehicular Systems
Mississippi State University, Mississippi State, MS 39762, USA

Keywords: Solidification modeling, porosity, aluminum A356

Abstract

A quantitative prediction of the amount of gas microporosity in alloy castings is performed with a continuum model of dendritic solidification. The distribution of the number and size of pores is calculated from a set of conservation equations that solves the transport phenomena during solidification at the macro-scale and the hydrogen diffusion into the pores at the micro-scale. A technique based on a pseudo alloy solute which is transported by the melt is used to determine the potential sites of pore growth, subject to considerations of mechanical and thermodynamic equilibrium. Two critical model parameters are the initial concentration of the pseudo solute and the initial size at which pores start to grow. The dependence of the model prediction on these parameters is analyzed for aluminum A356 plate castings and the results are compared with published experimental data.

Introduction

Although porosity occurring during alloy solidification is very harmful to the mechanical properties of alloys, the control or elimination of this defect is still a formidable task in modern foundries. The formation of microporosity in particular is known to be one of the primary detrimental factors controlling fatigue lifetime and total elongation in cast aluminum components [1]. Microporosity refers to pores which range in size from micrometers to hundreds of micrometers and are constrained to occupy the interdendritic spaces near the end of solidification. The micropores can form due to microshrinkage produced by the pressure drop of interdendritic flow or because of the presence of dissolved gaseous elements in the liquid alloy. In this work, we focus on the calculation of gas microporosity and, unless noted otherwise, we use the term porosity to mean gas-induced microporosity. In the case of aluminum alloys, hydrogen is the most active gaseous element leading to gas porosity [2]. Many efforts have been devoted to the modeling of porosity formation in the last 20 years, particularly in aluminum alloys [3-8] and, in a lesser degree, to nickel superalloys [2,9] and steels [10,11]. More recently, rather sophisticated models have been developed to include the effect of pores on fluid flow (three-phase transport) [12], multiscale frameworks that consider the impingement of pores on the microstructure [13], and new mechanisms of pore formation based on entrainment of oxide bifilms [14]. A recent review on the subject of computer simulation of porosity and shrinkage related defects has been published by Stefanescu [15].

Mathematical Model

The model presented here is based on a robust and well-tested multicomponent solidification program which calculates macrosegregation during solidification of a dendritic alloy with many solutes [16]. The model (named MULTIA) solves the conservation equations of mass, momentum, energy, and each alloy component within a continuum framework in which the mushy zone is treated as a porous medium of variable permeability. In order to predict whether microporosity forms, the solidification shrinkage due to different phase densities, the concentration of gas-forming elements and their redistribution by transport during solidification were later added to the model [2]. In this form, the model was able to predict regions of possible formation of porosity by comparing the Sievert's pressure with the local pressure, but it lacked the capability of calculating the amount of porosity. This model has already been presented in detail in Refs. [2] and [16] and references therein, and only the main assumptions and governing equations are presented here. In particular, we discuss a method that adds to MULTIA the capability of calculating the volume fraction of porosity and the pore size distribution during solidification. In this method, the growth 6pores is simulated with a micro - scale growth model that is coupled to the macro-scale governing equations in MULTIA. The criterion for the formation of pores is based on equilibrium conditions between the pores and the alloy and on the transport of a pseudo-solute that represents inclusions or impurities dissolved in the alloy. The pore volume fraction as well as the pore size distribution can be determined from the evolution of the population and size of pores during solidification.

Governing Equations

The following assumptions are invoked: the liquid is Newtonian and the flow is laminar; the Boussinesq approximation is made in the buoyancy term of the momentum equation; the solid phase is stationary; the gas phase does not affect the transport equations (twophase model); and the densities of solid (ρ_s) and liquid (ρ_l) are different but constant. Additional assumptions are made at appropriate places in the article. With these assumptions, the conservation equations can be written as [2]:

Mass and momentum:

$$\nabla \cdot \mathbf{u} = \beta \frac{\partial g_l}{\partial t} \tag{1}$$

$$g_l \frac{\partial}{\partial t}\left(\frac{\mathbf{u}}{g_l}\right) + \mathbf{u} \cdot \nabla \left(\frac{\mathbf{u}}{g_l}\right) = -\frac{g_l}{\rho_l}\nabla p + \frac{\mu}{\rho_l}\nabla^2 \mathbf{u} - \frac{\mu}{\rho_l}\frac{g_l}{\mathbf{K}}\mathbf{u} + \frac{\mu\beta}{3\rho_l}\nabla\left(\frac{\partial g_l}{\partial t}\right) + \frac{\rho g_l}{\rho_l}\mathbf{g} \tag{2}$$

Energy:

$$\overline{\rho c}\frac{\partial T}{\partial t} + \rho_l c_l \mathbf{u} \cdot \nabla T = \nabla \cdot \overline{\kappa}\nabla T - \rho_s \left[L + \left(c_l - c_s\right)\left(T - T^H\right)\right]\frac{\partial g_l}{\partial t} \tag{3}$$

Solutes:

$$\frac{\partial \overline{\rho C}}{\partial t} + \rho_l \mathbf{u} \cdot \nabla C_l = \nabla \cdot \overline{\rho D \nabla C} - \beta \rho_l \frac{\partial g_l}{\partial t} C_l \tag{4}$$

In the above equations, $\mathbf{u}$ is the superficial velocity, g_l is the volume fraction of liquid, t is time, ρ is density, β is the shrinkage coefficient $\beta = \left(\rho_s - \rho_l\right)/\rho_l$, p is pressure, μ is viscosity, $\mathbf{g}$ is gravity, $\mathbf{K}$ is the permeability, T is temperature, c is specific heat, κ is the thermal conductivity, L is latent heat, T^H is a reference temperature, C is the solute concentration in weight per cent, and D is solute diffusivity. The subscripts "s" and "l" refer to solid and liquid, respectively, while a bar over a variable means a volume average of the variable over the solid plus liquid mixture; for

202

example, $\bar{\rho} = g_l\rho_l + g_s\rho_s$, where g_s is the volume fraction of solid. Several equations (4) are solved in the model, one per each solute. A particular solute is hydrogen, which can precipitate in gas form when its dissolved concentration in the liquid exceeds the solubility at the local temperature and pressure. The energy and solute equations are rearranged in modified form depending whether the solute is assumed to have negligible or complete diffusion in the local solid (like hydrogen). The reader is referred to references [2] and [16] for more specific details on the model regarding additional rearrangement of equations and numerical solution procedures.

Calculation of Porosity

A pore growth model is implemented at the microscopic scale together with a criterion for nucleation of pores. The term nucleation is here used in the general sense to refer to the origination of pores, without necessarily implying any particular mechanism of classical nucleation. We assume that, dispersed in the liquid, there is an initially known distribution of microscopic inclusions. These can be oxide bifilms that were entrained during melt pouring, old oxide bifilms that existed in the melt before pouring, or other impurities that serve as possible nucleation sites for hydrogen pores. We call $n(\mathbf{x},t)$ the number of these inclusions per unit volume of alloy, where $n(\mathbf{x},0)$ is known. The inclusions are transported with the velocity field $\mathbf{u}$ of the liquid and they can partition to the solid like the other solutes of the alloy. For implementation purposes, the inclusions are treated as another alloy solute with negligible diffusion.

We assume that hydrogen pores can nucleate and grow only at places where the following two conditions are met:

$$n > 0 \tag{5a}$$

$$p + \frac{2\sigma}{r} < p_S \tag{5b}$$

In which p_S is the Sievert pressure [17], r is the pore radius and σ is the surface tension of the pore-liquid interface. Most of the mechanisms that have been proposed for the nucleation of pores are based on the size of interdendritic cavities and the theory of heterogeneous nucleation on non-wetted surfaces. These ideas have been challenged by John Campbell [14,18,19] and others, who propose a nucleation-free mechanism for pore formation based on the concept of double oxide films or bifilms. In this scenario, during pouring in a casting process, the liquid surface of the alloy can fold upon itself. Because the liquid surface is covered by an oxide film, the folding action leads to bifilms, which are entrained into the bulk melt as a pocket of air enclosed by the bifilm. In effect, the bifilm with its air pocket is the beginning of a pore. After entrainment, the turbulence causes the bifilm to convolute and contract. Posterior pore growth can occur by the simple action of unfurling of the bifilms which in turn can be caused by mechanical action of the surrounding melt and by the aid of hydrogen diffusion into the pore.

It is interesting to note that if the bifilm theory is correct, then it follows that Eq. (5b) is irrelevant because there is no direct contact between gas and liquid and hence no surface tension is involved. However, the unfurling of bifilms is probably affected by the diffusion of hydrogen into the bifilm through the oxide layer. Consequently, a threshold amount of hydrogen in the liquid may be necessary to produce sufficient unfurling.

In this work, we keep (5b) as a criterion for pore origination and test the modeling results thus obtained against experimental data. When conditions (5) are met, we assume that a concentration

n of spherical pores form with a known average initial radius, r_0. If the pores are in a supersaturated environment, they will grow by hydrogen diffusion. Assuming that the pores maintain the spherical shape during growth in the liquid, the mass rate of hydrogen entering the pore by diffusion from the liquid is given by:

$$\frac{dm^H}{dt} = 4\pi r_P^2 \rho_l D_H \frac{\partial C_l^H}{\partial r}\bigg|_{r=r_p} \tag{6}$$

Where r_P is the pore radius, r is the radial coordinate measured from the center of the pore, and D_H is the diffusivity of hydrogen in the liquid. We assume that the hydrogen gas inside the pore behaves as an ideal gas and that the partial pressure of other gases in the pore is negligible compared to that of hydrogen (this is reasonable in Al alloys given the high diffusivity of H compared to other gases). In this case, the rate of increase of the volume of the pore can be calculated as:

$$\frac{dV_P}{dt} = \frac{R_H T}{p_S} \frac{dm^H}{dt} \tag{7}$$

where R_H is the hydrogen gas constant. The radius of the pore is then obtained as:

$$r_P = \left(\frac{3}{4\pi} V_P\right)^{\frac{1}{3}} \tag{8}$$

To estimate the radial derivative in Equation (6), we follow Yin and Koster [20] and consider the thickness of the diffusion boundary layer around the pore:

$$\frac{\partial C_l^H}{\partial r}\bigg|_{r=r_p} \cong \frac{C_l^H - C_P}{\delta} \quad ; \quad \delta = 4\sqrt{D_H t} \tag{9}$$

Where C_P is the solubility of hydrogen at the local pressure and temperature, given by Sievert's law [17], and t is the time measured since pore nucleation. We must keep in mind that the pore growth model exists at the microscopic scale; there are no actual pores that are part of the geometry of the macroscopic model (the radial direction has no meaning in the macroscopic model). The pore radius calculated in Equation (8) should be interpreted as the average radius of pores in a location x where there are $n(x, t)$ pores per unit volume.

Equation (6) is valid for pores that grow in the liquid. For pores growing in the mushy zone, the diffusion flux is taken as an average for liquid and solid [7] and the pore area is multiplied by a shape parameter, α, in order to account for the distortion of the pores as they impinge into dendrites, with:

$$\alpha = \frac{r_P S_V}{3} \tag{10}$$

Where α is the pore shape parameter and S_V is the specific area of the pore (ratio of pore area to pore volume). For spherical pores, $\alpha = 1$, while $\alpha > 1$ for pores distorted by dendrites.

The pores grow while there is liquid remaining around them and lock in size after complete solidification. The total fraction of porosity in the casting as a function of time can be calculated as:

$$f_P(t) = \frac{1}{V} \int_V n(\mathbf{x},t) V_P(\mathbf{x},t) d\mathbf{x} \qquad (11)$$

where V is the volume of the casting.

To close the model, we need to provide some mechanism by which the concentration of dissolved hydrogen in the bulk liquid around the pore decreases to compensate the hydrogen provided to the pore (otherwise the pore will continue to grow indefinitely). That is, the transport equation for hydrogen needs to be modified to include a sink term. In the liquid, this equation is:

$$\frac{\partial C_l^H}{\partial t} = D_H \nabla^2 C_l^H - \mathbf{u} \cdot \nabla C_l^H - n C_l^H \frac{dV_P}{dt} \qquad (12)$$

Where the last term in the right hand side represents the amount of hydrogen entering the pores from the liquid by diffusion. Because MULTIA is a two-phase code (liquid and solid), the gas phase is not included in the transport equations. Therefore, the validity of the proposed model needs to be restricted to small volume fraction of porosity, which is reasonable for the usual level of hydrogen microporosity measured in aluminum castings (< 1%). In this case, we can assume that the presence of the pores does not considerably affect the transport of other quantities like energy and momentum.

Model Application and Discussion

The solidification model is discretized in space and integrated in time using a finite element algorithm that is described by Felicelli *et al.* [2,16]. Aluminum A356 alloy is solidified by simulation in a bottom-cooled two-dimensional mold. The two-dimensional simulated casting has dimensions of 26 mm in width and 300 mm in height. Gravity acts downwards. In addition to the alloy solutes in A356 (Si and Mg), the gas-forming element, hydrogen, is considered. The computational domain is the casting; the top boundary is left open in order to allow for liquid flow to feed shrinkage. A no-slip condition is used for velocity at the bottom and two vertical boundaries, a stress-free condition is used on the top open boundary, and solute diffusion flux is set to zero on all closed boundaries. The thermal boundary conditions utilized in Poirier *et al* [7] are used here, which are extracted from a measured thermal history in the plates cast by Fang and Granger [3]. The simulations start with an all-liquid alloy of the nominal composition initially at a uniform temperature of 958 K, which is 70 K of superheat. The thermodynamic and transport properties of the alloy, including the alloy elements and hydrogen, are the same as the ones in Ref. [7], with the exception of the partition coefficient of hydrogen, for which we used the developments of Poirier and Sung [21] to include the effect of the high eutectic fraction in A356.

We performed simulations for the following values of initial hydrogen content: 0.11, 0.25, and 0.31 cc/100g (Note: 1 wt% = 1.12 x 10^4 cc/100g), for which measured volume fraction of porosity and pore diameters were reported in the experiments by Fang and Granger [3]. In their work, these represented three different castings with the same geometry.

For all the calculations presented in this work, we used an initial pore diameter of 3μm and a density of inclusions of 2x10^{11} m^{-3}. Taking the density of alumina as 4000 kg/m^3 and spherical inclusions of size equal to the initial pore size, this inclusion density corresponds to a concentration of approximately 5 ppm . This selection was guided by the work of Simensen and Berg [22], who found that the smallest alumina particles in aluminum and aluminum alloys ranged from 0.2 to 10 μm, while the concentration of oxides varied between 6 and 16 ppm.

Figure 1 shows the variation of pore volume fraction and pore diameter vs. cooling rate in the solidified casting for all three values of initial hydrogen content. In this figure, the pink dots are calculated values that span all the casting; each dot represents the pore volume fraction or pore diameter calculated at a mesh node in the casting.

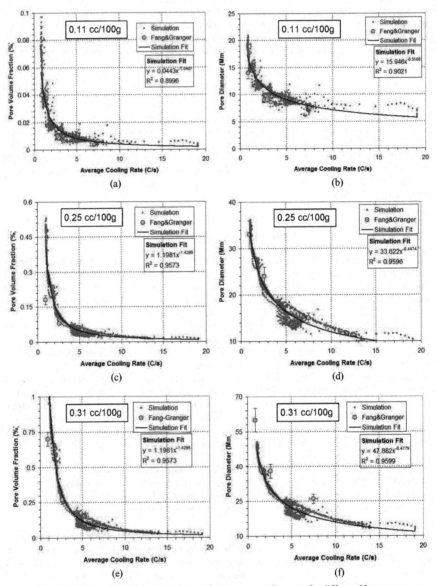

Figure 1. Pore volume fraction and pore diameter vs. cooling rate for different H contents

A least square fit of the calculated values is also shown as a solid black line. The experimental data of Fang and Granger [3] are indicated as green dots; these were taken by manual reading from their paper, so bars estimating possible reading error are added. The experimental green dots represent average values measured at a certain section of the casting, while the simulation shows the space variation within the entire casting. Certainly, the pore volume fraction and diameter are affected by other solidification variables in addition to cooling rate, but an average trend can be identified which is that they both decrease for higher cooling rates.

The quantitative agreement of simulated results with the experimental data is reasonable, considering that we are using a relatively simple two-dimensional continuum model. As previously mentioned, all the results in Figures 1 were obtained using the same values of the initial pore diameter ($d_0 = 3$ μm) and concentration of inclusions ($n = 2 \times 10^{11}$ m^{-3}). Although the selected values fall in the experimentally measured range reported in Ref. [22], it is possible that pores originate with a range of sizes and that the concentration of inclusions may differ from one casting to another in the experiments of Fang and Granger [3]. A closer agreement with the experimental data can be obtained if the parameters d_0 and n are individually adjusted for each level of hydrogen content, but this approach was not pursued. In this sense, it is interesting to note that a same set of parameters works rather well for all three castings.

In addition to d_0 and n, a pore shape parameter $\alpha = 1$ was used to obtain the results for the castings with 0.25 and 0.31 cc/100g of hydrogen content, indicating that the growth of pores in these castings was apparently not significantly affected by impingement of the pores on dendrites. However, we needed to use $\alpha = 4$ to reproduce the results of the 0.11 cc/100g casting, probably indicating that in this casting the pores were significantly distorted during growth. This observation is supported by the calculated fraction of liquid at which pores activate in each casting: 0.45, 0.75 and 0.85, for the 0.11, 0.25 and 0.31 cc/100g castings, respectively. We observed in the simulations that once activated, pores grew very fast, indicating that in the 0.25 and 0.31 cc/100g castings, the pores developed most of their size at high fraction of liquid and were not significantly affected by dendrite impingement. In contrast, in the 0.11 cc/100g casting, pores started to grow at an already high fraction of solid and were most probably largely distorted by dendrites during growth.

Conclusions

A continuum solidification model was extended to calculate the volume fraction of porosity and the pore size distribution during solidification of aluminum alloys. The formation and growth of individual pores are calculated with a new hydrogen-diffusion technique in which the inclusions are treated as an additional alloy solute and subject to transport equations in the liquid and mushy zone. The method requires two parameters, the initial pore size and the concentration of inclusions, which are of physical nature and can be linked to measured data. The simulations show that the same set of these parameters is able to reproduce with reasonable agreement experimental data of different castings with varying levels of hydrogen content. A limitation of the method occurs when pores grow at high fraction of solid, which happens for the lowest level of hydrogen content. In this case, the growth of pores is highly affected by impingement on dendrites. Although the experimental data can still be reproduced through the use of a pore shape factor, a micro-model that links the pore shape parameter to physical quantities in the mushy zone would be desirable.

Acknowledgements

This work was partly funded by the National Science Foundation through Grant Number CTS-0553570. The authors gratefully appreciate the financial support provided by the Center for Advanced Vehicular Systems (CAVS) at Mississippi State University.

References

1. Major J F: Porosity control and fatigue behavior in A356T61 aluminum alloy, *AFS Trans.* Vol. 105 (1997), 901-906
2. Felicelli S D, Poirier D R and Sung P K: A model for prediction of pressure and redistribution of gas-forming elements in multicomponent casting alloys, *Metall. Mater. Trans. B* Vol. 31B (2000), 1283-1292
3. Fang Q T and Granger D A: Porosity formation in modified and unmodified A356 alloy castings, *AFS Trans.* Vol. 97 (1989), 989-1000
4. Han Q and Viswanathan S: Hydrogen evolution during directional solidifications and its effect on porosity formation in aluminum alloys, *Metall. Mater. Trans. A* Vol. 33A (2002), 2067-2072
5. Lee P D and Hunt J D: Hydrogen porosity in directional solidified aluminum-copper alloys: a mathematical model, *Acta Mater.* Vol. 49 (2001), 1383-1398
6. M'Hamdi M, Magnusson T, Pequet Ch, Amberg L and Rappaz M: Modeling of micropososity formation during directional solidification of an Al-7%Si alloy, *Modeling of Casting, Welding and Advanced Solidification Processes X*, D.M. Stefanescu, J. Warren, M. Jolly, and M. Krane (Eds.), The Minerals, Metals, & Materials Society, 2003, 311-318
7. Poirier D R, Sung P K and Felicelli S D: A continuum model of microporosity in an aluminum casting alloy, *AFS Trans.* Vol. 109 (2001), 379-395
8. Zhu J D, Cockcroft S L, Maijer D M and Ding R: Simulation of microporosity in A356 aluminium alloy castings, *Int. J. Cast Met. Res.* Vol. 18 (2005), 229-235
9. Guo J and Samonds M T: Microporosity simulations in multicomponent alloy castings, *Modeling of Casting, Welding and Advanced Solidification Processes X*, D.M. Stefanescu, J. Warren, M. Jolly, and M. Krane (Eds.), The Minerals, Metals, & Materials Society, 2003, 303-311
10. Carlson K D, Zhiping L, Hardin R A and Beckermann C: Modeling of porosity formation and feeding flow in steel casting *Modeling of Casting, Welding and Advanced Solidification Processes X*, D.M. Stefanescu, J. Warren, M. Jolly, and M. Krane (Eds.), The Minerals, Metals, & Materials Society, 2003, 295-302
11. Sung, P K, Poirier D R and Felicelli S D: Continuum model for predicting microporosity in steel castings, *Modelling Simul. Mater. Sci. Eng.* Vol. 10 (2002), 551-568
12. Sabau A S and Viswanathan S: Microporosity prediction in aluminum alloy castings, *Metall. Mat. Trans. B* Vol. 33B (2002), 243-255
13. Lee P D, Chirazi A, Atwood R C and Wang W: Multiscale modeling of solidification microstructures, including microsegregation and microporosity, in an Al-Si-Cu alloy, *Mater. Sci. Eng. A* Vol. A365 (2004), 57-65
14. Yang X, Huang X, Dai X, Campbell J and Tatler J: Numerical modeling of entrainment of oxide film defects in filling of aluminium alloy castings, *Int. J. Cast Met. Res* Vol. 17 (2004), 321-331
15. Stefanescu D M: Computer simulation of shrinkage related defects in metal castings – a review. *Int. J. Cast Met. Res.* Vol. 18 (2005), 129-143
16. Felicelli S D, Heinrich J C and Poirier D R: Finite element analysis of directional solidification of multicomponent alloys, *Int. J. Numer. Meth. Fluids* Vol. 27 (1998), 207-27
17. Flemings M C: *Solidification Processing* (New York: McGraw-Hill), pp 203-10, (1974)
18. Campbell J: Castings 2nd Edition – The New Metallurgy of Cast Metals. Butterworth-Heinemann, Oxford, UK, 2003
19. Campbell J: Entrainment Defects, Materials Science and Technology Vol. 22, n 2, (2006), pp 127-145
20. Yin H. and Koster J., "In-situ observed pore formation during solidification of aluminium", *ISIJ International*, vol. 40(4), pp. 364-372 (2000)
21. Poirier D R and Sung P K: Thermodynamics of hydrogen in Al-Si alloys, *Met. Mater. Trans. A* Vol. 33A (2002), 3874-3876
22. Simensen C J and Berg G, A survey of inclusions in aluminum, *Aluminium*, Vol. 56, No. 5, pp. 335-340 (1980)

Shape Casting: The 2nd International Symposium *Edited by Paul N. Crepeau, Murat Tiryakioğlu and John Campbell*
TMS (The Minerals, Metals & Materials Society), 2007

REDESIGN OF AN INDUSTRY TEST FOR HOT TEARING OF HIGH PERFORMANCE ALUMINIUM CASTING ALLOYS USING CASTING SIMULATION SOFTWARE

ACM Smith[1] and MR Jolly[1]

[1]The University of Birmingam, Edgbaston
Birmingham, B152 TT, UK

Keywords: hot tearing, simulation, modeling, mould design

Abstract

Hot tearing propensity in aluminium alloys is commonly measured using dog-bone (or "I" beam) and ring tests. Hot tearing occurs as a result of a number of factors including; level of stress and strain, hot spots and nucleation sites. This paper presents the results of a study to redesign a dog-bone type hot tear test using casting simulation software to ensure that the location of the tearing was always in the same location to improve the reproducibility of the test.

In the simulation of the original five fingered die both the stress and strain were sufficiently high for hot tearing but there was no defined hot spot implying that the random hot tear locations would result depending upon suitable nucleation sites. A number of design iterations were carried out to produce more focussed hot spots and to ensure that the die was easy to manufacture and use, and was economically viable.

Introduction

Previous work at Birmingham has investigated the redesign of tensile test moulds [1]. In this work moulds for measuring hot tearing susceptibility are redesigned to give more consistent predictions. Hot tearing is a phenomenon that occurs during the solidification of a cast material when the stress created generated by the thermal contractions (both solidification and linear) become greater than the inherent strength of the material. There is a tendency for them to occur in hot spots within the geometry as this will be the weakest material. Hot tear tests have been developed over the years so that engineers can determine the susceptibility of alloys to hot tearing and to investigate the effect of trace elements for example. Despite much work on hot tearing over the last several decades there is still no consensus on the mechanism of the nucleation of hot tears. There is almost no doubt that they are initiated on pre-existing defects [2] but there have been a number of mechanisms proposed for the growth of the crack. Pellini [3] proposed a theory of hot tearing based on the accumulation of strain which must fulfill the following criteria: cracking occurs in the hot spot, hot tearing is controlled by the level of strain occurring within the hot spot and finally that the accumulated strain in this region depends upon the strain rate and a time factor. This has been further developed by Clyne and Davies [4].

Rapaz [5] and previously Prokhorov [6] have suggested that it is the strain rate which is the critical factor for controlling hot tearing. This is justified by the assuming the strain rate during solidification is limited to the rate at which fracture can occur. A third approach assumes that failure occurs at a critical stress with the remaining liquid around the solidifying grains acting as a stress raiser.

The final theory is that hot tearing occurs because there is not enough feed metal to supply the hot tearing region [78,9]. Foundries will often grain refine their alloys in order to promote better feeding so that hot tearing doesn't occur. Katgerman [10] has summarized these mechanisms as presented in Table I.

Table I: Possible hot tearing mechanisms (after Katgerman)

Temperature range fraction of solid	Nucleation of crack	Propagation of crack	Fracture Mode
$T_{rigidity} < T < T_{coherency}$ $F_s = 50-80\%$	Grain boundary covered with liquid; shrinkage or gas pore	Liquid film rupture Filled gap	Brittle, intergranular Healed crack
$T < T_{rigidity}$ $F_s - 80-99\%$	Pore, surface of particle or inclusion, liquid film or pool, vacancy clusters	Plastic deformation of bridges Liquid film rupture, liquid metal embrittlement of solid bridges	Brittle, intergranular Plastic deformation of bridges possible
Close to solidus $F_s = 98=100\%$	Pore, particle or inclusion, segregates at grain boundary, liquid at stress concentration point, vacancy clusters.	liquid metal embrittlement Plastic deformation of bridges, creep	Brittle transgranular propagation possible Macroscopically brittle or ductile, transgranular propagation possible

Hot spots occur at section increases and at intersections in the casting so that any die or mould should be designed to have uniform section thickness to equalize cooling [11] in order to reduce the occurrence of hot tearing.. Where this is not possible, chills should be used to alter the cooling rate. Any potential stress raisers should have a gradual change in cross section. Moulds for predicting the susceptibility of metals and alloys to hot tearing are deliberately designed with features that act as stress raisers but few of them create localized hot spots.

Hot tear test methods

A number of different tests have been developed to demonstrate the susceptibility of cast metals to hot tearing. Some of these are described in the following sections.

The "I" Beam
There are many variants to this method but they all involve casting a bar with resistance to contraction at both ends (Figure1). The resistance form the end geometry increases the stress and strain in the material promoting hot tearing. A common variation based on the "I" beam test involves casting fingers of differing lengths, from one runner. The amount of strain available in each finger is proportional to the length of the beam implying that the longest finger should fail first by hot tearing. The more fingers that fail, the more susceptible the material is to hot tearing.

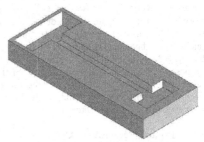

Figure 1: Simple "I" beam mould for identifying hot tear susceptibility

The Ring Test

The ring test involves pouring liquid material into the area between the inner and outer regions of a steel ring shaped die, (Figure 2), producing a 'ring' shaped casting.

The cast material cools where it contracts onto the inner section of the die whilst the inner core of the die expands slightly at the same time. This produces the constraining forces, which will initiate transverse hot tears in a susceptible material. It is an unusual test as there is no specific area of strain concentration or a hot spot, yet it still produces notable consistency.

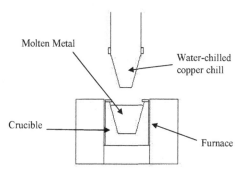

Figure 2: The mould for a ring test used for identifying hot tear susceptibility

Figure 3: The cold finger test [12]

The cold finger test

The cold finger test, developed by Warrington and McCartney, consists of a steel crucible [13] contained within an open furnace, holding the molten metal being tested (Figure 3). Above the crucible is a copper chill which is also water cooled. Both the steel crucible and the copper chill have angled sides of 17.5 degrees, which allow an exact match when the chill is lowered into the melt. It is lowered a set distance to produce a casting with a predetermined 10 mm thick wall. The melt cools and solidifies with a tear being initiated in the surface where the restraint stress is at its highest.

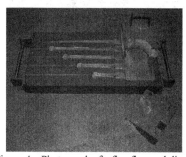

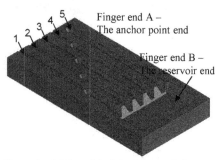

Figure 4a: Photograph of a five fingered die

Figure 4b: CAD model of the 5 fingered die

211

Hot tear test design for current work

The current work was based around a design from N-Tec, The geometries is expected to create hot tears in a systematic way. The geometry is shown in Figures 4 a&b.

The mould is essentially a multiple finger "I" beam test that produces five cast fingers of increasing length, which are connected to a single reservoir. There are slight differences between the actual mould and the CAD due to manufacturing issues. The fingers are all of the same depth so should all start to fill at the same time. The liquid metal is poured into the reservoir where it subsequently flows down these fingers, filling the mould and solidifies. If the test material is susceptible to hot tearing, the fingers will tear upon solidification. The idea is that the more susceptible to hot tearing the alloy is then the more fingers will show hot tears. Those with lower susceptibility will only show tearing in the longer fingers whereas highly susceptible alloys will show tearing even in the shortest finger.

The fingers and the reservoir have a draft angle on the depth for easy removal from the mould after solidification. The reservoir was originally triangular in shape but after testing and previous modeling at the University of Birmingham, it was found that a rectangular reservoir allowed the fingers to fill quickly and more evenly. The mould contains vents at the end of each finger, ensuring there is no backpressure build-up during filling. One of the most important aspects of this design are the cones located at the end of each finger. These downward pointing cones fill with the liquid material and solidify providing an anchor point allowing stress and strain to build in the fingers.

The mould produced by N-Tec does induce hot tears in the cast material but there is inconsistency with the location of the failure. Figures 5 a&b show castings produced from the mould. It is clear that despite using the same alloy and mould, the hot tears have occurred at different locations on each test.

 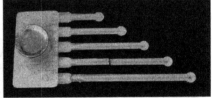

Figure 5 a&b: Test castings showing the random location of cracks in the hot tear test

Table 2: Experimental parameters for hot tearing

Variable Experimental Parameters	Value
Cast Alloy Type	AlSi7Mg [~LM25]
Pouring Temperature	715 °C
Permanent Die Material	Grey Cast Iron - Grade 250
Permanent Die Initial Temperature	300 °C
Pouring Sleeve	Foseco Kalpur insulating sleeve
Filter	10 ppi

Experimental Parameters

Table 2 gives the initial experimental parameters used in the initial simulations which were run using Magmasoft casting simulation software running on a Dell PC running an Intel Dual Xeon 3.06 GHz processor and 4 Gb RAM.

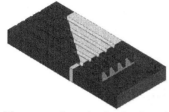 

a) large ceramic section insulated mould   b) cut-away section insulated mould
Figure 7: Two of the design iterations considered in the research work.

Simulation results

The simulation results for the unmodified die are shown in Figure 6. The figure shows the order of solidification on the left hand pictures and indicates there is a progressive solidification front which moves back towards the reservoir. This would be an ideal scenario if we wanted to avoid hot spots and the associated hot tearing. However the premise for this test is that materials that are susceptible to hot tears will tear within the test section. The right hand pictures show the development of stresses in the direction along the fingers. Hot tearing is indicated by a region where the maximum difference of stress occurs. During the solidification the only region where this appears to be of significance is in the centre of the middle finger. However the differences of stress levels are not high. It would appear that the location of any tears in the other fingers will be totally reliant on the existence of a defect to initiate the tear. Thus this backs up the results obtained from the experimental work where hot tears appeared in a random fashion. The highest level of stress predicted from the modeling was at the junction between the shoulder and the reservoir (Fig 6e).

Mould redesigns based on initial simulation results

One obvious conclusion from these results is that a focused hot spot would be a method of concentrating where the hot tears should occur. This would be the equivalent in mechanical testing of providing a notched test sample for toughness. It was decided that providing a section of insulation on the mould which would retard the solidification in that area. Figures 7 a&b show two of the iterations. Figure 7a depicts a mould with a ceramic section replacing the top part of the mould. Another feature which was incorporated into all the moulds was a cooling fin which replaced the conical anchor of the original mould and an additional cooling fin just after the shoulder at the reservoir end of the mould. The fins perform two functions; they promote a rapid solidification from each end of the gauge length by having lower thermal modulus than the conical sections they replaced, thus trapping liquid metal in the centre of the gauge. They also function as the anchor point to ensure stress build up within the gauge length of the test pieces. Fig 7b is a section through the cut-away mould. In this case the thinnest section was 10 mm. Figure 8 shows the location of the hot spots predicted from these design iterations. Although both of these designs worked reasonably effectively there some issues with each one. It was felt that the large ceramic insert would be difficult to use without damage and the large differences in thermal expansion coefficients might give problems during the use of the mould. Although not complex in shape it was also felt that this would be an expensive mould to manufacture. The cut-away mould didn't give as precise a hot spot as the ceramic insulated mould (Figure 8). An extreme cut-away mould was modeled with thinnest section being only 1 mm and although this gave a more controlled result it was felt that the mould would be prone to distortion over time.

The final design was adapted from the large ceramic insert and consisted of 5 ceramic fiber inserts 15x 25x30 mm with the cutout of the finger cross section in them (Figure 9a), positioned close to the reservoir end fins (Figure 9b). The mould was developed to be practical, cheap and effective in producing localized hotspots in each test finger.

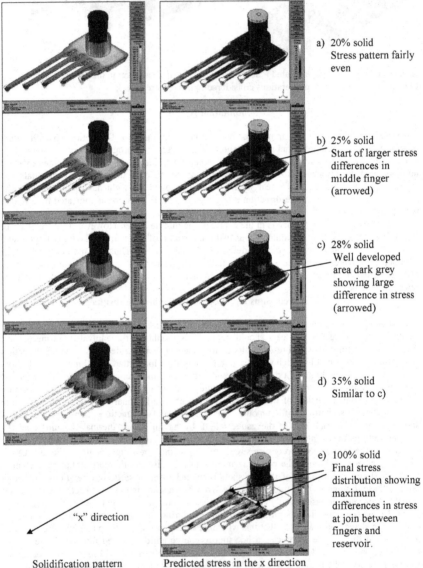

a) 20% solid
   Stress pattern fairly even

b) 25% solid
   Start of larger stress differences in middle finger (arrowed)

c) 28% solid
   Well developed area dark grey showing large difference in stress (arrowed)

d) 35% solid
   Similar to c)

e) 100% solid
   Final stress distribution showing maximum differences in stress at join between fingers and reservoir.

"x" direction

Solidification pattern   Predicted stress in the x direction

Figure 6: Predicted solidification sequence for the original 5 fingered die and stresses developed within the fingers at different times during the solidification

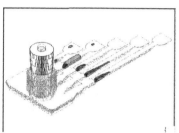

a) large ceramic section mould   b) cut-away section mould

Figure 8:Hot spot prediction from two of the design iterations.

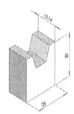
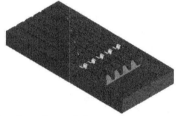

a) ceramic fiber insert design   b) location of ceramic inserts

Figure 9: Final design of mould to promote highly localized hot spot

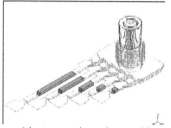
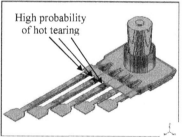

a) Hot spot prediction showing the extremely localized hot spot produced.   b) MAGMA prediction using a bespoke hot tear criterion indicating high probability hot tearing in the middle and two longest fingers

. Figure 10: Simulation results for the final design using small ceramic fiber inserts near the chills at the reservoir end of the mould.

Final design simulation results

Results from the simulations of the final design are shown in Figure 10. The predicted hot spots generated by using the small ceramic fiber inserts are extremely localized and can be see in Figure 10a.

The maximum predicted level of strain was of the order of 0.1 in the Al7%Si alloy and using the proprietary hot tear criterion cracking was predicted in the three longest fingers.

Summary

A hot tear mould that creates localized hot spots has been designed using computer numerical simulation experimentation. It is perceived by the users of the hot-tear test that this would give more reproducible results. The optimum design, considering efficacy of creating hot spots, ease of use and manufacture, consisted of a five fingered cast iron mould which incorporated chill fins at each end of each finger and a disposable ceramic fiber insert close to the reservoir end of the die. Designs that incorporated monolithic ceramic inserts or cut-way sections of mould produced less focused hot spots and were perceived to be less easy to use and more expensive to manufacture.

Acknowledgments

The authors would like to acknowledge the support given by staff at N-Tec.

References

1 Gebelin J-C. Lovis M. & Jolly M.R. "Simulation of tensile test bars: Does the filling method matter?", **Symposium on Simulation of Aluminum Shape Casting Processing**, TMS2006 March 2006, TMS, Warrendale, PA, Eds Q. Wang, M.J.M Krane and P.D Lee

2 Campbell J., *Castings,* Oxford, U.K.: Butterman-Heinemann Ltd., 1991

3 Pellini WS, Foundry, **80**, 1952, pp 125-199

4 Clyne TW, Davies GJ., British Foundrymen **68,** 1975, p 238

5 Grasso P.D., Drezet J.M., and Rappaz M., "Hot Tear Formation and Coalescence Observations in Organic Alloys", *JOM-e*, Jan, 2002

6 Prokhorov NN, Russian Castings Production, **2**, 1962, p 172

7 Rappaz M., Drezet J.-M., & Gremaud M., *Met. and Mater. Trans. A*, 30A, 1999, p. 449.

8 Niyama E. Uchida T., Morikawa M. & Saito S., "A method of shrinkage prediction and its application to steel casting practice", Int. J. Cast Metals Res., 7, 1982, pp 52–63

9 Feurer U., *Giessereiforschung, Heft 2* (Neuhausen am Rheinfall, Schweiz, **28**, 1976, p 75

10 Katgerman L., Post-graduate lectures, University of Birmingham, September 2006

11 M'Hamdi M., & Mo A., "On modelling the interplay between microporosity formation and hot tearing in aluminium direct chill casting", Materials Science and Engineering A, **413-414**, 2005, pp 105-108

12 Warrington D. & McCartney D.G., Cast Metals, **2(3)**, 1989, p 134

13 Eskin D.G., Suyitno & Katgerman L., "Mechanical properties in the semi-solid and hot tearing of aluminium alloys", *Progress in Materials Science*, **49**, 2004, pp.629-711

PREDICTING THE TORTUOUS THREE DIMENSIONAL MORPHOLOGY OF MICROPOROSITY IN ALUMINUM ALLOYS

Junsheng Wang[1], Robert C. Atwood[1], Ludovic Thuinet[2], Peter D. Lee[1*]

[1]Department of Materials, Imperial College London, South Kensington Campus, Prince Consort Road, London SW7 2AZ, UK;
[2]Laboratoire de Métallurgie Physique et Génie des Matériaux, USTL, CNRS UMR 8517, Bât C6, F-59655 Villeneuve d'Ascq Cedex, France
*corresponding author: p.d.lee@imperial.ac.uk

Keywords: Simulation, Microporosity, Aluminum Alloys, Casting, X-ray Micro-Tomography.

Abstract

A microscale model is presented to predict the complex interaction of pores and the developing solid microstructure. The 3D simulation results illustrate how the pore growth is influenced by both the grain and dendritic structure, forming highly tortuous and irregular shapes. Methods to quantify the 3D structures are proposed and used to qualitatively compare the predicted size and tortuosity of the porosity to experimental observations using x-ray micro-tomography. The predicted average and extreme sizes closely matched those observed in Al12Cu wedge castings.

Introduction

The dramatic increase in the application of aluminum alloys in a variety of industries calls for the significant improvement of mechanical properties of castings. Many components experience cyclic loading, where casting defects such as microporosity, intermetallic particles and oxide films can be detrimental to the fatigue behavior [1]. Therefore, in recent years a number of studies have focused on the prediction of these defects, in particular porosity formation [2-6]. However, most of these studies predict only the percentage porosity or the average radius assuming the pores are spherical. This study aims to extend the predictions and experimental observations to include the tortuous three dimensional morphologies that pores can form during solidification.

Methods to quantify the complex range of shapes and sizes of pores are required if quantitative models for relating their presence to final component fatigue life are to be developed [7]. In most microporosity models, the morphology of porosity is not predicted due to simplification of microstructure evolution during microporosity formation [2, 4, 6]. Growth of microporosity depends upon diffusion of hydrogen, shrinkage, and the interaction of the expanding pores with the surrounding dendrites. Therefore, the prediction of the morphology of porosity should consider the restriction of bubble growth by the solid phase.

In this paper, a three dimensional mathematical model is used to predict the shape, size, and distribution of porosity in a dendritic structure. The model is validated by quantitatively comparing the size and shape of the pores to three dimensional images of experimental castings obtained from a high resolution x-ray micro-tomography (XMT). This comparison shows a good agreement between numerical and experimental results and clarifies the mechanisms of pore shape development.

Mathematical Model

An in-house mesoscale model was developed to simulate the nucleation and growth of the primary phase, the hydrogen gas diffusion, and pore formation. This model aims to apply the relevant physics as much as feasible, using stochastic submodels for nucleation of the primary phase and porosity and solving the continuum equations describing the diffusion of solute and gas.

Assuming a local equilibrium at the S/L interface, both the hydrogen and the solute are partitioned between the solid and the liquid if there is no pore formation. Any change in the solute concentration affects the degree of undercooling and hence the nucleation and growth processes. Variations of solute concentration and temperature change the local gas solubility and thus provide the opportunity for porosity nucleation and growth. Once the supersaturation is sufficient to overcome the nucleation barrier, a pore nucleates.

In this multiphase system, the extent of each phase is explicitly tracked on a regular cubic grid. The orientation of solid grains is tracked using a modified decentred octahedron algorithm [8]. Expansion of gas bubbles is limited by the presence of the solid phase. The simulated microstructure develops in response to applied conditions without the use of imposed penalty factors [9, 10], deterministic factors [11] or limitation variables [6] that have been used in prior models to enforce the behavior. Details of the model excluding the new features have been published previously [5, 8]. Therefore, only a summary of the model is presented below, with special emphasis on the growth of microporosity.

Solidification Model

In the model, a cell may have three different states: (i) liquid; (ii) solid; and (iii) growing/melting (a mixture of liquid and solid), and may contain all or part of a pore. By assuming thermodynamic equilibrium in the mushy cells, the dissolved gas is partitioned between solid and liquid and thermodynamic equilibrium condition is maintained at the S/L interface. The propagation of the solid crystal is modeled by a cellular automaton algorithm including a decentred octahedron algorithm [12] modified by Wang et al [8]. A linearized Al-Cu phase diagram was used, with eutectic temperature, T_{eu}=548°C, eutectic composition, C_{eu}=32, liquidus slope, m=-3.39, and solute partition coefficient, k_P=0.173 [4]. The propagation of each gas bubble is governed not only by the diffusion of gas but also by the dendritic structure around it. Therefore, the model not only predicts the percentage porosity but also predicts the detailed morphology of pores.

Nucleation of the Primary Phase. To account for the stochastic nature of grain nucleation, a grain nucleation function previously developed by Charbon, Rappaz and Gandin is implemented in the model [13, 14] in a similar manner to that presented previously [8]. A population of N_{max} nucleation sites whose critical undercooling, ΔT_{crit}, obeys a Gaussian distribution about a mean undercooling ΔT_{μ} with standard deviation ΔT_{σ} is randomly distributed within the domain, with no more than one nucleus per grid cell. In order to activate the nucleation event in a cell, a nucleation site must be present and the undercooling in the cell, ΔT_{cell}, must exceed the critical undercooling, ΔT_{crit}, for that nucleation site, which can be written as [15]:

$$\Delta T_{Cell} = \Delta T_C > \Delta T_{Crit} \tag{1}$$

where ΔT_C is the constitutional undercooling.

Growth of the Primary Phase. By applying a lever rule to partition the solute between primary phase and liquid, the local solute conservation can be established in each cell, described by:

$$\frac{\partial C_E}{\partial t} = \nabla \cdot (D_E \nabla C_L) \tag{2}$$

218

where C_E and D_E are the equivalent solute concentration and the effective diffusion coefficient respectively. The solute concentration in the liquid, C_L, is derived from linearised phase diagram:

$$C_L = C_0 - \frac{\Delta T_C + \Delta T_R}{m_L} \qquad (3)$$

where C_0 is the initial solute concentration, m_L is the liquidus slope on the linearized phase diagram, and ΔT_R is the curvature undercooling, which is calculated by the correlation introduced by Nastac [11].

Porosity Formation

The models previously presented by Lee and Hunt[16] and Atwood and Lee [5] were modified as described below. The solubility of hydrogen in the liquid at one atmosphere pressure, S_l, was given by the formula reported by Ransley and Neufeld [17] with an adjustment for copper content given by Doutre[18]:

$$\log S_L = -\frac{2760}{T} - 0.0269 C_{Cu} \quad \text{(ml STP/100g)} \qquad (4)$$

The nucleation of porosity depended on a Gaussian distribution law of the hydrogen supersaturation ratio, C^H/S_l, where C^H is the hydrogen concentration.

Porosity Growth.

As in the previously reported model, restricted curvature was applied to the pore growth. However, in this study, the minimum dimension of the pore at the previous timestep was used as an estimate of the restricted curvature of the pore instead of the solid fraction. The restricted curvature was then used to calculate the pressure inside the pore, and therefore the volume occupied by n mols of gas, using the ideal gas law.

Methods

Casting experiment

A wedge was cast from Al-12 wt% Cu after melting commercial purity aluminum (99.7wt.%Al) with Al-50wt.%Cu in an induction furnace. The alloy was held at 715 ℃ for 15 minutes and then cast into an instrumented permanent steel wedge mould as described previously [1].

From a location at the middle of the wedge in width, and 100 mm from the bottom in height, two cylindrical samples, 8 mm and 2 mm in diameter, were analyzed using XMT. The X-ray tube conditions were 80kv for acceleration and 80 μA for tube current. Images were obtained with voxel resolutions of 9.25 μm and 4.25 μm for the large and small samples, respectively.

Simulation Methods

The averaged cooling rate during solidification, as measured using a 'K' type thermocouple at the sample location, was 2.9°C/s and this value was used for the simulations. No temperature gradient was applied in the simulations, allowing development of an equiaxed grain structure to compare with microstructure found in experiments. A value of 0.24 ml/100g STP was used for the initial hydrogen content throughout the simulations because the hydrogen level analyzed from wedge casting experiment was 0.23-0.25 ml/100g STP.

Two domain sizes were used for the simulation: i. a small cube of 0.425 mm on each side, with a cell size of 8.5 μm; and a larger cube, 1.2 mm on each side with a cell size of 12.5 μm. Three runs were carried out in the large domain and statistical analysis was performed, and one run on the small domain for high-resolution image generation.

219

<u>Image Analysis</u>

To extract the pores from the reconstructed x-ray tomographic volume, the following procedures were carried out using commercial image analysis programs: median filter with a 3×3×3 voxel[1] kernel, binary threshold with internal hole closure[2], and automatic quantification of defect particles[3]. In order to evaluate the tortuous pores in three dimensions, the bounding box diagonal L was taken as the characteristic length. A dimensionless sphericity factor, ε, defined as the ratio of the measured surface area to the equivalent spherical surface area was used to express the deviation of three dimensional porosity from spherical shape [19]:

$$\varepsilon = \frac{S}{\sqrt[3]{36\pi V^2}};$$ (5)

where V is the pore volume.

Results and Discussion

The size and 3D morphology of pores were revealed by applying image analysis techniques to the XMT images, allowing qualitative and quantitative comparison of experimental results and simulations. Fig. 1 shows the typical results from an XMT scan. In Fig. 1 (a) a single slice from the scan is shown, similar to a traditional optical micrograph – even here the pores are clearly not spherical. In Fig. 1 (b) the full 3D reconstructed pores are shown illustrating that most of the pores are roughly plate-like in shape, but branched and twisted. Only a few may be described as rod- or string-like. It appears that most have formed by spreading out in two dimensions, between grains, rather than growing in one dimension as observed in directional solidification [20]. A small cubic volume including one pore is shown in Fig. 1 (c), in which the branching sheet structure lying between the solid grains may be seen more clearly. The restriction of microporosity growth by the dendritic structure must be considered if the pore shape as well as maximum dimension is to be predicted accurately, as required to predict the fatigue life of casting components [1].

At a 4.25 µm resolution the sample size and hence total number of pores observed is too small to gain good quantitative data on the whole population of pores. The data from the larger 8 mm diameter sample provides an ample population size albeit with reduced detail. The normalized pore volume fraction distribution is presented in Fig. 2, and the size ranges are highlighted and color coded to match Fig. 1 (b). The distribution illustrates the range of pore sizes found in this type of cast aluminum. However, pores of similar volume may have very different shapes, as seen in Fig 1 (b), and the length of pores of similar volume can vary significantly. It is expected that compact, nearly-spherical pore would have less deleterious effect on the fatigue properties of the casting than a long, thin pore having the same volume [19].

However, the pore shape is not taken into account in Fig. 2; this distribution alone does not adequately describe the pores visible in Fig. 1. Therefore, Equation 5 was used to quantify the shape factor information for characterizing the microporosity in both the simulations and experiments.

---

[1] Amira, Mercury Computer Systems **SAS**…PA Kennedy 1 - BP 50227 F-33708 Merignac Cedex France
[2] ImageJ 1.37s, Public Domain Java-based image processing program, National Institutes of Health, USA
[3] VgStudioMax Volume Graphics GmbH Wieblinger Weg 92a 69123 Heidelberg Germany

Comparison of Experiment and Simulation

The quantitative results from the experiments and simulations are shown in Table I, including percentage porosity (%P), number density (N_v), maximum pore dimension (L), equivalent diameter (D), and sphericity (ε). The average values (L_{avg}, D_{avg} and ε_{avg}) are also listed. Since only one 8 mm sample was analyzed in the XMT experiment, no experimental conclusions about casting-to-casting variation may be made at this stage. From three high resolution simulations, some idea of random variation in the pore features may be inferred. The difference between the simulations was only in the random value used to assign locations and thresholds to the pore and grain nucleation sites.

The model predicts percentage porosity with reasonable accuracy, considering that there is significant uncertainty in the true conditions of the experimental casting, and little knowledge at all about the nucleation events other than those inferred from prior experiments [21]. The volume fraction of porosity is higher in simulations than that calculated from experiments because a much smaller domain (one hundredth of the sample volume in experiment) was used in simulations. This is reasonable because any volume fraction calculation based on a small 'sample' will overestimate the results.

The number density in the simulations is lower than in the experiment and almost half of the observed value. This should be expected, since number density includes pores of all sizes, and the observation of the smallest pores is highly dependent on the resolution. In all of those simulations, the cell size was kept 12.5 μm whereas it was 9.25 μm in experiments.

The maximum pore length gives good agreement with the XMT experiment, though the range of pore length includes a greater number of smaller pores in the experiment than in the simulation. It can be seen that maximum pore size varies less than 10%

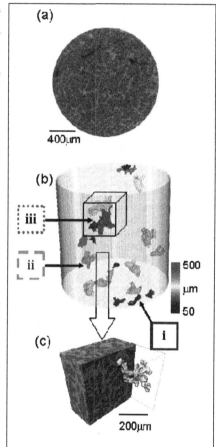

Fig. 1. X-ray micro-tomography characterization of microporosity in a 2 mm cylindrical sample from the wedge mould casting, with 4.25 μm per voxel resolution. (a) A 2D slice from a reconstructed volume. (b) microporosity visualization with (i) small, (ii) medium and (iii) large pores shaded according to their size. (c) 3D rendering of the microstructure and microporosity morphology in a 425 μm cubic sub-volume.

around the value from experiment. That is acceptable because the mean value should be reached once statistical study is done on sufficient simulations. However, the sphericity of pores between experiment and simulations give very good correlation both in maximum values and the mean value. Although the model exhibits promising capabilities to predict the pore size as well as morphology quantitatively, it is only validated against one cooling rate. A systematic comparison using different cooling rates and alloy compositions is ongoing.

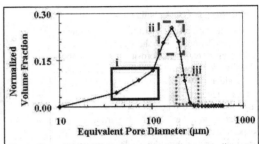

Fig. 2. Normalized distribution of equivalent pore diameter quantified from the 8 mm diameter cylindrical sample imaged at 9.25 µm/voxel. The outlines (i), (ii), and (iii) indicate the size ranges shown in Fig. 1 (b).

Fig. 3 shows the morphology of microporosity quantitatively by comparing the results from the 8 mm cylindrical sample with simulations of a cubic 1.2 mm³ domain. Although fewer pores were quantified in the simulations compared to the experiments, a very close correlation shows the model's ability to predict the change in pore morphology as pores grow. The trend in these graphs demonstrates that larger pores deviate from sphericity (larger ε) more than the smaller ones, both in experiment and simulations. The pores must grow within a developing dendritic structure, spreading between dendrite arms, therefore forming flattened and possibly branched shapes as they become larger.

Table I. Quantitative comparison of microporosity from experiments and simulations

	%P	N_V	L_{max}	D_{max}	ε_{max}	L_{avg}*	D_{avg}*	ε_{avg}
		mm⁻³	µm	µm		µm	µm	
XMT	0.42	13.7	863	240	4.7	102/+140/-59	44/ +39/ -20	2.0 ±0.6
Model	1.03	7.5	979	203	4.2	482/+254/−168	118/+56/−38	2.4 ±1.0
Model	1.05	5.8	933	253	4.0	349/+250/−146	120/+80/−48	2.4 ±0.8
Model	1.13	5.2	787	202	3.7	445/+206/−141	153/+34/−28	2.5 ±0.7

*Parameters analyzed as log-normal distributions, and the reported ranges correspond to plus or minus one standard deviation of the log of the parameter.

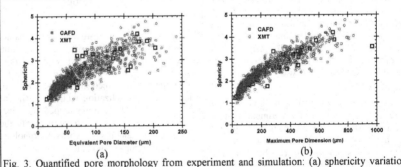

(a)   (b)

Fig. 3. Quantified pore morphology from experiment and simulation: (a) sphericity variation with maximum pore dimension, and (b) sphericity changes with equivalent pore diameter.

The influence of microstructure on the shape of microporosity can be observed from a comparison of typical individual pores from the experiment and simulations, Fig 4. The shape of pores observed in the casting, Fig 4 (a) and (c) shows a very similar flattened and tortuous fashion to those from the simulation Fig 4 (b) and (d). The pores have formed by expanding into thin interdendritic spaces, thus forming these bent sheet-like shapes, and this process has been reproduced by the simulation. However, there is a limitation of the resolution and domain size due to computational cost. This 3D mathematical model allows investigation of evolution of the morphology of microporosity as well as the primary phase during solidification.

Conclusions

Microporosity formation in 3D is studied experimentally and computationally. By comparing the simulations with experiments, the following points have been successfully achieved. (1) X-ray micro-tomography combined with image analysis is a promising approach to study the morphology and size of microporosity. (2) The mesomodel successfully predicts the observed 3D morphology of pores qualitatively and quantitatively. (3) Good correlation between the mathematical model and experiments demonstrate that the diffusion of gas and the impingement of microstructure are the key factors affecting the growth of pores. (4) The morphology of pores is not well described by volume alone. The pore shapes observed in casting are highly variable and complex. The model allows the subsequent use of the porosity simulation to predict large tortuous pores which dominates the crack initiation and propagation in castings.

In summary, a robust model is available to predict the 3D microporosity morphology as well as size distributions. Further investigations need to be done to compare the simulations with x-ray tomographic results using different processing conditions.

Acknowledgements

The authors would like to acknowledge support from Dorothy Hodgkin Postgraduate Awards, the funding from Ford Motor Company, and the EPSRC platform grant EP/C536312.

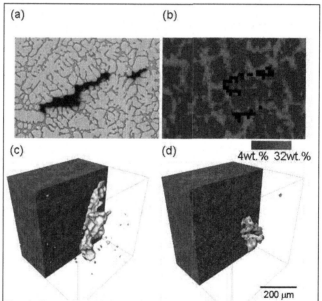

Fig. 4. Comparison of the morphology of microporosity in 2D and 3D from experiment and simulation: 2D images from (a) optical micrograph, (b) 2D slice from simulation showing the solute concentration profile and microporosity morphology, (c) XMT measured 3D grain and microporosity morphology; and (d) mesomodel predicted grain and pore morphology. (Note, in (c) and (d) the primary phase is rendered blue, eutectic/inter-dendritic liquid in green and pores in gold.) [22].

223

References

[1]   J.Z. Yi et al., "Scatter in fatigue life due to effects of porosity in cast A356-T6 aluminum-silicon alloys," *Metall. Mater.Trans. A*, 34A (9) (2003), 1879.

[2]   J. Huang et al., "Simulation of microporosity formation in modified and unmodified A356 alloy castings," *Metall. Trans. B*, 29B (6) (1998), 1249.

[3]   Q.T. Fang and D.A. Granger, "Prediction of Pore Size Due to Rejection of Hydrogen During Solidification of Aluminum Alloys," Light Metals, Las Vegas, Nevada, USA, 27 Feb.-3 Mar. 1989: TMS, 1989, p.927.

[4]   D.R. Poirier et al., "A Thermodynamic Prediction for Microporosity Formation in Aluminum-Rich Al--Cu Alloys," *Metall. Trans. A*, 18A (11) (1987), 1979.

[5]   R.C. Atwood and P.D. Lee, "Simulation of the three-dimensional morphology of solidification porosity in an aluminium-silicon alloy," *Acta Mater.*, 51 (18) (2003), 5447.

[6]   J.D. Zhu et al., "Simulation of microporosity in A356 aluminium alloy castings," *Int. J. Cast. Metals Res.*, 18 (2005), 229.

[7]   Q.G. Wang, "Fatigue behavior of A356-T6 aluminum cast alloys. Part I. Effect of casting defects," *J. of Light Metals*, 1 (1) (2001), 73.

[8]   W. Wang et al., "A model of solidification microstructures in nickel-based superalloys: predicting primary dendrite spacing selection," *Acta Mater.*, 51 (10) (2003), 2971.

[9]   C.C. Pequet et al., "Modeling of Microporosity, Macroporosity, and Pipe-shrinkage Formation during the Solidification of Alloys Using a Mushy-Zone Refinement Method: Applications to Aluminum Alloys," *Metall. Mater.Trans. A*, (7) (2002), 2095.

[10] J. Ampuero et al., "Modelling of Microporosity evolution during the solidification of metallic alloys," Materials Processing in the Computer Age, 17-21 Feb. 1991, New Orleans, Louisiana, USA: TMS, 1991, p.377.

[11] L. Nastac and D.M. Stefanescu, "Modeling of Growth of Equiaxed Dendritic Grains at the Limit of Morphological Stability," In: Piwonka TS, Voller V, Katgerman L, editors, MCWASP and Advanced Solidification Processes. VI,, Palm Coast, Florida, USA,, 1993, p.209.

[12] C.A. Gandin and M. Rappaz, "A 3D Cellular Automaton algorithm for the prediction of dendritic grain growth," *Acta Mater.*, 45 (5) (1997), 2187.

[13] C.H. Charbon and M. Rappaz, "3D Probabilistic Modelling of Equiaxed Eutectic Solidification," *Modell. Simul. Mater. Sci. Eng.*, 1 (4) (1993), 455.

[14] M. Rappaz and C.-A. Gandin, "Probabilistic Modelling of Microstructure Formation in Solidification Processes," *Acta Metall. Mater.*, 41 (2) (1993), 345.

[15] W. Kurz and D.J. Fisher, *Fundamentals of Solidification*, 1998).

[16] P.D. Lee and J.D. Hunt, "Hydrogen porosity in directionally solidified aluminium-copper alloys: A mathematical model," *Acta Mater.*, 49 (8) (2001), 1383.

[17] C.E. Ransley and H. Neufeld, "The solubility of Hydrogen in Liquid and Solid Aluminium," *J. Inst. Metals*, 74 (1948), 599.

[18] D. Doutre, *Alcan Int. Ltd, KRDC Lab., Internal report*, (1991).

[19] A. Chaijaruwanich et al., "Pore evolution in a direct chill cast Al–6 wt.% Mg alloy during hot rolling," *Acta Mater.*, on-line proof (2006).

[20] P.D. Lee and J.D. Hunt, "A model of the interaction of porosity and the developing microstructure," Welding and Advanced Solidification Processes VII, London, UK: TMS, 1995, p.585.

[21] P.D. Lee and J.D. Hunt, "Measuring the nucleation of hydrogen porosity during the solidification of aluminium-copper alloys," *Scripta Mater.*, 36 (4) (1997), 399.

[22] P.D. Lee et al., "Microporosity Formation during the Solidification of Aluminum-Copper Alloys," *JOMe (USA)*, (Accepted) (2006).

224

MODELING THE INFLUENCE OF MULTI-COMPONENT ALLOYING ADDITIONS ON MICROSTRUCTURE AND PORE FORMATION IN CAST ALUMINUM ALLOYS

Peter D. Lee, Junsheng Wang, Robert C. Atwood

Department of Materials, Imperial College London, South Kensington Campus, Prince Consort Road, London SW7 2AZ, UK, p.d.lee@imperial.ac.uk

Keywords: Porosity; Modeling; Microstructure; Aluminum alloys; X-ray Tomography.

Abstract

A three dimensional (3D) microstructural model was developed to simulate the nucleation and growth of grains and pores formed during the solidification of multi-component aluminum alloys. The model solves for the diffusion limited growth of the solid phases, combined with diffusion and shrinkage driven growth of pores. The interaction of the developing solid and gas phases was simulated and the resulting composition dependent curvatures were found to control the morphology of the pores formed. The predictions were compared to both *in situ* transmission radiographic observations of the kinetics of pore formation and 3D x-ray micro-tomographic observations of pore shape.

Introduction

Increasing demand in the aerospace and automotive industries for high strength to weight ratio components at low cost has made aluminum-based alloy castings an increasingly attractive solution. However, defects formed during the casting process have limited their application [1]. One of the most detrimental defects to the mechanical properties, especially to the fatigue life and total elongation, is microporosity formed during the solidification. These pores are formed due to the combined action of hydrogen segregation and solidification shrinkage [2].

Defect formation during the solidification of aluminum alloys involves many complex physical processes occurring over a wide range of spatial, temporal and energy scales. Multiscale modeling is a powerful methodology for understanding such complex systems [3]. The use of this methodology and its application to the study of solidification processes in metallic alloys promises to be an important factor for the improvement of alloy and process design for casting industries.

An existing in-house micromodel of the nucleation and growth of both grains and pores was extended to incorporate multi-component alloys. The pore-grain impingement was also solved in order to predict the final pore size, distribution and morphology [4, 5]. The temperature and pressure fields used in the micromodel may be interpolated from the solution of the heat and fluid flow calculations from macroscale simulations or applied as analytical functions.

In this paper, this in-house mathematical model was used to predict the shape, size, and distribution of porosity and the columnar/equiaxed dendritic structures in which they form. The model predicted size and shape of the pores was validated quantitatively by comparison to high resolution x-ray micro-tomography (XMT) analysis of experimental castings. The kinetics of pore formation was validated by comparing the predicted growth rates to those measured using *in situ* radiography. A good agreement between numerical and experimental results was found, clarifying, under some conditions, the mechanisms controlling the rate of pore growth and its complex morphology.

Experimental Methods

One technique which has been used for many decades for *in situ* observation of the development of the solidification structures in metals is radiography [6-8]. Lee and Hunt combined real-time microfocal radiography together with a temperature gradient stage [9] to produce a system with independent control of the growth velocity and thermal gradient. This system, termed an XTGS, allows *in situ* quantification of solidification microstructures including porosity [10]. With the recent availability of high flux synchrotron and high resolution laboratory sources, greater temporal and spatial resolution is now possible and several recent *in situ* studies have taken advantage of this technique [11-13].

In this study a high resolution microfocal source (Phoenix|X-ray nano-tome, Germany) was used to study the solidification of two alloys, Al-4 wt% Cu and Al-12 wt% Cu. In each digital frame the individual pores were characterized using the maximum pore dimension, L_{max}, calculated from the radiographs.

In addition to the *in situ* observations, x-ray microtomography was used to characterize the final solidification microstructure in three dimensions. Therefore, the same two alloys were cast into an instrumented permanent mould wedge casting rig using the methodology described in [14]. From a location at the middle of the wedge in thickness, and 100 mm from the bottom, a 2 mm diameter cylinder was machined and analyzed using the same micro-focus source as described above, but in tomography mode. An x-ray tube current of 80 µA and acceleration of 80 kV was used, obtaining a pixel size of 4.25 µm. The resulting data sets were quantified using VGStudioMax and visualized in Amira with L_{max} set equal to the diagonal of the rectangular box bounding each pore.

Mathematical Model

An in-house mesoscale model was used which incorporates stochastic rules for the nucleation of nuclei and pores, combined with partitioning-diffusion growth laws, is used to simulate pore and grain formation. The model comprises of:

1. Stochastic grain density based on local undercooling.

2. Stochastic pore density based on local supersaturation (the ratio between local hydrogen concentration in the liquid phase and the solubility limit of hydrogen gas in liquid).

3. Grain growth based on local constitutional and curvature undercoolings [15].

4. Pore growth based on hydrogen partitioning and diffusion.

5. Interaction between the grain and pore growth based on solute concentration-dependent hydrogen solubility and solid-restricted pore growth.

6. Pressure within the pore determined by the local metallostatic pressure (set via an analytic equation or through coupling to a macroscopic code) and a surface tension curvature term (which is set as a function of the fraction solid, dendrite arm spacing and pore shape).

These features are all solved in 3D, with details of most of the algorithms are given in prior publications [4, 5]. In the simulations reported here, a calculated constant cooling rate and constant decrease in pressure over time were used to simulate a small region within the experimental wedge casting.

226

Results and Discussion

Experimental Observations

The evolution of the solidification microstructure as observed in the XTGS for both the Al-4wt%Cu and Al-12wt%Cu alloys is shown in Fig. 1. The sequence of four time frames shows how the microstructure evolves as a function of time while the temperature decreases, as marked in the graph in Fig. 1c. The α-aluminum columnar dendrites are the lighter grey color and the copper enriched inter-dendritic liquid partitioned between the dendrites is darker grey. The primary and secondary dendrites are clearly visible in the Al-12wt%Cu (Fig. 1b) due to the large fraction of Cu enriched eutectic formed. The extent of Cu segregation, and hence x-ray attenuation variation, is much lower in Al-4wt%Cu. Therefore, in this alloy (Fig. 1a) the Cu depleted primaries are just visible and the secondary arms almost indistinguishable from the background noise.

Regions of even lighter grey (lower x-ray attenuation) are also visible (outlined in color). These are pores that have formed in the interdendritic regions and grow with time. Image analysis routines (ImageJ, NIH, Bethesda US and ImageMagick, ImageMagick Studio LLC, Landenberg, PA) were used to measure and outline these pores, and the results are shown as a graph of maximum pore length versus temperature in Fig. 1c. The color plotted matches the color used to outline the pores in Figs. 1a and 1b. In Al-4wt%Cu there are three pores that nucleate at a temperature of approximately 640°C. These pores initially grow slowly, then the pore outlined in red shows a sudden spurt of growth at approximately 570°C. This growth is almost entirely in a direction towards higher fraction solid (down the page in Fig. 1), suggesting it is expanding to feed the volume contraction of eutectic.

One of the pores (1, highlighted in green) can be seen to expand and contract in its maximum dimension as shown by the

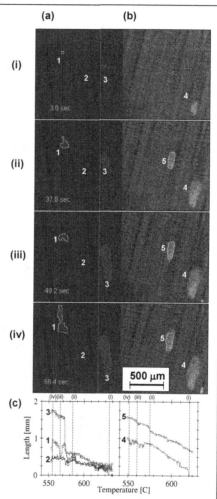

Fig. 1. Series of x-ray transmission images at four different times/temperatures (i– iv) showing the dendritic structure and pore growth during solidification in (a) Al-4%Cu and (b) Al-12%cu. (c) Graph of the maximum pore length versus time/temperature, with the indicated temperature points (i – iv) corresponding to the radiography frames. The colors and numbers of the pores match the colors and numbers of the lines.

227

oscillations in the green curve (1) in the graph. This pore (1, highlighted in green) appeared to split, rejoin and then split again, although this behavior is not captured with only four frames. This pore grows upwards towards the region of lower fraction solid, unlike the largest pore. This upward growth (towards lower fraction solid) was observed to be the most common during several similar experiments.

In the Al-12wt%Cu alloy the two pores shown in Fig. 1b nucleate at a much lower temperature (ranging from 610-595°C) but grow more quickly than in Al-4wt%Cu. All the pores expand in size, primarily in the low f_s (upward) direction, growth more regularly than the pores in Al-4%Cu. The growth appeared to occur in smaller spurts than in Al-4wt%Cu. The spurts were on the same scale as the secondary dendrite arm spacing - perhaps the pores are constrained by the secondaries, building up pressure in the pore until it can push through to the next gap. The size of the spurts was larger in the more constricted Al-4wt%Cu microstructures.

When comparing pore growth in the two alloys, it is surprising that although the nucleation temperatures and rates of pore growth are very different, the final pore sizes are similar. Examining the radiographs more closely in Fig. 1, it can be see that the attenuation contrast between the pores and alloy is quite different in the two cases. This difference is greater than can be explained by the increased attenuation of the additional copper, suggesting that the size of the pores in the through-thickness direction is different.

Three dimensional imaging via x-ray microtomography was performed on samples form wedge castings of both the Al4Cu and Al12Cu alloys with typical results shown in Fig. 2. The pores in

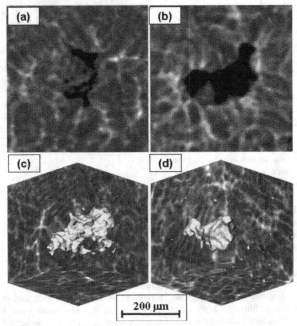

Fig. 2. XMT imaging of the Al-4wt.%Cu and Al-12wt.%Cu alloys. 2D slices from the reconstructions of (a) Al-4wt.%Cu and (b) Al-12wt.%Cu. 3D renderings showing the microsegregation on cut planes at the back of the volumes and the pores (gold) in the foreground for (c) Al-4wt.%Cu and (d) Al-12wt.%Cu.

228

Al-4wt%Cu are thin in cross-section (Fig. 2a) and highly tortuous (Fig. 2c), while the pores Al-12wt%Cu are thicker (Fig. 2b) and more rounded (Fig. 2d). In Al-12wt%Cu there is significantly more eutectic and hence the interdendritic spaces are wider and the pores can grow larger before they start spreading out between grains. In the Al-4wt%Cu the interdendritic spaces are narrow and hence the pore has to grow in the thin and tortuous interdendritic channels producing higher curvatures and hence higher surface tension components of the pressure in the pores. The high curvatures/interfacial energies may also explain the erratic growth of the green pore in Fig. 1a, as the resulting forces may be sufficient to move the solid dendrite surrounding the pore, and allow the pore to round out thus decreasing its projected length.

Model Results

The micromodel of grain/pore nucleation and growth was run to simulate the conditions and composition in the Al-12wt%Cu wedge casting. Fig. 3 shows the results of the model predictions compared to a pore observed by XMT in the Al-12wt%Cu wedge. The model (Fig. 3d) predicts the solute segregation and complex morphology of the pores caused by their interaction with the solid phases. The model correctly predicts the more rounded and thicker cross-sectional area in this alloy (compare Fig. 3 a and b). The fine detail of the secondary arms was not resolved in the

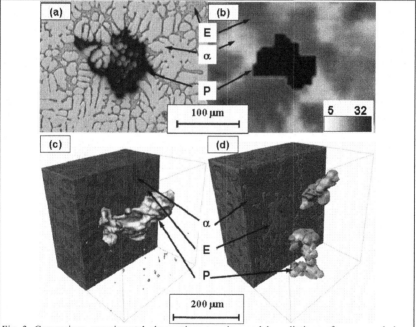

Fig. 3. Comparison experimental observations to micromodel predictions of pore morphology and microstructure for Al-4wt%Cu solidified at 2.9°C/s and a hydrogen level of 0.23 ml/100g STP. (a) Optical micrograph, showing pores (P) primary phase (α) and eutectic (E) (b) 2D slice from simulation showing the solute segregation. In both (a) and (b) the pore (P) is black. (c) XMT image of dendrites (α, blue), eutectic/inter-dendritic liquid (E, green) and pores (P, gold). (d) Micromodel prediction using the same coloring as (c).

model since a relatively coarse mesh was used. However, the complexity in pore shape is still present despite the averaging caused by the coarse mesh.

The XTGS quantified rate of pore growth is compared to micromodel predictions in Fig. 4 for a few pores in both the experiments and model. The shape of the curve, as well as the final sizes, are in reasonable agreement, illustrating that the model can predict the kinetics of defect formation (Fig. 4) as well as final size and morphology (Fig. 3). This model incorporates both the gas and shrinkage effects in order to obtain a reasonable match in the kinetics. Other factors such as the curvature or interfacial energy terms, which are only calculated via a first order approximation in the micromodel, could account for the minor variations at low fractions solid.

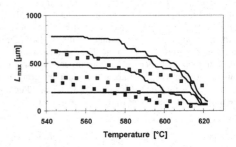

Fig. 4. Micromodel predicted (solid lines) and XTGS measured (squares) maximum pore length as a function of temperature during the solidification of Al-12wt%Cu.

<u>Influence of Ternary Alloy Additions on Pore Formation</u>

In order to investigate the influence of multicomponent additions on pore formation with this micromodel, simulations were run on an Al-3wt%Cu binary alloy and an Al-7wt%Si-3wt%Cu ternary. The simulation conditions were a cooling rate of 5°C/s and a hydrogen content of 0.23 ml/100g STP. The results are shown in Fig. 5.

The addition of 7wt% silicon to Al-Cu alloy results in a greater amount of eutectic phase, i.e. a greater fraction of liquid persisting to a lower temperature. This allows pores to expand in the liquid before eutectic reaction and develop rounded shapes since they are not restricted by the solid as soon as those in the Al-3wt%Cu alloy. In the latter case (Fig 5a), the fraction solid is greater by the time the pores begin to form, and the pores have to grow into narrow spaces between dendrites. In the former case (Fig 5b) the pores are able to take on nearly rounded shapes.

Moreover, the solubility limit of hydrogen in the ternary alloy is altered, decreasing the amount of H segregation necessary to form pores and therefore causing pores to form at even smaller fraction solid. These combined effects result in pores that are more tortuous in the Al-3%Cu binary alloy than in the Al-3%Cu-7%Si ternary alloy, even though the same cooling rate and hydrogen concentration has been applied in both simulations.

A similar effect was observed experimentally in Fig. 1 and 2 and in the simulations of Fig. 3 due to the increase in copper concentration, which also increases the volume of eutectic and alters the hydrogen solubility. These results highlight the importance of alloying elements on pore nucleation and growth and the capability of available models to predict this behavior.

Conclusions

A microscale model was developed to simulate the nucleation and diffusion controlled growth of grains and pores. The predicted pore size, morphology and rate of growth all compare well to experimental observations made using *in situ* real-time radiography observation and *ex situ* x-ray micro-tomography. The strong influence of the alloying content on the morphology of the pores was illustrated both experimentally and via the simulations.

Acknowledgements

The authors would like to acknowledge partial support for this research from an EPSRC platform grant (EP/C536312) and for one of the author's (JW) from Ford Motor Company and EPSRC through a Dorothy Hodgkin Postgraduate Award.

References

[1] J. Campbell, *Castings*, 2 Ed. (London: Butterworth-Heinemann, 2003), 337.

[2] P.D. Lee et al., *J. Light Metals*, 1 (1) (2000), 15.

[3] H. Rafii-Tabar and A. Chirazi, *Phys. Reports*, 365 (3) (2002), 145.

[4] R.C. Atwood and P.D. Lee, *Acta Mater.*, 51 (18) (2003), 5447.

[5] W. Wang et al., *Acta Mater.*, 51 (10) (2003), 2971.

[6] J. Forsten and H.M. Miekk-oja, *J. Institute of Metals*, 95 (1967), 143.

[7] E.W.J. Miller and J. Beech, *Metallography*, 5 (--3--) (1972), 298.

[8] M.P. Stephenson and J. Beech, Proc. Int. Conf. Sol. Casting Metals, 18-21 July 1977 Metals Society, 1977, p.18.

[9] J.D. Hunt et al., *Review Sci. Instrum.*, 37 (6) (1966).

[10] P.D. Lee and J.D. Hunt, Mod. Casting, Welding and Adv.Sol. Processes VII, London, 10-15 Sept. 1995: TMS/AIME, Warrendale, PA, 1995, p.585.

[11] R.H. Mathiesen and L. Arnberg, *Acta Mater.*, 53 (2004), 947.

[12] R.H. Mathiesen and L. Arnberg, *Mat. Sci. Eng. A*, 413-414 (2005), 283.

[13] O. Ludwig et al., *Metall. Mater.Trans. A*, 36 (A) (2005), 1515.

[14] J.Z. Yi et al., *Metall. Mater. Trans. A*, 34A (9) (2003), 1879.

[15] W. Kurz and D.J. Fisher, *Fundamentals of Solidification*, 1998).

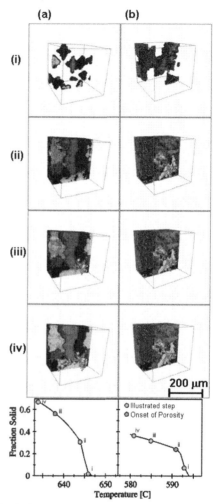

Fig. 5. Comparison of porosity and microstructure in binary and ternary alloys: (a) Al-3wt.%Cu and (b) Al-3wt.%Cu-7wt.%Si. Frames (i-iv) are for different times during solidification simulation at the points indicated by circles in (c), the graphs of solid phase evolution. Solid grains are colored individually and the pore isosurface is shaded in gold. The metal in the near half of the domain is rendered transparent.

231

Shape Casting: The 2nd International Symposium *Edited by Paul N. Crepeau, Murat Tiryakioğlu and John Campbell*
TMS (The Minerals, Metals & Materials Society), 2007

A MODEL FOR PREDICTION OF SHRINKAGE DEFECTS AS A RESULT FROM THE REDUCTION IN PRESSURE DURING FEEDING

A. Reis[1], Zhian Xu[2], Rob Van Tol[2], A.D.Santos[1], E. Wettinck[3], A.B. Magalhães[1]

[1] FEUP – Faculdade de Engenharia da Universidade do Porto
R.Dr.Roberto Frias s/n, 4200-110 Porto, Portugal e-mail: arlr@fe.up.pt
[2] WTCM Foundry Center, Research Center of Belgium Metalworking Industry9052
Zwijnaarde, Gent, Belgium
[3]UGent – University of Ghent, Technologiepark Zwijnaarde 903, Gent, Belgium

Keywords: Casting process, Numerical Modeling, Shrinkage Defects.

Abstract

A 3-D feeding flow model based on the pressure drop concept has been developed to evaluate shrinkage defects for casting alloys. By combining Darcy's law, which governs fluid flow in the mushy zone, with the equation for Stokes flow, which governs the motion of a slow flowing liquid, it is possible to derive a momentum equation that is valid everywhere in the solution domain. A pressure equation is then derived by combining this momentum equation with a continuity equation. The model solves the coupled macroscopic conservation equations for mass, momentum and energy with phase change.

It is assumed that the degree of feeding depends on the pressure drop at any location of a solidifying region. The higher the pressure drop, the greater the possibility of porosity formation.

It must be noted here that the present model doesn't deal with mechanisms of pore formation. However the value of pressure drop is associated with pore formation and the probability of pore formation can be inferred. An adapted free surface algorithm is used to model surface defects "sinks" and also for internal pores. A finite volume mesh is used to solve the equations numerically.

The model was applied to the study of a simple geometry to illustrate the basic physical phenomena involved and to test the ability of the model to predict the progression from internal porosity to the external sinking as it goes from short to long freezing range alloy.

Introduction

Shrinkage related defects in shape casting are a major cause of casting rejections and rework in the casting industry. The most important defects that arise from shrinkage solidification are:

- External defects– pipe shrinkage and caved surfaces
- Internal defects –macroporosity and microporosity

Shrinkage related defects result from the interplay of several phenomena such as heat transfer with solidification, feeding flow and its free surfaces, deformation of the solidified layers and the presence of dissolved gases. There have been many attempts to model shrinkage related defects. An extensive review on these models has been made by Lee [1] and later Stefanescu [2].

The initial effort of casting simulation was to develop codes that only analyzed the solidification behavior by heat conduction models a solution of the energy transport equations. For defect prediction they use criteria functions: empirical models for evaluation of shrinkage porosity based on some relationship to the local temperature gradient. The most well known is the Niyama Criterion [3], based on finding the last region to solidify as most probable to present shrinkage defects. These and other functions have been summarized by Berry et al. [4], Spittle [5] and later evaluated by Taylor [6].

Some models were based on solving the heat transport and mass conservation to predict the position of the free surface and macro shrinkage cavity. Trovant [7] proposed a model to account for shrinkage and consequently determine the shrinkage profile resulting from phase and density change. Beech et al. [8] presented a method for macro-shrinkage cavity prediction based on a continuum heat transfer model which determines when an area will be completely cut off from sources of liquid metal (such as risers) where a void will form to account for volume deficit. Its size is calculated through the mass conservation equation.

A more complex approach is to consider feeding flow analyses. The first model that took into account feeding flow dates back to the early 1D analytic work of Piwonka & Flemings [9]. This early analytical work formed the basis of a later category of models based upon Darcy's law which relates the flow through a porous medium to the pressure drop across it. Kubo & Pehlke [10] were the first to couple Darcy's law to the equations of continuity estimating the fluid flow in a 2-D numerical model. Later other 2-D models were developed by Zhu [11] and Huang [12].

In terms of 3-D models, Bounds et al. [13] presented a model that predicts macroporosity, misruns and shrinkage pipe in shaped castings. Later Sabau & Viswanathan[14], Pequet et al.[15] and Carlson et al. [16] also presented 3-D models that included the concept of pore nucleation and growth.

In the present model the volume deficit due to shrinkage can only be compensated by two phenomena: depression of the outside surface or by creating internal pores. The model is tested in a simple test case using a short and long freezing alloy. The objective is to test the ability of the model to predict the progression from internal porosity to the external sinking as it goes from short to long freezing range alloy.

Model description

The model calculates the pressure distribution and feeding flow during the solidification and couples it to the formation of shrinkage porosity defects, internal porosity and external sinking.
The coupled macroscopic conservation equations for mass, momentum, and energy are solved. Movement of the free surface, both external and internal, is considered to ensure volume conservation and to model the shrinkage defects. A Finite Volume method is used to solve the equations numerically employing a pressure correction algorithm to couple the mass and momentum continuity equations.

Continuum equations

The shrinkage induced flow is described by the term $\partial \rho / \partial t$. Liquid flow will be driven toward the volume element when the average density increases with time, which is the case of most alloys:

234

$$\frac{\partial \rho}{\partial t} + \nabla(\rho \mathbf{V}) = 0 \qquad (1)$$

where $\mathbf{V}$ is the velocity vector and ρ is the density.

Neglecting the influence of convection and diffusion during solidification, the feeding velocities in the casting are determined from the momentum equation:

$$\frac{\partial}{\partial t}(\rho \mathbf{V}) = -\nabla P + \rho g + S \qquad (2)$$

where P is pressure and ρg the body force. It is assumed that when solidification starts the solid particles adhere and form a cohesive immobile network. Liquid flow through the network and can be approximated as a Darcy flow through a porous medium. This is represented in the model by adding a momentum loss "S". The Darcy term is represented by:

$$S = -K\mathbf{V} \qquad (3)$$

where K is the permeability of the solid skeleton. K reflects the resistance of the already solidified structure to the feeding flow. K depends on the amount of solid fraction, fs, and can also account for the solidification morphologies in the form of C_{drg} coefficient.

In this model a Kozeny-Carman relation for permeability was used:

$$K = C_{drg} \frac{f_s^2}{(1 - f_s)^3} \qquad (4)$$

where C_{drg} [1/s] is a characteristic dimension, typically function of the secondary dendrite arm spacing [17].

The solid fraction represents the amount that is solid in a cell and is obtained through the internal energy of that cell. It is reasonable for many cast alloys to assume that the fraction solid is derivable from the solution of the energy conservation equation that can be expressed in terms of temperature as:

$$\frac{\partial}{\partial t}(\rho c_p T) + \nabla(\rho c_p \mathbf{V} T) = \nabla(k\nabla T) + \frac{\partial}{\partial t}(L\rho f_s) \qquad (5)$$

where L is the latent heat of fusion of the alloy, c_p is the specific heat and k is the thermal conductivity, the evolution of solid fraction is calculated with:

$$f_s = \frac{T_l - T}{T_l - T_s} \qquad (7)$$

where T_l and Ts are the liquidus and solidus temperature, respectively.

The free surface of the liquid is located by means of an adapted fluid function F [18] which represents the fraction of the element filled with liquid. The value F is 0 for empty elements, 1 for full elements and between 0 and 1 for surface elements. The fluid function F is satisfied by the conservation equation:

$$\frac{\partial F}{\partial t} + \nabla F\mathbf{V} = 0 \qquad (8)$$

Shrinkage defects formation

The governing equations are solved to give the pressure and feeding velocity through the casting. When solidification starts the casting will be fed by the highest point, usually the feeder. The pressure in the liquid begins to drop due to the solid crystals that form and as soon as the pressure in other surface element, besides the top of the riser, reaches the atmospheric pressure they become also free surface elements that can feed the solidifying casting to compensate metal contraction. This will result in external shrinkage defects also called "surface sink".

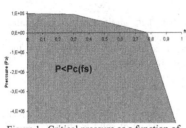

Once the feed path becomes obstructed by solid or partially solid material the liquid pressure quickly drops and when an element reaches a critical value, $P_c(fs)$, Figure 1, it becomes a free surface element resulting in a pore that grows to account for the mass deficit., determined from the continuity equation.

Figure 1 –Critical pressure as a function of fraction solid

Numerical method

The finite difference method is used to discretize the governing equations. A 3-D structured orthogonal mesh is employed to discretize the mold and the casting. A staggered grid serves to discretize the flow fluid equations

The prediction of fluid flow coupled with heat and mass transfer, requires a particular algorithm to obtain the solution from the transport equations. The Semi-Implicit-Method for Pressure Linked Equations, SIMPLE method, described in detail by Patankar [19] is used to handle the velocity coupling for the momentum equations. The velocity field is calculated from the momentum and continuity equations at each time step using the updated properties of each volume element, such as solid fraction, density and permeability. Within each time step, the temperature distribution used to calculate solid fraction was obtained by solving the energy equation. Next, the properties of each volume element were updated again. Iteration is continued until the convergence criterion is satisfied. The movement of the free surface is evaluated using the velocity field. In the beginning only the top riser is a free surface.

After each time step, the condition $P<P_{atm}$ in any of the outside layer volume elements, is checked and if satisfied this element becomes a free surface element as well.

At each time step the condition for pore nucleation $P<P_c(fs)$ is tested. If this happens this volume element will be treated as a free surface element and the amount of porosity that forms is determined from the continuity equation.

Results

To illustrate the features of the developed model, a simple geometry was simulated. This geometry is a block (35 x 35 x 35mm) surrounded with sand mould (15mm), Figure 2. Two alloys with a binary eutectic phase, a short freezing and a long freezing alloy, were analyzed.

Table 1 shows the calculation parameters for the two alloys. The parameters, were determined by comparing material data for different AlSi alloys.

Table 1. Calculation properties and parameters

Properties and parameters	Short freezing	Long freezing
Liquid density (kg/m³)	ρ_l = 2600 (kg/m³)	ρ_l = 2600 (kg/m³)
Solid density (kg/m³)	ρ_s = 2700 (kg/m³)	ρ_s = 2700 (kg/m³)
Liquidus Temperature	T_l =580 (°C)	T_l =580 (°C)
Solidus Temperature	T_s =565 (°C)	T_s =530 (°C)
Initial melt Temperature	T_0=590 (°C)	T_0=590 (°C)
Conductivity	=181 W(m.K)	=181 W(m.K)
Heat capacity	=1150(W/(kg.K)	=1150(W/(kg.K)
Cdrg (permeability constant)	10E6	10E6

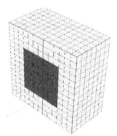

Figure 2. Mesh of finite volume sand+casting (middle cut)

Short freezing material

At the beginning of solidification unrestrained feeding occurs and the top level will sink. Once solid crystals begin to form the flow will be restricted.

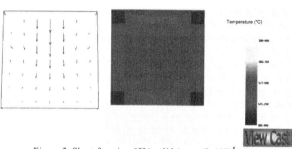

Figure 3. Short freezing 68% solid (v_{max}= 7. 10E^{-5}m/s)

The surface shape depends on the remaining surface area open to the liquid. During solidification because the corners solidify faster the open surface is reduced and sinking velocity increases, Figure 3. This mechanism results in a steep slope in the top as illustrated.

As solidification evolves the feeding is restrained resulting in a pressure drop as shown in Figure 5. The pressure distribution and the solid fraction are plotted for the same amount of solidification 68%. At this point, a cell fulfils the conditions for a pore to open: P<P_c(f_s). A pore will form and grow to account for the need of feeding as shown in Figure 7 for 90% of solidification.

Long freezing material

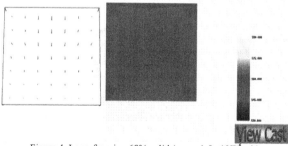

Figure 4. Long freezing 68% solid (v_{max}= 1.5x 10E^{-5}m/s)

In the long freezing material the top feeding velocity is much more uniform, Figure 4, resulting in a flatter top surface.

The condition for a pore to open isn't met, Figure 6. The principal mechanism for feeding is carried by the outside layers. As shown in detail in Figure 8 for 90% already solidified. Only at the end (98% solid) some cells will fall under the criteria P<P_c(f_s) and very small pore will form as shown in the final shape, Figure 9.

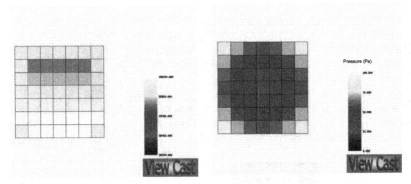

Figure 5.Pressure distribution (68% solid) short freezing

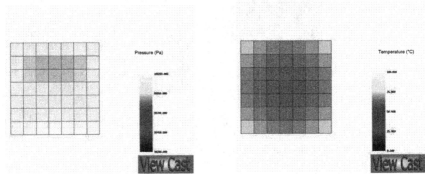

Figure 6. Pressure distribution (68% solid) long freezing

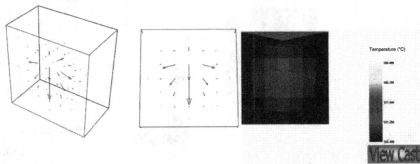

Figure 7. Short freezing, 90% solid (v_{max}= 15x 10E^{-5}m/s)

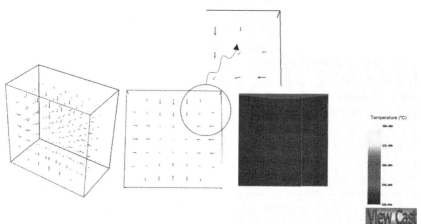

Figure 8. Long freezing, 90% solid ($v_{max} = 0.5 \times ^{-5}$m/s) and detail of side movement

Figure 9 shows the resulting geometry for both cases (middle section). They illustrate how the freezing range was found to have a pronounced effect on the type, internal or external, distribution of porosity. The short freezing range alloy exhibits maximum internal shrinkage porosity and the top shape is much more sharp, where the long freezing range alloy exhibits maximum external porosity and practically zero internal porosity.

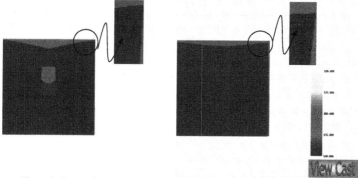

Figure 9. Comparison between the final in a short and a long freezing material

Conclusions

A numerical model was presented, that can qualitatively describe the movement of the melt surface as well as the internal porosity formation during solidification.

The calculated examples, short a long freezing alloy show a promising agreement with expected results.

The model parameters still need to be estimated to quantify the effects.

References

1. P. D. Lee, A. Chirazi, D. See, "Modeling microporosity in aluminum-silicon alloys: a review", Journal of Light Metals, 1 (2001), 15-30.

2.   D. M. Stefanescu, "Computer simulation of shrinkage related defects in castings – a review", Foundry Trade Journal, (2005), 189-194.

3.   E. Niyama et al., "Method of shrinkage prediction and its application to steel castings practice" AFS Casting Meals Research Journal, (1982), 52-63.

4.   R. A. Overfelt, R. P. Taylor and J. T. Berry, "Dispersed Porosity in Long Freezing Range Aerospace Alloys", Proceedings of the 4th Decennial International Conference on Solidification Processing, ed J. Beech and H. Jones, (1997), 248-250.

5.   J. A. Spittle, S. G. R. Brown and J. G. Sullivan, "Application of criteria Functions to the Prediction of Microporosity Levels in Castings", Proceedings of the 4th Decennial International Conference on Solidification Processing, ed J. Beech and H. Jones, (1997), 251-255.

6.   R. P. Taylor and J. T. Berry, "A critical evaluation of criteria functions for use with shaped-casting modeling", Modeling of Casting, Welding and Advanced Solidification Process VIII, ed. B.G. Thomas and C. Beckermann, (The Minerals and Materials Society, 1998), 1055-1062.

7.   M. Trovant and S. A. Argyropoulos, "Mathematical modeling and experimental measurements of shrinkage in the casting of metals", Canadian Metallurgical Quarterly, 35 (1996), 77-84.

8.   J. Beech et al "Computer modeling of the formation of macroshrinkage cavities during solidification", Modeling of Casting, Welding and Advanced Solidification Process VIII, ed. B.G. Thomas and C. Beckermann, (The Minerals and Materials Society, 1998), 1071-1078.

9.   T. S. Piwonka and M. C. Flemings," Pore formation in solidification", AIME Met. Soc. Trans., 236 (1966), 1157-1165.

10. K. Kubo and R. d. Pehlke, "Mathematical modeling of porosity formation in solidification", Metallurgical and Materials Transactions B, 16B (1985), 359-366.

11. J. D. Zhu and I. Ohnaka, "Computer simulation of interdentritic porosity in aluminum alloy ingots and casting", ", Modeling of Casting, Welding and Advanced Solidification Process V, ed. M. Rappaz, M. R. Ozgu and K. W. Mahin (The Minerals and Materials Society, 1991), 435-442.

12. J. Huang, T. Mori and J. G Conley, "Simulation of microporosity formation in modified and unmodified A356 alloys castings", Metallurgical and Materials Transactions B, 29B (1998), 1249-1260.

13. S. Bounds, G. Moran, K. Pericleous, M. Cross and T. N. Croft, " A computational model for defect prediction in shape castings based on the interaction of free surface flow, heat transfer, and solidification phenomena", Metallurgical and Materials Transactions B, 31B (2000), 515-527.

14. S. Sabau and S. Viswanathan, "Microporosity prediction in aluminum alloy casting", Metallurgical and Materials Transactions B, 33B (2002), 243-255.

15. Ch. Pequet, M. Gremaud, M. Rappaz, "Modeling of microporosity, macroporosity, and pipe-shrinkage formation during the solidification of alloys using a mushy-zone refinement method: application to aluminum alloys", Metallurgical and Materials Transactions A, 33A (2002), 2095-2106.

16. K.D. Carlson, Z. Lin, R. A. Hardin, C. Beckermann, G. Mazurkevich, M. Schneider, "Modeling porosity formation and feeding flow in steel casting", Modeling of Casting, Welding and Advanced Solidification Process X, ed. D. M. Stefanescu, J. Warren, M. Jolly and M. Krane, (The Minerals and Materials Society, 2003), 295-302.

17. M.C. Shneider and C. Beckermann, "A Numerical Study of the combined effects of microsegregation, mushy zone permeability and flow, caused by volume contraction and thermosolute convection, on macrosegragation and eutectic formation in binary alloys solidification", Int. Journal Heat Mass Transfer, 38 (1995), 3455-3473

18. Z. Xu,"Fluid flow and thermal analysis during mould filing and solidification castings" (ph.D. thesis, Ghent University, 1993)

19. S. V. Patankar: Numerical heat transfer and fluid flow", (Washington, DC: Hemisphere Publishing, 1980).

240

Shape Casting: The 2nd International Symposium *Edited by Paul N. Crepeau, Murat Tiryakioğlu and John Campbell*
TMS (The Minerals, Metals & Materials Society), 2007

RELATIONSHIP BETWEEN HTC EVOLUTION, GAP FORMATION AND STRESS ANALYSIS AT THE CHILL INTERFACE IN ALUMINUM SAND CASTING

A.Meneghini[1], L.Tomesani[1], G. Sangiorgi Cellini[1]

[1]University of Bologna, Dept. DIEM; Viale Risorgimento 2; Bologna, 40136, Italy

Keywords: Stress analysis, Heat Transfer Coefficient, Sand casting, Gap formation

Abstract

The effect of the interface pressure on the metal-chill Heat Transfer Coefficient (HTC) during the cooling of a A356 aluminum alloy in a gravity sand casting process is analyzed. HTCs are evaluated by inverse analysis in different processing conditions, by varying both chill material (copper, aluminum and cast iron) and size (30, 60 and 90 mm depth). All the experimental conditions are reconstructed by numerical analysis with the aim of evaluating the increase of contact pressure between casting and chill before the formation of an air gap. The experimental heat flow as determined in each experiment was used as a boundary condition for the calculations. The evaluated interface pressures were put in relation to the particular HTC evolution in the different experiments in order to point out correlations.

Introduction

During the solidification process in gravity casting of aluminum alloys, complex phenomena take place at the chill-material interface as a consequence of heat flow: cooling, solidification and shrinkage of the casting material, heating and expansion of the chills [1,2,3,6]. Therefore, the global heat flow between cooling material and chill is determined by the particular history of the interface at each position; it is therefore evident that both local temperature and pressure, together with physical variables related to processing materials and surface conditions, will all affect the HTC evolution.

The importance of the formation of an air gap due to the casting shrinkage is widely known; Htcs are frequently modeled with two distinct formulations, before and after the insurgence of an air gap, respectively. In gravity sand casting, the particular position of the chill surface and the casting contraction behavior determine different heat transfer conditions on different chills. The way the two surfaces separate the one from the other plays an important role as well in determining the HTC evolution; this is confirmed by the concave or convex shape of the casting surfaces once cooled. The air gap, then, does not form instantly, but develops over a period of time during which two different heat transfer phenomena are simultaneously active [9,10]. A third factor, the nature of the contact at the still adhering surfaces, is complex. This is composed of both actual contacting surfaces and entrapped gas. The type of gas, together with the distribution and dimensions of the entrapped bubbles (dependent on the surface roughness and topography) will determine the HTC on a particular surface.

Despite the relevant research aimed at HTC modeling, data in the literature are still to be considered as representative of the particular experimental condition; to use the data in practical conditions is still considered unacceptable. [12,13,14]. This is particularly true in gravity sand casting, which is a sophisticated process aimed at high performance components such as engine heads, and other automotive and aerospace products. Here, great variations arise in different chills, owing to their different dimension [15], material [16], orientation and metal head [17]. A little-investigated matter in HTC modeling in the gravity sand casting process is that of the effect of the interfacial pressure. During cooling, two distinct phenomena are observed: the casting shrinkage and the chill thermal expansion. Depending on the chill variables (material, dimension, location, surface) and in particular casting shrinkage behavior, a temporary increase in interface pressure may develop, before the eventual air gap formation.

The aim of the present work is to describe the history of this temporary interface pressure, to try to quantify it by a mechanical point of view, and, if possible, to correlate such pressure increment to the HTC evolution. With this aim, HTC data obtained from sand casting experiments in different cooling conditions were used to reconstruct by numerical analysis, the interface cooling history and to evaluate the development of a temporary pressure on it. Then, the comparison between the HTC experimental behavior and the interface pressure is performed in order to become aware of possible correlations. It must be noted that this pressure contribution is different from what happens in squeeze casting and die casting, as it is not the result of an external process parameter but is produced inside the cooling system to a different extent at each point.

Experimental

An A356 alloy was cast and cooled in an assembly described in [15] and represented in figure 1. HTC values were evaluated in a side arm (20x20 mm section), where a chill end ensured a dominant unidirectional heat flow during cooling. The surfaces of the chills, made of cast iron, copper and aluminum in different experiments, were those typically used in foundry practice, having a mean Ra value of 0.8 μm and with a square grid 10 mm side of about 1 mm depth grooved on the surface. The particular surface design is aimed at improving the casting-chill contact time.

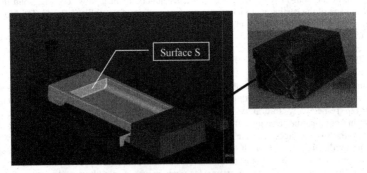

Figure 1. CAD model of the experimental equipment and chill geometry (left) and a photograph of cooling interface (right).

Two K-type thermocouples of 1 mm diameter were inserted in the chills at 6 and 12 mm from the interface, respectively. Two more thermocouples of 0.5 mm diameter were placed in the casting at 6 and 100 mm from the cooling surface. Temperature data, which were taken every for 120 s of cooling, were used for HTC evaluation at the casting-chill interface.

The alloy, melted in an electric furnace, degassed with argon and modified with strontium, was kept still until a temperature of 730 °C was reached; it was then poured into the mold. A silica sand used in the casting experiments. Before casting, the assembly was heated with a C_2H_2 flame for 10 min from the basin.

Figure 2. Casting and chill at the end of the cooling;
thermocouple locations are visible as well as the interface gap.

To solve the inverse heat conduction problem under the one-dimensional heat flow assumption, a non-linear estimation method was used [7,8,11] by considering the phase change and temperature-dependent thermal properties of the casting and the chill. The inertia in the thermocouples during the temperature measurement is also considered. The HTCs at the interface are evaluated with the following method: the two chill temperatures (T_2 and T_3) are first used to determine the heat flux Q at the chill interface S:

$$Q = -S \cdot \lambda \cdot (T_2 - T_3) \tag{1}$$

where λ is the chill conductivity. Based on the general Fourier equation for heat transfer, the latent heat of solidification and also considering the temperature T_1 measurement, the surface temperature of the chill Tb_{chill} and the casting temperature $Tb_{casting}$ at the casting-chill boundary can be evaluated [9]. Finally, the chill and casting temperatures at the boundary interface S are used to evaluate the HTC on that surface:

$$HTC(t) = S \frac{T_{bcasting}(t) - T_{bchill}(t)}{Q(t)} \tag{2}$$

The estimation technique assumes that the heat flux is either a constant or a linear function of time within a given time interval, then determines the heat flux in that period by solving the least-squares problem of the following function:

$$F(q) = \sum_{t}^{N1} \sum_{j}^{N2(N3+1)} (T_{ij} - Y_{ij})^2 \tag{3}$$

where T_{ij}, Y_{ij} are the calculated and measured temperatures at location i and time instant j respectively, N1 is the number of internal points in temperature measurement (excluding those used as boundary conditions), $N2$ is the number of temperature measurements per time interval, $N3$ is the number of "future" time intervals considered for heat flux calculation in each time interval (the future temperatures at a certain location are the measured temperatures ahead of each time interval, which are used for solving the heat flux or the temperature at the current time interval). By applying $\partial F/\partial q = 0$ for minimization, we have:

$$\sum\sum\left(Y_{ij}-T_{ij}\right)\frac{\partial T_{ij}}{\partial q}=0 \tag{4}$$

Then, using the Taylor series expression of $T_{ij}\left(q\right)$, we have:

$$T_{ij}\left(q_{l+1}\right)\approx T_{ij}\left(q_{l}\right)+\Theta_{ij}\delta q_{l+1} \tag{5}$$

where $\Theta_{ij}=\partial T_{ij}/\partial q$ can be solved by a numerical method; the following iterative expression for solving the heat flux at each time interval is then obtained [3,4]:

$$\partial q_{l+1}=\frac{\sum\sum\left(Y_{ij}-T_{ij}\left(q_{l}\right)\right)\Theta_{ij}}{\sum\sum\Theta_{ij}^{2}} \tag{6}$$

By repeatedly applying the expression $q_{l+1}=q_{l}+\delta q_{l+1}$ to correct the heat flux at location $q=q_{0}\left(l=0\right)$, the heat flux q at the time interval can be found when $\delta q_{l+1}/q_{l}$ is small enough. The HTC as a function of time with copper, aluminum and cast iron chills of the experiments are shown in figure 3.

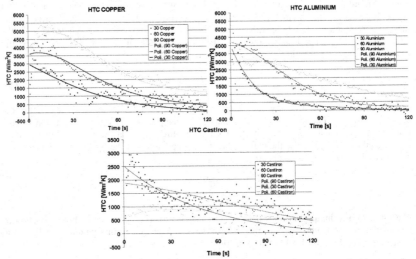

Figure 3a,b,c. Evolution of HTC with time for A356 alloy cast against copper, aluminum and cast iron chills

Numerical analysis

The casting processes were reconstructed by complete thermal-stress numerical analyses. The casting was modeled by means of 1,300,000 tetrahedral elements, the mold with 1,500,000 elements, the two chills with 40,000 and 20,000 elements, respectively. Nodal distance was 1 mm; chill and casting meshes were coincident. Only the cooling phase was considered, with a initial temperature of 710 °C. Heat transfer coefficients at the chill-casting interfaces were those obtained from the inverse analysis experiments as a function of time. In the early stages of cooling, no data were available from the inverse analysis and HTCs were adjusted in order to fit experimental temperature at the thermocouple locations. HTCs at the casting-sand and chill-sand interfaces were assumed to be constant at 500 $Wm^{-2}K^{-1}$. Thermal properties of the casting alloys and of the chill materials were obtained from the database supplied by the software. Simulation runs were interrupted when the temperature difference at the casting-chill interface was below 80 °C. ProCAST was the commercial code used for the calculations.

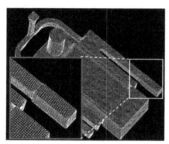

Figure 4. Mesh of the casting system

Nodal temperatures at the thermocouple locations in the casting (6 mm from the interface) and in the chill (6 and 12 mm from the interface) were used for validation. The agreement between the experimental and calculated temperatures was considered acceptable when they were within 20 °C. Boundary conditions for the chills displacements were given only at the face opposed to the casting interface, by locking horizontal displacements at each node (Figure 5). In this way, lateral thermal expansion of the chills are allowed, owing to the poor strength of sand to compression loads. Boundary conditions for the casting displacements were given at the surface S of figure 1, by locking horizontal displacements; no other constraints were considered for the mold during the casting shrinkage ("vacant mold" option in the code preprocessor).

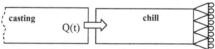

Figure 5. Representation of displacement and heat boundary adopted for the numerical simulations

Results and discussion

In Table 1 the summary of the main simulation results are shown. The linear expansion of the chills were always evaluated at the center of the interface. However it was found that the separation starts at the periphery of the interface and ends at its center. Thus, the interface center point is always the last to detach and its separation time can be considered as the end of the casting-chill contact.

The maximum chill expansion is not reported here because it always occurs after the air gap is formed. Instead, the chill displacement at a specific time is a good parameter for showing behavior of different cooling systems. The chill displacements at 100 s increase with increasing the chill dimension. Clearly, the chill material has a great effect: copper chills had the highest linear expansion (0.099-0.134-0.183), aluminum chills gave intermediate values (0.062-0.111-0.131) while cast iron ones had the lowest expansion properties (0.016-0.022-0.085).

245

The casting-chill separation time (which represents the end of the separation phase) also depends on the chill material, always increasing with the chill dimension. Very high separation times were found for copper chills (>100 s); aluminum chills obtained separation time between 100 and 125 s, cast iron chills had it in the range 39-55 s. These results can be related to the HTC as a function of time of figure 3: there, the "high HTC" zone with aluminum and copper chills is

Table I. Results of numerical analyses

	Chill expansion [mm] at 100 [s]	Casting-chill separation time [s]	Maximum Pressure [MPa]
Copper 90 mm	0.183	>125	5
Copper 60 mm	0.134	>125	5
Copper 30 mm	0.099	100	5
Aluminum 90 mm	0.131	>125	6
Aluminum 60 mm	0.111	125	6
Aluminum 30 mm	0.062	100	6
Cast iron 90 mm	0.085	55	1.5
Cast iron 60 mm	0.022	50	1.5
Cast iron 30 mm	0.016	39	1.5

more extended than with cast iron chill. Furthermore, for any class of chill material, the larger the chill dimension (and thus, the higher the separation time), the longer the "high HTC" zone. The effectiveness in time of the chill is then related to its expansion capabilities, which can prolong the contact.

Another question arises as to whether some pressure is present at the interface during the evaluated contact times and, if so, what effect it could have on the cooling process. If the HTC are represented as a function of the temperature difference between casting and chill (figure 6a-6b), it emerges that, especially in the case of copper chills, there is a very clear tendency of the HTC to increase (up to 5,500 $Wm^{-2}K^{-1}$ for 90 mm copper) instead of the decreasing temperature difference, which would be an absurd, if some other effect was not considered. This "increasing HTC" zone is present, in particular, when the two surfaces are still in full contact. In fact, in figure 6a, the beginning and the end of the separation procedure of the casting from the chill are shown and the regular decrease in HTC, as it is observed in those regions, is the effect of the decrease in contact area, as explained. Thus, HTC steadily increase before separation begins and it is interesting to note that HTC values are almost coincident, in that phase, for all three chill dimensions. Clearly, the smaller the chill dimension, the earlier the beginning of separation and the HTC decrease. The effect of a transient interface pressure is the only reason for an HTC increase when temperature difference is decreasing.

Maximum pressures at the interface were found to be higher for copper and aluminium chills (5 and 6 MPa respectively) and very low with cast iron chills, where they never exceeded 1.5 MPa. With all chill materials, the maximum interface pressure was not sensitive to the chill dimension. In Table 1 pressure values were approximated the nearest 0.5 MPa owing to the uncertainty of the evaluation: the pressure distribution at the interface was not constant and the reported values are calculated as a mean on the whole interface.

These results are in general agreement to the HTC behaviour as it is observed in figures 3 and 6. The generally high value of pressure with copper and aluminium chills interface is coherent to the high HTC values in those cases (except with 30 mm aluminium chill, where HTC drop very quickly to zero in few seconds). The poor pressure development in cast iron chills seems to be very well related to always poor HTC on those boundaries.

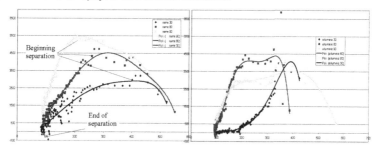

Figure 6a-6b. a) HTC for copper chills as a function of ΔT casting-chill;
b) HTC for aluminum chills as a function of the ΔT casting-chill.

In all cases the pressure did not exceed the value of 6.7 MPa, which was the sand compression strength. However, it is doubtful that such interface pressure, which is theoretically evaluated, can be fully supported by the still thin shell of solidifying material. In fact, such pressure should be transferred to (see figure 7) the sand surrounding the solidifying shell and the hydrostatic pressure of the casting at the liquid-semisolid interface. Closer observation of the "rising HTC" zone in copper and aluminium, shows a sequence of rises and falls which could be related to pressure increase and then to shell backward displacement.

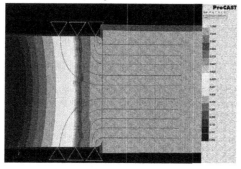

Figure 7. Model showing the extent of the solid shell and mushy zone when the maximum interface pressure is applied to the interface (percent of fraction of solid)

Conclusions

The HTC at the interface with chills of different materials and dimensions were evaluated during the solidification of the A356 aluminium alloy in a sand casting process. The experimental conditions were reconstructed by numerical analysis in order to evaluate the chill expansion effect and the local conditions of pressure between the chill and the growing solid phase. A clear and consistent effect of a transient interface pressure was found; this effect is responsible for great HTC increases with decreasing temperature differences at the interface.

Acknowledgements

The authors would like to thank Modelleria Meccanica CPC (Modena, Italy) for the help in the experimental activities.

References

1. J. Campbell, *Casting 2nd Edition*, (Oxford, Butterworth-Heinemann, 2003).

2. W.D. Griffiths, "A Model of the Interfacial Heat-Transfer Coefficient during Unidirectional Solidification of an Aluminum Alloy", *Metal. and Mat. Trans. B*, 31B (2000) 285-295.

3. C.P. Hallam and W.D. Griffiths, "A Model of the Interfacial Heat-Transfer Coefficient for the Aluminum Gravity Die-Casting Process", *Metal. and Mat. Trans. B*, 35B, (2004) 721-733.

4. S. Shen, A "Numerical Study of Inverse Heat Conduction Problems", *Computers and Mathematics with Applications*, 38 (1999), 173-188.

5. J.Liu, "A Stability Analysis on Beck's Procedure for Inverse Heat Conduction Problems", *J. of Computational Physics*, 123 (1996), 65–73.

6. E. Velasco et al., "Casting-Chill Interface Heat Transfer during Solidification of an Aluminum Alloy", *Metal. and Mat. Trans. B*, 30B (1999), 773-778.

7. J. V. Beck and B. Blackwell, "Comparison of some inverse heat conduction methods using experimental data", *Int. J. Heat Mass Transfer*, 39 No.17 (1996), 3649-3657.

8. Mythily Krishnan and D.G.R. Sharma, "Determination of the interfacial heat transfer coefficient h in unidirectional heat flow by Beck's non linear estimation procedure", *Int. Comm. HeatMass Transfer*, Vol. 23, No. 2 (1996), 203-214.

9. C.A. Santos, J.M.V. Quaresma and A. Garcia, "Determination of transient interfacial heat transfer coefficients in chill mold castings", *Journal of Alloys and Compounds*, 319 (2001), 174–186.

10. K.N. Prabhu, W.D. Griffiths: "One-dimensional predictive model for estimation of interfacial heat transfer coefficient during solidification of cast iron in sand mould", *Mat. Sci. and Tech.*, 18 (2002), 804-810

11. J.V. Beck, B. Blackwell, *Inverse Heat Conduction – Ill-Posed Problems*, (Wiley-Interscience Publication, 1985).

12. M. A. Gafur, M. Nasrul Haque et al., "Effect of chill thickness and superheat on casting/chill interfacial heat transfer during solidification of commercially pure aluminum", *Journal of Mat. Proc. Tech.*, 133 (3) (2003), 257-265.

13. F. Lau, W.B. Lee et al. "A study of the interfacial heat transfer between an iron casting and a metallic mould", *Journal of Mat. Proc. Tech.*, 79 (1998) 25–29.

14. F. Lau, W.B. Lee et al. "A study of the interfacial heat transfer between an iron casting and a metallic mould", *Journal of Mat. Proc. Tech.*, 79 (1998) 25–29

15. A. Meneghini and L. Tomesani, "Chill material and size effects on HTC evolution in sand casting of aluminum alloys", *Journal of Mat. Proc. Tech.*, 162-163 (2005), 534-539

16. Meneghini, L. Tomesani, "Chill Boundaries HTC in sand casting of A201 and A356 alloys" (Paper presented at AMPT06 Conference, Las Vegas 2006)

17. A. Meneghini, L. Tomesani "The Influence Of The Metal Head On The HTC Evolution In The Sand Casting Of A357" (Paper presented at 6th MATEHN Conference, Cluj-Napoca, Romania, 2006)

SHAPE CASTING:
2nd International Symposium

Applications/Novel Processes

Session Chair:
David StJohn

A NOVEL TECHNIQUE FOR MELTING AND CASTING SUPERALLOYS

Sanjay B. Shendye[1] and Blair King[1]

[1]Metal Casting Technology, Inc.; 127 Old Wilton Road, Milford, NH 03055, USA

Keywords: Counter-gravity, Investment Casting, Superalloys, Inert Atmosphere

Abstract

Superalloys are traditionally melted and cast under vacuum to eliminate the formation of inclusions in the casting that can reduce their mechanical properties. Counter-gravity Low-pressure Inert-gas (CLI) process is a unique method of melting and casting superalloys under inert atmosphere which precludes the use of vacuum during the melting and casting process. In addition, the CLI casting process fills the mold through a single-use fill pipe inserted into the center of the melt thus avoiding the oxides that often congregate along the periphery of the molten alloy.

A strict inert atmosphere control during the CLI melting process ensures a very low level of molten alloy contamination. Oxygen levels of under 50 ppm are typically maintained during CLI melting and casting. Resultant gas content in the casting is typically less than 50 ppm oxygen and nitrogen. Such low levels of oxygen and nitrogen do not adversely impact the mechanical properties.

Introduction

Vacuum induction melting (VIM) process is one of the processes used to melt Ni-base superalloys. The investment casting of Ni-based superalloys involves vacuum induction melting and pouring the molten alloy into ceramic molds in a vacuum atmosphere [1]. The CLI process uses a different approach, in which the alloy is melted in a refractory crucible protected by inert argon gas at slightly above atmospheric pressure as opposed to vacuum, and the alloy is filled into the mold by creating a partial vacuum in the mold cavity. Shown in Figure 1 is the principle of the CLI melting and casting process [2].

In the CLI process, as soon as the molten alloy is ready to be cast, a preheated ceramic fill pipe is placed on the bottom opening of the mold chamber. As in any other investment casting process, a preheated ceramic mold is transferred from the oven and placed on top of the fill pipe. A small thin ceramic blanket acts as a seal between the mold and the fill pipe. Support media which essentially consists of mullite sand is packed around the preheated mold. The mold chamber containing the hot mold surrounded by sand is then transferred to the melting furnace. The fill pipe at the bottom of the chamber is then inserted into the argon atmosphere above the molten alloy. Vacuum thus created on the mold draws argon into the mold cavity and displaces air in the mold cavity. The fill pipe is then inserted deep into the molten alloy and the vacuum in the mold cavity is increased at a controlled rate, enabling the mold filling. The level of vacuum in the mold cavity depends upon the alloy to be cast and the total height of the mold. Typically a partial vacuum (1/3 of an atmosphere) is sufficient to fill the mold. Controlling the rate of change to this vacuum level enables control of mold filling which is typically accomplished in 2 to 5 seconds.

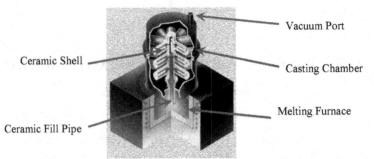

Ceramic Shell

Ceramic Fill Pipe

Vacuum Port

Casting Chamber

Melting Furnace

Figure 1: Schematic illustration of the CLI melting and casting process. Support media packed around the mold is not shown in this illustration for the purpose of clarity.

After the mold cavity is completely filled, it is held under vacuum for a specific amount of time. During this time interval, the fill pipe is kept immersed into the molten alloy and the castings with in-gates are solidified. At the end of the mold fill cycle, the mold chamber is returned to atmospheric pressure, which returns the molten alloy in the mold center sprue back into the crucible. The mold chamber containing the filled mold is lifted up and the ceramic fill pipe is raised above the molten alloy. The mold containing the solidified castings is then dropped into a sand bucket and air-cooled to room temperature. The furnace is then recharged with alloy, if needed, for casting the next mold.

One of the concerns in using the CLI process is the possibility of contamination of the alloy. Certain reactive elements such as Al and Ti in superalloys can react with gases such as O_2 and N_2 within the argon atmosphere, and form inclusions in the castings [1-3]. Another concern is that since a significant amount of liquid alloy in the center sprue is returned to the crucible after each mold is filled, contaminants such as loose refractory material in the mold cavity or the alloy-ceramic fill pipe and/or the alloy-shell material reaction product could enter the melt, and the mold, during the subsequent molds cast.

Several studies were undertaken to evaluate the effect of such processing variables on the chemical composition, cleanliness of the alloy, and tensile and stress rupture properties for Ni-base alloys such as IN713C, Haynes 230 and IN718. Described in this paper are the results of a recent trial that was conducted to assess the cleanliness of the alloy after it was cast and to determine the microstructure across different wall thicknesses on the experimental samples. Specifically, results pertaining to IN713C alloy melted and cast using the CLI process are discussed. For comparison purposes, chemical composition and microstructure of vacuum induction melted and vacuum gravity-cast IN713C are also discussed in this paper.

Experimental Procedure

A total of 5 molds were cast using IN713C alloy ingot in the CLI process. The first and the last molds cast in the casting campaign were test bar molds with chemical test coupons and several test bar blanks (12.5 mm x 100 mm long) per mold. Other molds consisted of experimental samples varying in length and width up to 200 mm and in wall thickness from 1.5 mm to 25 mm.

All the molds were made using the standard investment casting shell building techniques. Each mold was preheated per the standard procedures to 1093 °C in a gas fired oven and transferred to the mold chamber and cast in approximately 150 seconds. Alloy charge consisted of ingot and all molds were cast in succession in approximately 15 minute intervals. All molds were cast at 1540 °C alloy temperature.

As part of the production process, a solid state oxygen sensor capable of reading oxygen levels up to 1000 ppm is placed on top of the molten alloy surface in the CLI furnace to monitor the level of oxygen. According to the standard operating procedure, all molds were cast when the oxygen levels in the atmosphere were measured below 50 ppm [4]. One alloy charge of approximately 23 kg was made to the furnace after the 4th mold was cast to ensure there was sufficient metal in the furnace to cast the remaining mold.

Processing conditions such as the rate at which metal was drawn into the mold cavity, the maximum vacuum achieved in the mold, mold preheat temperature, alloy temperature, mold transfer time from preheat oven to cast, and argon gas flow rate over the molten alloy in the furnace were all held constant. Dwell time – the time for which the ceramic fill pipe is held in the molten metal after the mold is filled - was also the same for all the molds. Small amount of dross which typically forms on the molten alloy surface during melting was removed manually with a spatula before casting a mold.

After the molds were cooled to room temperature, chemical test coupons were removed from the molds and all castings were inspected using the fluorescent penetrant and x-ray inspection techniques. Test coupons from the first and the last mold were sent to Sherry Laboratories, IN for the determination of chemical composition per AMS 5391F specification [5]. In addition, two samples of vacuum-melted and cast IN713C sample (gravity-cast in a different foundry, not Hitchiner) were also sent to Sherry Laboratories, IN for chemical analysis. The foundry conditions used to cast the vacuum gravity-cast parts were unknown but their size in terms of overall dimensions and wall thickness was comparable to the experimental shapes cast using the CLI process. Test bars could not be excised from vacuum gravity-cast parts, and therefore only the chemical composition and microstructure of those parts could be analyzed. Microstructure of the CLI-cast experimental shapes was also analyzed.

IN713C test bars were not tested for tensile or stress rupture (SR) properties. Those properties were tested in a prior study [6] and are reported in this paper for reference purposes. Results of mechanical property testing from another study conducted in 2003 on IN713LC were reported in an earlier paper [7].

Results

Shown in Table I is the chemical composition of the CLI-cast and vacuum gravity-cast IN713C samples. Also shown for comparison purposes is the chemical composition of the ingot used to CLI-cast all the molds. For reference purposes, the mechanical properties of IN713C tested per AMS 5391F specification are shown in Figure 2. Typical SR properties of CLI-cast bars tested at 982 °C temperature and 151.7 MPa load per AMS 5391F specification are shown in Table II. These properties were generated from test bars cast in an earlier study [6].

There were no rejectable fluorescent penetrant and x-ray indications on the samples. Shown in Figures 3 to 5 is the typical grain-size observed in the microstructure of the CLI-cast and vacuum gravity-cast samples of different wall thicknesses. The carbide distribution in the microstructure is shown in Figures 6 and 7 in two different wall thicknesses – 1.5 mm and 5 mm.

Table I: Chemical composition (wt% or ppm) of the IN713C test coupons, ingot and the vacuum gravity-cast samples.

	C	Cr	Al	Mo	Nb	Ti	Si	B	Zr	H (ppm)	N (ppm)	O (ppm)
First mold	0.13	12.92	6.02	3.94	2.12	0.78	<0.01	0.012	0.07	9	<50	39
Last mold	0.12	12.98	6.02	3.88	2.04	0.74	<0.01	0.013	0.06	4	<50	50
Ingot (CLI)	0.14	13.14	6.03	3.98	2.27	0.79	<0.01	0.013	0.08	3	<50	17
Vacuum gravity-cast (1)	0.12	13.53	6.20	4.22	2.27	0.87	<0.01	0.013	0.06	12	20	60
Vacuum gravity-cast (2)	0.11	13.50	6.07	4.49	2.22	0.86	<0.01	0.013	0.07	<5	10	20
AMS 5391F spec	0.08 to 0.20	12.00 to 14.00	5.50 to 6.50	3.80 to 5.20	1.80 to 2.80	0.50 to 1.00	0.50 max	0.005 to 0.015	0.05 to 0.150			

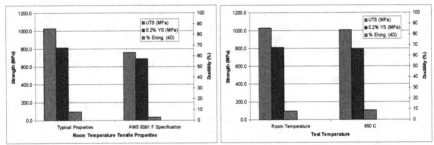

Figure 2: Typical room temperature and elevated temperature tensile properties of CLI-cast IN713C bars.

Table II: Stress rupture properties of CLI-cast IN713C test bars machined from ½" diameter blanks (6).

	Stress Load	Stress Rupture Life (hours)	Elongation in 4D (%)
Typical Properties	Up to 193 MPa to achieve rupture	48.5	14
AMS 5391 F Specification	151.7 MPa	30	5

Figure 3: Optical micrographs of the mounted and etched x-section from a 25 mm thick wall of
a) CLI-cast and b) vacuum gravity-cast sample showing the grain size.

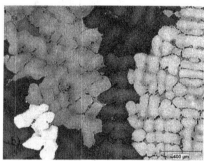

Figure 4: Optical micrographs of the mounted and etched x-section from a 5 mm thick wall of
a) CLI-cast and b) vacuum gravity-cast sample showing the grain size.

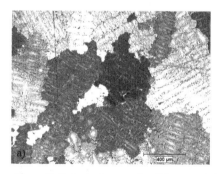
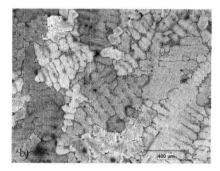

Figure 5: Optical micrographs of the mounted and etched x-section from a 1.5 mm thick wall of
a) CLI-cast and b) vacuum gravity-cast sample showing the grain size.

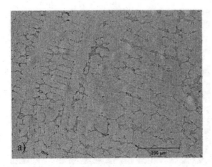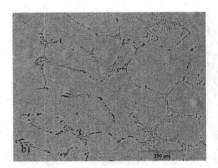

Figure 6: Optical micrographs of the mounted and polished x-section from a 5 mm thick wall of a) CLI-cast and b) vacuum gravity-cast sample showing the carbide distribution.

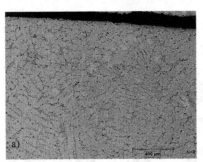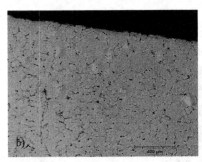

Figure 7: Optical micrographs of the mounted and polished x-section from a 1.5 mm thick wall of a) CLI-cast and b) vacuum gravity-cast sample showing the carbide distribution.

Discussion

Ti and Al in the IN713C alloy are known to react with gases such as O_2 and N_2 at high temperature and form inclusions, which can adversely affect the mechanical properties of the cast component [1-3]. In addition, such gases also tie up these reactive elements that form the gamma prime (γ') phase, which is the primary strengthening phase for many Ni-base superalloys including IN713C [1, 3]. Such gases must, therefore, be excluded from the environment where reactive superalloys are melted.

Oxygen level during CLI melting, as measured by a probe inserted in the argon atmosphere right above the molten alloy, is typically below 50 ppm. This is a requirement for casting any superalloy product using the CLI process in accordance with the standard operating procedure [4]. These levels of oxygen may be considered to be higher than those encountered in a vacuum melting and vacuum gravity-casting environment. However, as shown in Table I, the chemical composition of the CLI-cast alloy showed a very minor increase in oxygen and nitrogen compared to the starting ingot material. Similar results were reported in an earlier study conducted on CLI-cast IN713LC and IN718 [7, 8].

Since the slag on top of the molten alloy is typically skimmed off in the CLI process before casting, and the ceramic fill pipe at the bottom of the mold is plunged deep inside the melt before filling the mold cavity, gas content of the solidified CLI-cast sample is expected to be comparable to that of the vacuum-cast starting alloy ingot or a vacuum gravity-cast sample. As shown in Table I, the chemical composition of the CLI-cast samples was comparable to the vacuum-melted and vacuum gravity-cast samples. The ability to skim off the slag and to fill the mold cavity with clean alloy below the molten surface is a major advantage of the CLI process.

Inherent to the CLI process is the back and forth transport of about 23 kg of molten alloy through a relatively colder ceramic mold, when several molds are cast in succession. One might surmise that such back and forth transport of liquid alloy from the mold to the melting furnace would alter the chemical composition, specifically the oxygen content of the alloy. This process could also add ceramic oxide inclusions which can adversely affect the mechanical properties [1-3]. However, chemical composition of the cast product shown in Table I, indicated an insignificant change in the overall alloy composition from the first to the last mold cast. Tensile and SR properties shown in Figures 2 and 3 and in Table II also indicated that those properties exceed the specifications. There were no rejectable fluorescent penetrant and x-ray indications on the CLI-cast samples. These results confirm that the CLI process does not introduce any deleterious phases into the cast product that can adversely affect the alloy composition and the mechanical properties. Similar results have been observed with other alloys such as IN718 and Haynes 230 cast using the CLI process [7, 8].

Insignificant micro-shrinkage was observed in both the CLI and the vacuum gravity-cast samples. However, the grain-size in the two samples was different as shown in Figures 3 to 5. In the heavier sections such as 25 mm thick, both the CLI and the vacuum gravity-cast samples had a comparable grain-size. However, the grain-size in the 5 mm thick section of the CLI-cast product was smaller – ranging from 0.75 mm to 1.25 mm – compared to the same wall thickness in vacuum gravity-cast product – ranging from 0.75 mm to 2 mm. The grain size in the 1.5 mm wall thickness was also smaller ranging from 0.40 mm to 0.70 mm in the CLI-cast samples vs. 0.50 mm to 1 mm in the vacuum gravity-cast samples.

The finer grain-size in the thinner walls in the CLI-cast samples can be explained based upon the chilling effect of the backup mullite sand that is packed around the hot mold before filling the mold with molten alloy. Backup sand, which is at room temperature when it is filled in the casting chamber, takes away the heat from the mold thus reducing its temperature. This reduction in the mold temperature results in a temperature gradient such that the mold is hottest at the center and colder toward the outer edges of the mold. This effect is found to be beneficial and enhances directional solidification.

The microstructure of both the CLI and the vacuum gravity-cast products consisted of the Nb, Mo, Zr and Ti rich carbides and borides, typically observed in IN713C alloy [9]. In addition, a large volume fraction of cuboidal γ' and a small volume fraction of primary eutectic γ' was also observed as reported in an earlier study [7]. No other secondary phases were observed in the microstructure.

The distribution of carbides in the microstructure was found to be different in the two samples as shown in Figures 6 and 7. Carbides which form along the grain boundaries and in the inter-dendritic spaces [9] were more widely dispersed in the case of vacuum gravity-cast samples, than in the CLI-cast samples in both the 5 mm and 1.5 mm thick walls. This can also be attributed to the chilling effect of the backup sand on the CLI-cast mold. A reduction in the mold temperature produces a large number of chill grains, which in turn produce a large number of dendrites that grow from the casting surface into the body of the casting. A larger number of dendrites reduce the dendrite arm spacing, which results in a finer distribution of carbides.

Conclusions

The CLI process is a novel process that uses inert argon atmosphere for melting and casting superalloy components. The CLI process, similar to the vacuum melting and gravity-casting process, does not result in an increase in the oxygen and the nitrogen levels in the alloys such as IN713C. Consequently, the microstructure and mechanical properties are not adversely affected. The CLI process also offers other unique advantages such as the ability to fill thin-wall components and the ability to control the mold filling rates to reduce turbulence and to promote thermal gradients. Stringent customer requirements for chemical composition, mechanical properties and microstructure are routinely met and exceeded for alloys such as IN713C. As a result, the CLI process is currently used in production to cast components in IN713C.

References

1. C. T. Sims, N. S. Stoloff and W. C. Hagel ed., *Superalloys II: High Temperature Materials for Aerospace and Industrial Power*, (New York, NY: John Wiley & Sons, 1987) 387-439.

2. Hitchiner Manufacturing Co., Inc. web site http://www.hitchiner.com.

3. Matthew Donachie and Stephen Donachie, *Superalloys: A Technical Guide* (Metals Park, OH: ASM International, 2002).

4. Blair King, Metal Casting Technology internal communication, August 2005.

5. Aerospace Material Specification 5391F - Nickel Alloy, Corrosion and Heat Resistant, Investment Castings UNS N07713 (Warrendale, PA: SAE International, January 2005).

6. Blair King, Metal Casting Technology internal communication, July 17, 2003.

7. Sanjay B. Shendye, Blair King and Paul McQuay, *Counter-gravity Investment Casting of Ni-base Superalloys Using Inert Atmosphere* (Near-Net-Shape Technologies: Proceedings of the MS&T'05 Materials Science & Technology Conference and Exhibition, Pittsburgh, PA September 25-28, 2005), 45.

8. Sanjay B. Shendye, Blair King and Paul McQuay, *Mechanical Properties of Counter-gravity Cast IN718* (Proceedings of the Superalloys IN718, 625, 706 and Various Derivatives Conference, ed. E. A. Loria, Pittsburgh, PA October 2-5, 2005), 123.

9. C. T. Sims, N. S. Stoloff and W. C. Hagel ed., *Superalloys II: High Temperature Materials for Aerospace and Industrial Power*, (New York, NY: John Wiley & Sons, 1987) 97-133.

THE QUEST TO CAST THE PERFECT ALUMINIUM SAND CASTING

Vian F. Coombe

Ferrari Spa. (Gestione Sportiva, Direzione Tecnica - Metallurgia);
Via Ascari 55/57, Maranello, 41053, (Mo), Italy.

Keywords: Campbell's 10 Casting Rules, Water Cooled Chills, Formula 1 V8 Cylinder Head.

Abstract

Modern Formula One racing rule changes have certainly not made the life of the foundryman any easier. The engines must run for two races (doubling the previous requirement). The 3.0 liter V10 changed to 2.4 litre V8. The designers expected the same ~250 BHP/litre at ~18,500 rpm, but with increased casting reliability. This paper describes how John Campbell's 10 Rules of Casting were used for the foundry methods and process for a new Ferrari F1 V8 sand cast cylinder head. With casting simulation and only the simplest of foundry equipment, the running system produced a "casting right the first time". Later, after a little innovation, cut samples taken from the casting exhibited a Q-index of 526 MPa, closer towards the perfect casting.

Introduction

Cylinder head castings required for Formula 1 are not in the same league as a casting for an everyday car petrol engine. In the latter stresses of the cold start are probably the severest, rather than the stress of extreme racing power. Secondly, the life expectancy of automotive engines is higher, reaching thousands of kilometers. According to latest F.I.A rules the F1 engine must perform 2 races, approximately 1,200 kilometers (including test and shake down) or the car is penalised 10 positions on the starting grid. The 2.4 litre engine produces ~650 BHP (~250 BHP/liter) at ~18,500 rpm. Under such loading the smallest of defects in the casting can cause fatigue failure. Therefore the castings are required to have zero defects (a perfect casting).

With *Casting Practice – the 10 Rules of Casting* by John Campbell [1] as a guide, casting methods for the 2006 season Ferrari F1 V8 cylinder head[1] incorporated thin runners and downsprue, tangential/transverse placed filters (chicanes[2]), simulation and some added interpretations to aid solidification. Since the quality of the final casting depends on the quality of the metal used, the process included degassing, fluxing and casting with minimum turbulence to eliminate the formation of bifilms and thus achieve an optimum quality product for the F1 team.

This paper describes the author's interpretation of the 10 rules, plus some 32 years of foundry experience (and a little black art) were combined into a practical methods system in an attempt to make the perfect cylinder head casting. The prior 2005 season V10 cylinder head methods followed philosophy taken primarily from an earlier publication *Castings* by John Campbell [2], producing sound and perfectly usable 357-T6 castings and achieving Q-values of 475 MPa (Q =

---

[1] Because the cylinder head is used in a current Formula 1 racing engine, the head design cannot be shown.
[2] A chicane is a series of tight S bends used to slow down racing cars at the end of a fast straightaway. The tangential/transverse placed filters produce exactly the same effect for the liquid velocity.

UTS + 150 log %EL). To evaluate precisely the effect of the new methods for the V8 heads the same casting alloy was used.

Our F1 engine designers are not generous with their timetables: the pattern maker and foundry get 8 weeks to deliver a pair of metallurgical qualified cylinder head castings to the machine shop. Therefore the casting methods must be right first time.

Application of the Rules

Flexibility of design, rapid delivery and thin wall capability (2.75mm average) make chemically bonded precision sand the preferred moulding technique. In the complex cylinder head casting, the fire face and combustion chambers cannot have defects. Feeding the highly stressed upper sections of the head is of paramount importance. One trick is to leave the spark plug towers solid and to feed other parts of the casting through them. Another is to attach ingates to the main bolt bosses to the block. The combustion chamber is normally rapidly solidified with chills moulded into the sand. Chills or cast fins may be added elsewhere depending on design and section thickness.

The Running and Feeding System

One of the greatest challenges to the methods engineer using a gravity poured system is to guarantee that the critical velocity of 0.5 m/s is not exceeded. Where this cannot be avoided, particularly in the running system, the metal liquid front must be constrained to remain intact without splashing. Hyperbolic downsprues, thin rectangular runner bars and fan ingates limit surface turbulence and filters built into the system reduce the velocity of the metal, thus producing quiescent filling within the casting.

To enable optimum filling, the mould is made of 3 primary parts, a normal cope and drag plus a lower 3rd part that houses the exit of the downsprue, filters and runner bars. The ingates can be placed anywhere on the lower surface of the casting without the restrictions of division joints, core prints and the like. This satisfies the no-fall requirement and allows the runner bar to be fully optimised to the ingates.

The running and feeding system was developed as follows:

Position and number of ingates. The filling distance (50 x average wall thickness is used), lowest point of entry, heat distribution and T-section rule are all considered. An ingate section thickness is derived; the length is ascertained during the final calculation. All ingates incorporate a ceramic foam filter that helps guarantee laminar flow into the casting.

Position of downsprue and runner bar(s). The path of the runner bar(s) with 8 mm max thickness is established allowing for space for primary ceramic foam filters. These filter(s) are oriented horizontally creating a natural double chicane, Figure1 – inset. The main flow path of the runner is offset to the ingates. Space is also allocated between the primary filter and the 1st ingate of the system to smooth liquid metal flow before entry into the casting. The terminus of the runner bar is extended beyond the last ingate and includes a fin to trap any oxide produced by the initial filling liquid metal front.

Position and size of feeders. Directional solidification is calculated by modulus methods.

Average flow rate to fill the casting. For a cylinder head a rate between 1 - 2 kg/s will maintain a continuous moving liquid front without cold shuts. (This where a bit of the black art is needed). We now have approximate parameters of total casting weight, total mould height, fill rate and therefore fill time.

Cross sectional area (CSA) of various parts of the system. Thanks to M. Cox's Internet Downsprue Calculator [3], CSA and section dimensions are easily calculated to obtain the correct liquid entry velocity into the casting (>0.25 m/s and <0.5 m/s). However, the 8 mm thick runner bar(s) commonly calculate to be 150 mm wide. The author uses a tee system to compact the runner into a more manageable width, e.g., tee horizontal 75 mm (filter width) and tee vertical 75 mm. The runner bar is reduced in area, equal to each ingate CSA in a smooth taper to balance ingate metal entry pressure, Figure 1.

Chills. The combustion chamber is formed with individual cast iron chills formed into the mold.

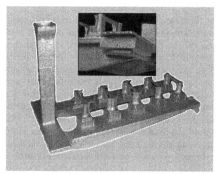

Figure 1: Final runner system. Inset: Filter detail.

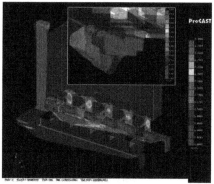

Figure 2: Procast system simulation result. Inset: Effect of filter chicane liquid velocity reduction.

Calculations complete, all the required parameters are now available to produce a casting simulation.

A Procast simulation programme was used to analyse the filling and solidification. Because of the tight delivery time the filling velocity parameters were analysed while the pattern was still being constructed. The filling simulation produced only two small velocity alarms. The first occurred at the exit of the primary filter; this was easily remedied by reducing the CSA under the filter and shortening the vertical drop of the "tee" section. The second was noted at the expansion from the filter/runner bar; this was remedied by enlarging the radius, thus eliminating the sharp angled transformation. Some predicted surface shrinkage was considered insignificant.

The simulation showed an interesting phenomenon: the metal filled the vertical of the tee and then cleanly filled the horizontal with an unbroken liquid front, finally ending at the fin trap at the terminus without any sign of a back wave. The first ingate filled slightly before the others but by the time the metal arrived vertically into the casting all ingates were balanced, Figure 2. The simulation also illustrated the effect of the filter chicane to reduce velocity, Figure 2 – inset.

The Moulding Process

The chemically bonded precision sand moulding process uses dry silica sand and a catalysed resin binder system. The process produces low out-gassing, is dimensionally stable and has good permeability. Primary cores and moulds are precisely fixed with consumable pins and bushes. The complete cylinder head mould assembly can have as many as 70 cores, requiring gluing and pasting of joints. Therefore, to reduce further the risk of gas blows, the assembled mould package is oven dried before casting.

261

The Metal

The quality of the casting depends heavily on the quality of the liquid metal. Time, if available is a major benefit to achieving a good quality metal. If the metal can be allowed to stand the impurities can settle and float, and are thus removed. The Ferrari melting cycle is as follows:

– Melting is performed in a 300 kg tilting electric resistance crucible furnace. Primary continuous cast ingot is melted under minimal superheat. The melt is covered with a deoxidising flux and held at 720°C for at least 3 hr after which chemistry is adjusted.

- The metal is degassed with Ar dispensed at low flow rate through a preheated rotary degasser. All surface dross and flux is removed, elemental analysis and RPT (reduced pressure test)

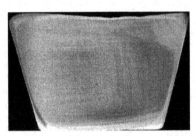

samples are taken and another layer of cover flux is added. The RPT liquid sample is taken directly from the melt in a crucible preheated to the melt temperature (permits slow solidification) and solidified at 70 mbar. The sample is checked for density and then cut and polished. The center should show no visible defects, Figure 3. The author has used this system for approximately 20 years but assumed that the defects observed were hydrogen, we know now that they are Bi-films.

Figure 3: Typical polished RPT sample.

– After meeting chemistry and RPT sample requirements the melt is settled 30 min between degassing and pouring during which time the melt is raised to the final pouring temperature. The moulds are transported to the furnace, grain refiner and modifier master alloy are carefully added and the castings are direct poured from the furnace.

The spout of the furnace has been modified to be exactly on the tilting pivot axis. The furnace spout to pouring basin is less than 25 mm. The offset pouring basin has also been redesigned in 2 parts to provide an undercut to reduce turbulence and spillage.

It is pleasing to watch the liquid metal front rise into the feeders and see a surface like a mirror – this means that everything has been done right.

Post Processing

The castings are cut and ground by normal processes and 100% inspected by real-time x-ray, not only to control for defects but to guarantee that all the internal core sand has been removed. Solution heat treatment is performed in a drop-bottom furnace certified to ± 3°C. The heads are positioned vertically and plunged into a polymer/water quenchant. After quenching castings are immediately checked for any distortion and proceed to high pressure jet washing to remove all traces of polymer. If the castings cannot be processed within 2 hours they are placed into a freezer. Aging is later performed to finish the treatment cycle.

The castings are subjected to shot blasting, 100% visual control and a further real-time x-ray to verify the absence of shot and residue. Test bars that are cast and processed with every batch are pulled and if all minimum requirements are met the castings are transferred to the machine shop.

When a new design or modification has been made to the tooling, the first castings are sectioned and subjected to a full x-ray. The sections are submitted for micrographic analysis to determine microporosity levels, Bi film presence, SDAS and degree of silicon modification. Further examination maybe performed by SEM.

Results

The first pair of metallurgically qualified V8 cylinder head castings was transferred to the machine shop 8 weeks from the final CAD design clearance, inclusive of some last minute modifications. As in previous cases the delivery time was met. The X-ray certificate reported Level 0, i.e. "no defects".

Test bars taken from the fire face sections exhibited averages as follows:
UTS 350 MPa, 0.2% YS 285 MPa, EL - 8%, Q 485 MPa, and SDAS 32 μm.

Results of test bars taken from other parts of the casting were somewhat lower but met design minimum requirements. The Rules Worked!

The results were an improvement over those from the previous V10 cylinder head. However the author was still not satisfied. Constant F1 design modifications give the opportunity to introduce additional methods improvements.

Research and Trials to Improve the Casting Method

One of the established methods to improve mechanical properties is to produce a finer structure by chilling. Copper and aluminium chills have a higher thermal conductivity than cast iron but a lower specific heat. The amount of liquid cast metal at the fire face requires a large heat sink to solidify the metal rapidly. What if there was and endless heat sink? Could the chills be chilled? How could this be achieved in a practical repeatable manner?

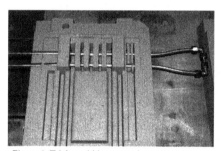

Figure 4: Trial mould incorporating water tube.

Moulds were prepared with identically sized chill materials of cast iron, copper and aluminium. In addition a mould was modified to incorporate an 8 mm diameter copper tube passing through fin extensions to introduce water cooling, Figure 4. The moulds were cast from the same melt at the same casting temperature of 730°C and the cooling rate of each casting was measured at 1 mm, 10 mm, 50 mm, and 150 mm from the chilled face with embedded thermocouples.

Table 1: Cooling rate versus chill material.

Chill Material	Max Cooling Rate, °C/s
Aluminium	11
Copper	12
Cast iron	8
Water cooled fins	15

Heavy chilling with water cooling and the like is not a new idea – gravity die/permanent moulds use the system frequently if only to increase production rates. We are working in sand. The water cooling gave a superior cooling effect, Table 1, and the cooling rate continued for a considerable time, Figure 5. Micrographic analysis measured the SDAS adjacent to the

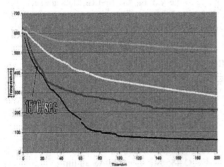

Figure 5: Cooling curves for water cooling tube.

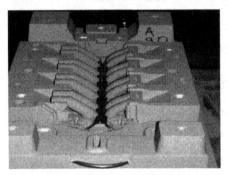

Figure 6: Copper tube cut into cylinder head mould.

Figure 7: Pure aluminium chill pack with cast embedded water tube. Inset: Traditional fire face cast iron chill.

cast-in tube at 9 μm. Even at 150 mm distance from the chill some effect was noted. The trial castings average wall thickness of 4 mm is somewhat smaller than a cylinder head fire face of 10 mm, but the results were considered sufficient to warrant further development.

The next task was to incorporate this method into the cylinder head mould to verify the results. A mould was prepared with a cut channel just below the fire face and an 8 mm copper tube embedded forming a U, Figure 6. Rubber hoses were attached and the cylinder head was cast with water flowing through the tube. The tube became encapsulated by the cast aluminium and formed part of the casting to be cut off later. The thought was that the cost of the copper tube to be insignificant to the possible advantage of increased mechanical properties. The resulting structure had a SDAS of 17 μm.

However, one major problem noted was that the casting bowed as a banana by 5-6 mm, caused by the rapidly cooling, rigid, thickened section along the length of the cylinder head. This was considered impractical to straighten without introducing stresses. (An attempt was made to straighten the cylinder head after quenching using a 20 ton press without success. The casting flexed like a spring). Back to the drawing board!

After various trials the best practical solution was to embed an 18 mm diameter, thin-wall stainless steel anti-vibration tube into the normal chill forms, forming a water-cooled chill package, Figure 7.

The corrugated form of the tube provided a large cooling surface area and more than compensated for the reduced coefficient of thermal conductivity in the stainless steel. A simple 4-impression pattern was produced with the chill spacing equal to that of the cylinder fire face tooling. The individual impressions were then cast with pure aluminium for optimum thermal conductivity. The combination chill/tube was easily mouldable, reusable and after casting could be simply adjusted if deformation occurred during knock-out. Rapid bayonet fittings were attached to the tube ends and slots made in the drag tooling sides so the tube ends extended beyond the finished mould.

The factory water system was piped to the casting furnace; and rubber hoses provided delivery and drain. The factory water pressure produced a flow rate of approximately 60 l/min.

The method proved to be simple to operate: the furnace man transports the ready mould to the casting furnace and attaches the hoses using the rapid connectors, starts the water flow and casts directly from the furnace as usual, Figure 8. The water cooling is continued during solidification, then closed, disconnected and readied for the next casting. The difference between water temperature at the inlet and outlet was just 2-3°C.

Figure 8: Direct casting from tilting furnace showing water cooling system. Inset: Modified offset basin.

Results

The flexibility of the anti-vibration tube allowed unrestrained natural contraction and the as-cast cylinder head remained straight to within 0.1 mm, within normal casting tolerances.

After the T6 heat treatment specimens were cut from various positions of the fire face and tensile tested. Uniformity of properties was ± 5%, Table 2. On average, Q was 529 MPa.

Table 2: Mechanical properties of specimens cut from the fire face cast with water-cooled chills

Test Bar Position	UTS MPa	0.2% Proof Stress MPa	Elongation %
Between cylinders	359.5	300.6	11.7
Across cylinder	363.7	305.1	12.4
In line cylinder	380.0	304.0	12.3
Average	367.7	303.0	12.1

The resulting microstructure (SDAS 27 μm, porosity 0.22%) was less refined than that from the fully embedded copper tube, but an improvement over that produced by the standard cast iron chill.

Further development to the chill pack has resulted in a double U-tube system producing a larger chill/fire face contact area, Figure 9, producing an SDAS of 22 μm, unfortunately the mechanical property results of this latest innovation were not available at press time.

Concluding Remarks

Only minor modifications were required to the methods on completion of the Procast simulation and these were easily introduced into the tooling without delay to the lead time. The castings produced were "right first time" by application of "The 10 Casting Rules". The first qualified cylinder heads were delivered in 8 weeks and met the programme targets. The mechanical properties were in specification.

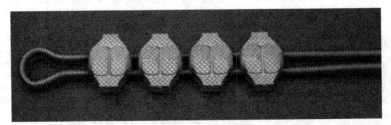

Figure 9: Modified 4-chill pack with U-tube.

The subsequent development of the water-cooled chill packs improved proof stress and UTS by 5%, and increased elongation by 50% in samples cut from the fire face. The properties from the casting are now better than the separately cast test bars and are nearing the A357-T6 material maximum.

The quality and mechanical values of the sand casting produced by simple casting equipment are at least similar to complex aerospace investment methods such as "Sofia" and "Lion" (Q 510 MPa – average results of investment castings supplied to Ferrari).

The latest U-tube chill pack has reduced the solidification time, further refining the structure (SDAS 22 μm) and therefore theoretically increasing fatigue life.

New F.I.A. Formula 1 rules are freezing engine design and material for the next 3 years which moves the burden from the designer to the foundryman to increase reliability for the cylinder head.

Provisions for further improvement of casting methods are already being studied, continuing the search to produce the "Perfect Casting".

References

1. J. Campbell, *Castings Practice - The 10 Rules of Castings*, Elsevier Butterworth-Heinemann, 2004.
2. J. Campbell, *Castings*, Butterworth and Heinemann, 1991.
3. M. Cox, "Work It Out Net - CASTAID – Downsprue Calculator".

INVESTMENT CASTING WITH ICE PATTERNS AND COMPARISON WITH OTHER TYPES OF RAPID PROTOTYPING PATTERNS

Chun-Ju Huang, Ming C. Leu, Von L. Richards
University of Missouri – Rolla, MO 65409
Keywords: investment casting, rapid prototyping, ice patterns

Abstract

The process of investment casting with ice patterns was studied with the objectives to reduce casting defects and increase dimensional accuracy of the cast metal part. Frequently observed casting defects included air bubbles and incomplete geometry of the cast part. These problems were addressed by optimizing the catalyst-to-binder ratio, vacuum treatment of water, pre-wetting the master mold, providing adequate mold support, controlling slurry pouring speed, and dressing the ice pattern before casting.

The dimensional accuracy and surface roughness of metal castings from ice patterns were measured compared with of metal castings from other rapid prototyping patterns. It was observed that the ice pattern has a contracted cast part while the Stereo Lithography (SLA) pattern has an expanded cast part. The surface roughness of metal castings from ice patterns is comparable to that of metal castings from Fused Deposition Modeling (FDM) patterns.

Introduction

The conventional method of investment casting is to first generate a mold from wax. The mold is then heated in an oven or autoclave to remove the wax. There are two problems with this method. First, melted wax patterns emit CO_2 and VOC's so ventilation and regulatory compliance may become issues. Second, the mold may crack because of the expansion of wax exerting stresses on the mold. To address these two problems, the use of ice patterns instead of wax patterns has been studied (Yodice, 1991; Yodice, 1998). Three benefits of using ice patterns instead of wax patterns are: (1) instead of expansion, ice shrinks when melted, thus avoiding the possibility of cracking the mold (2) the use of ice pattern makes the process environmentally benign, and (3) water is readily available and is more economical than wax.

The objective of the present study described in this paper was to investigate whether casting with ice patterns can be a viable alternative to the commercial approach with wax patterns. Investment casting with ice pattern has been studied with demonstrated successful examples of making intricate parts such as bolts and gears (Yodice, 1991; Yodice, 1998; Jose, 2005). The application of interface agent on the surface of ice pattern has also been studied (Liu, 2004; Pen, 2000; Wu, 2003). The main purpose of applying the interface agent is preventing ice from melting. In the current study, we investigate casting of dental bars with ice patterns and examine the difference between the metal bars obtained from ice patterns without application of an interface agent. We then measured the cast dental bars to determine the dimensional accuracy and surface finish of the generated metal castings.

The paper describes the experiments performed, the problems encountered, our approaches to address these problems, and the results obtained in investment casting with ice patterns. It also describes the result of an investigation that compares the dimensional accuracy and surface roughness of the metal castings obtained from patterns generated by rapid freeze prototyping and several other types of rapid prototyping processes used in the industry.

Experimental Procedure

Experiments were performed to study the dimensional accuracy of investment casting with ice pattern and to compare this process with investment casting with other RP patterns. The ice patterns were cast by us at the University of Missouri - Rolla, while the investment casting with other RP patterns was done by Dr. Steve Schmitt at Tel Med Technologies.

Patterns for investment casting with ice patterns in our study were made by a silicone rubber mold according to the process chart shown in figure 1. The template for the rubber mold was an SLA pattern. Before investment, the slurry ingredients including the ceramic powders, binder and catalyst were put in the freezer for many hours to lower their temperatures to between –18 and –20 °C before mixing. After putting these materials in the freezer, ice patterns and an ice tree were made by a silicone rubber mold. Before mixing slurry ingredients, the ice tree was connected to the base of a ring former. Several different catalyst-to-binder ratios have been investigated. Vacuum treatment was used to reduce air bubbles during the mixing of the ingredients to make the slurry at temperatures below –10 °C.

The slurry was then poured into the ring former and cured inside a freezer. The ring former was then taken out of the freezer to let the ice inside the mold melt at room temperature. After this, the mold was taken out from the ring former and allowed to set at room temperature. The mold was preheated to 1000°C before casting, and then put inside a centrifugal furnace (Bego Co.) to cast the dental bar.

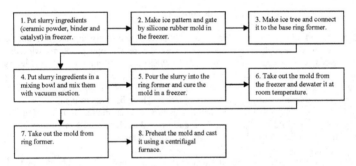

Figure 1 – Process Flow Chart for Casting with Ice Patterns

Problems of Investment Casting with Ice Patterns

In the beginning of the study, the investment slurry has 325 gm of Alumino-Silicate, 50 gm of gray matter (Nalcast fused silica powder with fiber), 7 ml of catalyst, and 130 ml of binder (the ratio of catalyst-to-binder is 5.4% and solid loading is 50.9% in volume).
The temperature of slurry was higher than –5°C. There are two main issues when the above ingredients are used in investment casting. The first problem is dimensional inaccuracy, which represents the differences in the dimensions of the metal castings from the dimensions of ice patterns. The second issue is casting defects. After casting, defects due to air bubbles on the metal part and incomplete geometry of metal parts were found (Figure 2).

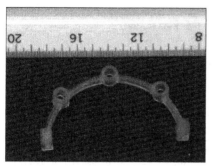

Loss of Pattern Cross Section Using the initial slurry from above as a starting point, the ice pattern is shown in Figure 2 and the finished metal part is shown in Figure 3. The difference between the ice pattern and the finished metal part can also be straightforward seen by visual inspection and comparing their dimensions. The measured differences in the width, height, and outer cylinder diameter between the pattern and the cast part were measured giving shrinkage allowance requirements of 15-30 percent.

Figure 2 -- Ice Pattern

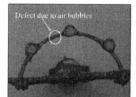

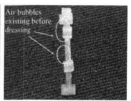

Figure 3 -Casting with Air Bubble Defect

Figure 4—Casting with Catalyst-to-Binder ratio 5.4%

Figure 5 – Ice Pattern before Dressing

Casting Surface Defects With the ratio of catalyst to binder at 5.4 %, there were mold defects caused by air bubbles, as shown in Figure 4. The air bubble defect is caused by the trapped air bubbles on the pattern when the slurry is poured around the ice tree. The air bubbles are attached to the ice pattern, and this is reproduced in the castings. Also, this ratio of slurry cannot gel in a short time; therefore, it may cause the ice pattern to melt or dissolve locally when the mold is curing in the freezer, forming an incomplete metal casting.

Process Development to Address Casting Problems

Improving Dimensional Accuracy. Both catalyst and binder are important ingredients in the slurry material. The catalyst in the slurry is used to shorten the gelling time, and the binder is to hold the refractory materials together to form the mold. When the solid loading is 51% volume, there is the shortest curing time and hence working time (Liu, 2006). Therefore, several different combinations of binder and catalyst were used in experiment. The ratio of catalyst-to-binder at 5.8 % (7 ml of catalyst with 120 ml of binder) with the solid loading of 52.8% volume did not allow the slurry to gel quickly enough. This caused ice pattern to melt prematurely in the mold and resulted in imperfect casting geometry. If gelling time was slow enough the water generated would damage the mold and cause the molten metal to flow out of the mold cavity during casting. In attempting to optimize the metal casting, solid loading was increased to 54.8% volume. For the ratio of catalyst to binder at 6.4 % (7ml catalyst and 110 ml binder, solid loading is 54.8% volume), this caused the ceramic mold to crack when casting, so it was not useable. Finally, the volume of catalyst was increased to shorten the gelling time of slurry and to keep the solids loading at 52 volume percent. A final combination of the ratio of catalyst-to-binder at 18.2 % (20 ml catalyst and 110 ml binder) with solids loading of 52.2% volume was chosen. This shortened the slurry gelling time and prevented the mold from cracking. The measured results from the castings obtained are listed on Table 1, which shows the range of shrinkage of four-to-eleven percent. However, on the sections other than the very thinnest where measurement is most difficult, the shrinkage was in the four-to-six percent range.

Eliminating Casting Defects. In order to eliminate casting defects get better casting results, five steps were used to control the quality of the ice part and the pouring of the slurry as follows:

1.) Vacuum treat the water to remove soluble gas before injecting water into the silicon rubber mold. This can remove air bubbles in the water and increase the quality of ice pattern.

2.) Pre-wet the master mold to decrease contact angle by injecting the water twice for each pattern. The first injection pre-wets the rubber mold.

3.) Provide adequate support; the soft silicon rubber mold was constrained by using a clamp to prevent the mold from warping during the freezing process. The clamp is removed after freezing.

4.) Implement good pattern room practices: the ice pattern was "dressed" before making the ice tree to ensure high quality of ice pattern during casting.

5.) Control pouring speed: Since the slurry is non-Newtonian fluid that exhibits shear thinning behavior, the shear rate during placement of the slurry around the pattern affects the viscosity immediately after placement. Consequently the slurry was poured with a controlled flow through a small nozzle will allow the slurry to conform to the pattern.

The positive results of following the above steps are illustrated in Figures 4 and 5. If the above five steps are not followed, there are some defects due to air bubbles observable after taking the ice pattern out from the silicon rubber mold, (Figure 5) The air-bubble defects were eliminated after following the above five steps (Figure 6).

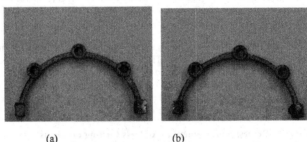

(a)   (b)

Figure 6. Metal castings from two ice patterns

Comparison with other Rapid Prototyping Patterns in Investment Casting

Patterns made from RP processes including Fused Deposition Modeling (FDM), Stereolithography (SLA), and ThermoJet were used in investment casting to compare with the results obtained from investment casting with ice patterns. Materials used by these RP processes are: DSM 9120-epoxy photopolymer for SLA, ABS for FDM, and TJ-88 wax for ThermoJet. The comparison was done using the average value of several measurements for each feature dimension.

The dental bar was made using the SLA, FDM, and ThermoJet RP processes by American Precision Prototyping in Tulsa, Oklahoma with the same CAD model of the dental bar.

Comparison of Dimensional Accuracy among Metal Castings The dental bars cast with SLA, FDM, ThermoJet were show in Figure 7. The dimensions of the metal castings were measured at several locations as were the four RP patterns. Figure 8 shows the dimensions that were measured. The results of measurement are given in Table 1. These results show that the differences in the dimensions of the metal castings from the patterns made by the four RP processes are from −0.030 mm to +0.076 mm. All of these values fall within the casting standard of "normal tolerance," which is ±0.007 inch as defined by the Investment Casting Institute (The Investment Casting Institute, 1980). Most of these values are within the casting standard of "premium tolerance," which is ±0.076 mm .

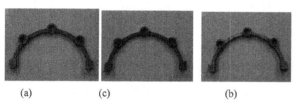

(a)   (c)   (b)

Figure 7. Metal castings from various types of RP patterns:
(a) SLA (b) FDM (c) ThermoJet

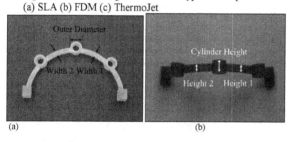

(a)   (b)

Figure 8. Measurement of dimensions

Table 1. Measured Dimensions of the Dental Bar (mm)

SLA	Width 1	Width 2	Height 1	Height 2	Outer Dia.	Cylinder Ht
Mean (Pattern)	1.6586	1.6861	2.50952	2.51291	4.73033	3.92091
Mean (Casting)	1.7454	1.7374	2.5582	2.54889	4.79891	3.96028
Mean (Ctg-Pattern)	0.0868	0.0512	0.0487	0.0360	0.0686	0.0394
Est.patternmakers shrinkage %	-4.972	-2.948	-1.903	-1.412	-1.429	-0.994
Pattern Std. Dev.	0.01418767	0.0238	0.02404	0.03191	0.06594	0.01972
Casting Std. Dev.	0.0210	0.0272	0.0400	0.0460	0.0245	0.0202
FDM	Width 1	Width 2	Height 1	Height 2	Outer dia	Cylinder Ht
Mean (Pattern)	1.508	1.499	2.400	2.415	4.457	3.872
Mean (Casting)	1.557	1.539	2.413	2.432	4.514	3.923
Mean (Ctg.-Pattern)	0.0487	0.0398	0.0123	0.0174	0.0567	0.0504
Est. pattern-makers shrinkage %	-3.127	-2.586	-0.509	-0.714	-1.257	-1.284
Pattern Std. Dev.	0.0165	0.0189	0.0396	0.0390	0.0292	0.0143
Casting Std. Dev.	0.0351	0.0116	0.0315	0.0321	0.0470	0.0440
ThermoJet	Width 1	Width 2	Height 1	Height 2	Outer Dia.	Cylinder Ht.
Mean (Pattern)	1.5697	1.5181	2.4668	2.4816	4.5038	4.0488
Mean (Casting)	1.581	1.582	2.545	2.541	4.548	4.105
Mean (ctg-pattern)	0.0110	0.0643	0.0779	0.0593	0.0440	0.0567

	Width 1	Width 2	Height 1	Height 2	Outer Dia.	Cylinder Ht.
Est. patternmakers shrinkage %	-0.696	-4.066	-3.061	-2.333	-0.968	-1.382
Pattern Std. Dev.	0.0426	0.1583	0.1318	0.0381	0.0271	0.0299
Casting Std. Dev.	0.0188	0.0193	0.0204	0.0350	0.0515	0.0481
RFP (ice)	**Width 1**	**Width 2**	**Height 1**	**Height 2**	**Outer Dia.**	**Cylinder Ht.**
Mean (Pattern)	1.659	1.679	2.565	2.538	4.703	4.028
Mean (Casting)	1.4707	1.4728	2.4168	2.4337	4.4928	3.8769
Mean (Ctg-Pattern)	-0.1888	-0.2057	-0.1486	-0.1041	-0.2104	-0.1511
Est. patternmakers shrinkage %	12.838	13.970	6.148	4.279	4.683	3.898
Pattern Std. Dev.	0.0407	0.0096	0.0732	0.0305	0.0610	0.0315
Casting Std. Dev.	0.0412	0.0273	0.0312	0.0149	0.0300	0.0222

When the dental bar was cast using an ice pattern, there is contraction on the metal casting part. Dimensional shrinkages for this dental bar are from four to eleven percent (Table 1). However the standard deviations are in the range between 0.015mm and 0.041mm, which is less than the casting standard of "normal tolerance," of ±0.178mm but suggests a process capability ($\pm3\sigma$) that is still below the standard on the very thinnest sections. The percent apparent shrinkage decreases with increasing pattern dimension, as shown in figure 9. One possible mechanism might be an effect of melting or dissolution of some of the ice pattern surface into the slurry for making the mold, this percent patternmakers shrinkage was also compared to thermal modulus (V/SA) of the sections that were measured. The correlation seems to be a little better for starting dimension than V/SA. One approach to limit this effect may be the use of an interface coating such as used by Liu et al (Liu, 2004), however the technology for uniform application of the coating needs development. An alternative is to make the pattrns oversize by a shrink rule that is a function of target dimension.

Comparison of Surface Roughness Among Metal Castings To compare the surface roughness of metal castings made from various RP patterns, the surface roughness of each metal casting was measured. The particle sizes of two main powders, alumino-silicate and gray matter (Nalcast fused silica powder) to make the slurry for investment casting with ice patterns were also measured. The mean particle size of alumino-silicate was 3.1μm and the mean particle size of "Gray Matter" was 3.3 μ m, both having an approximate log-normal distribution. Surface roughness was measured using a profilometer (Brown & Sharpe) on the metal castings made by the five types of patterns. The traverse length of measurement was set at 5mm, and the cutoff wavelength was set at 7.6mm. Thus the measured parts need to have length greater than 04.95 inch of a flat surface. Because of the geometry limitation, only the top surfaces of Section 1 and Section 2 in Figure 10 were measured. The measured surface roughnesses are given in Table 2

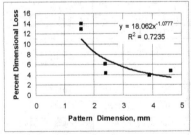

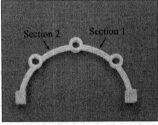

Figure 9. Percent dimensional loss (shrinkage) as affected by starting pattern dimension

Figure 10. Surface roughness measured sections

According to Table 2, metal castings made from the ThermoJet process patterns have the smoothest surfaces, ranging from 1.5 to 1.6 μm. Metal castings from FDM patterns have the largest surface roughness among the four RP processes, ranging from 3.3 to 3.5 μ m. Metal castings using ice patterns have surface roughness ranging from 3.1 to 3.2 μ m.

Table 2. Measured surface roughness of metal castings made from different RP patterns

	SLA		FDM		ThermoJet		RFP	
	Mean	Std.Deviation	Mean	Std.Deviation	Mean	Std.Deviation	Mean	Std. Deviation
Section 1(μ in)	2.81	1.08	3.30	.46	1.54	0.13	3.12	0.20
Section 2(μ in)	2.97	1.62	3.48	0.42	1.60	0.39	3.25	0.22

Data Analysis by ANOVA The ANOVA (Analysis of Variance) method was employed to assess whether the means of each group (SLA, FDM, ThermoJet, and RFP) from the measurements is statistically different from each other (Montgomery, 2005).

ANOVA Analysis of Dimensional Accuracy and Surface Roughness To investigate which RP pattern has the least difference in dimensions after casting, the ANOVA-test method is used again. When $\alpha = 0.01$ (confidence level is 99%), the critical F value ($F_{0.01, \, 4,25}$) is 4.18. All F_0 values of the measured dimensions (Width 1, Width 2, Height 1 Height 2, Outer Diameter, and Cylinder Height) are greater than $F_{0.01, \, 4,25} = 4.18$ (Table 3), thus we can say that the metal castings from the five different types of RP patterns have significant difference in the mean values of their dimensions. Then comparing the measurement difference between RP patterns and metal castings, we also find the F_0 of the standard deviations are greater than $F_{0.01, \, 4,25} = 4.18$ (Table 3). Thus, we can say that there are significant differences in the measurement differences. Then comparing the measurement values between RP patterns and metal casting, we find that the metal castings from SLA, FDM, and ThermoJet patterns expand (Table 1).

Table 3. Results of ANOVA test

	Width 1	Width 2	Height 1	Height 2	Outer-Diameter	Cylinder-Height
F_0 (Casting-Pattern)	27.33	14.37	8.65	13.75	14.92	21.54
F_0 (ΔCasting- ΔPattern)	0.002	0.0004	0.0012	0.0015	0.0011	0.0038

To compare the surface roughness of the metal castings from the four different patterns (SLA, FDM, ThermoJet, and RFP), ANOVA-test is used again. The critical F value ($F_{0.01, \, 3,32}$) in this case is 2.90. The two F_0 values of two surface roughness from the measured sections are 15.70 and 8.83. They are both greater than $F_{0.01, \, 3,32} = 2.90$. Thus we can conclude that the surface roughness of the metal castings made from the four different RP processes is significantly different. Note that the metal castings made with ice patterns have better surface finish than metal castings made with FDM patterns.

Conclusion

A study has been conducted to investigate the dimensional accuracy and surface roughness of investment casting with ice pattern. When increasing the ratio of catalyst-to-binder from 5.4 % to 18.2 % (20 ml catalyst and 110 ml binder), the dimensional accuracy of the metal castings from ice patterns has improved, from −0.44 to −0.21 mm. Other techniques to prevent casting defects and improve surface roughness are to dress the ice pattern before investment casting and to control the speed of pouring the slurry into the ring former. After applying these techniques, the surface roughness of metal castings from ice patterns has improved to a range between 3.2 and 3.4 μ m.

273

The dimensions of different dental bar patterns made using different RP processes, and the generated metal castings were measured. Metal castings made with SLA, FDM, and ThermoJet patterns showed expansion, while metal castings made with ice patterns showed contraction. All of the dimensional tolerances fall within ± 0.18 mm and most of these values are within ± 0.076 mm. When a dental bar is cast using an ice pattern, the range of dimensional differences is from -0.0041 to -0.0083 inch, indicating that there is contraction on the metal casting. The standard deviations are in the range between 0.18 and 0.091mm, which is within the casting standard of "normal tolerance." Metal castings made from the ThermoJet patterns have the smoothest surfaces, ranging between 1.55 and 1.60 μ m surface roughness. Metal castings from ice patterns have surface roughness ranging from 3.2 to 3.4 μm. This is comparable to the surface roughness of metal castings from FDM patterns. The study of dimensional accuracy and surface roughness of metal castings made from ice patterns in comparison to metal castings made from other RP patterns has shown that investment casting with ice patterns is a viable process for industrial application.

References

The Investment Casting Institute, "Dimensions, Tolerances Surface Texture," in Investment Casting Handbook, pp. 161-168 (1980).

Jose, H., "Solid Mold Investment Casting Process Using Ice Patterns," Master's Degree Thesis, University of Missouri – Rolla. (2005)

Liu, Q. Leu, M. C. Richards, V. and Schmitt, S. M., "Dimensional Accuracy and Surface Roughness of Rapid Rreeze Prototyping Ice Patterns and Investment Casting Metal Parts," International Journal of Advanced Manufacturing Technology, 2004, DOI: 10.1007/s00170-003-1635-9(2004).

Liu, Q. Richards, V. Daut, K. and Leu, M. C. "Curing Kinetics of Ceramic Slurry Used in Investment Casting with Ice Patterns," International Journal of Cast Metals Research, v 19 (3), P 195-200, June. (2006),

Montgomery D.C. ,, "The Analysis of Variance" in Design and Analysis of Experiments, sixth Edition, John Wiley & Sons, Inc., pp.63-75, Honoken, NJ, USA (2005).

Peng, X. Jiang, B. Yan, S. and Zhigang, Z. (2000), "Study on Rapid Prototyping Manufacturing in Ice-Pattern and Low Temperature Casting Technology," Tsinghua University, Beijing, China.

Wu, R. Yan, Y. and Feng, C.,,"Study on investment casting process based on ice-model," Tsinghua University, Beijing, China (2003).

Yodice, A., "Free Cast Process," US Patent 5,072,770 (1991).

Yodice, A. , "Freeze Cast Process Ready for Licensing," INCAST: International Magazine of the Investment Casting Institute, Vol. 11, No.12, 19-21 (1998).

EFFECT OF CASTING OVER-PRESSURE ON THE FATIGUE RESISTANCE OF ALUMINUM ALLOY A356-T6

M.A. Neri[1], D.R. Poirier[2] and R.G. Erdmann[2]

[1]Advanced Materials Research Center (CIMAV); Miguel de Cervantes # 120, Complejo Industrial Chihuahua; C.P. 31109, Chihuahua, México
[2]The University of Arizona; Department of Materials Science and Engineering, Mines Bldg.; Tucson, AZ, 85721-0012, USA

Keywords: Fatigue, Aluminum Casting Alloy, Microporosity, Pressure Casting

Abstract

The fatigue resistance of an Al-Si casting alloy (A356) solidified under pressures up to 20 atm and under a vacuum was investigated. Pressures of 10 and 20 atm mitigated microporosity in plate-castings, and this enhancement significantly improved fatigue resistance. Cooling rate during solidification, which affects the dendrite arm spacing (DAS), was also varied. Also studied were Sr-modification and the use of a flux to capture oxide-bifilms. Fatigue life increased as the maximum pore size decreased, which resulted from the over-pressures and/or a decrease in DAS. The use of flux in the mold and Sr-modification, especially when employed simultaneously, also increased fatigue life.

Introduction

Advances in degassing and fluxing, filtration, and design have improved the aluminum castings, but oxide-bifilms and porosity still persist as casting defects [1-3]. During solidification, hydrogen partitions into the intergranular liquid [4], where it diffuses to and expands the gas within the oxide-bifilms to make pores [1]. The advantages of applying 10 atm pressure in aluminum castings made by "pressurized lost foam" has been reported as "reduced porosity to near undetectable levels; increased elongation properties; and increased fatigue life" [5]. In a study of fatigue in A356.2 [6-8], it was found that cracks initiated at pores. When the DAS was < 30 μm and the pores were below a critical size of approximately 80-100 μm, however, the cracks initiated within large eutectic constituents or at oxide films. The effect of Sr-modification on the mechanical properties of Al-Si casting alloys has been studied many times [9-16]. On one hand, modification results in rounder and finer Si-particles after solution heat treatment. Hence, modification can improve the tensile strength and percent elongation [10-13, 15,16]. On the other hand, the added porosity associated with Sr-modification may nullify the beneficial effect of the improvement in the Si-morphology [15,16].

The porosity in plate-castings studied in this work were reported by Frueh et al. [4], who reported histograms of pore sizes in plates cast under 1, 10 and 20 atm. Porosity measurements were restricted to samples taken from plates where the cooling rate was 0.50 K s^{-1} with a DAS of about 55 μm. Most of the pores in the three cases were in the range of 0 to 30 μm. From their histograms, 38 pores > 80 μm and as large as 200 μm were found in the 1 atm - sample. With the exception of only one "outlier" at 130 μm, the largest pore in the plate cast at 10 atm was 70 μm. The largest pore in the plate cast under 20 atm was only 60 μm. For this work, the same plate-castings of A356 [4] were studied with emphasis on fatigue resistance. The vacuum/pressure vessel allowed both melting and casting under either vacuum or gas mixture. Here the effects of the over-pressure, DAS, Sr-modification and the application of flux to the mold cavity on the fatigue life of the aluminum alloy in the T6-condition were evaluated.

Materials and Methods

Casting and Post-Casting Processing

Details of special melting and casting chamber can be found in Frueh et al.[4]. Briefly, the alloy was induction-melted in a fixed alumina crucible, equipped with an alumina stopper rod. A ceramic-shell mold, set below the crucible, was used to produce a vertically cast plate (241 mm tall, 76 mm wide and 25 mm thick). The open-bottom mold was placed on a water-cooled chill. Directional solidification was abetted by flexible heating tapes, which were wrapped around the molds, for preheating (973 K). The molds with heating tapes were enveloped with silica sand and covered with insulation. Table I summarizes four groups of plates that were cast at four different pressures (vacuum, 1, 10 and 20 atm). Melting included an equilibration period under an atmosphere of argon and hydrogen. The fraction of hydrogen in the gas was set to get a concentration of hydrogen in the melt of 0.2 ppm, which ensured porosity in castings made under 1 atm. After equilibration, the chamber was evacuated to the vacuum-level, and the melt was poured; or the atmosphere was maintained during pouring. In some molds, a commercial flux was placed in the bottom of the mold cavity; temperatures during cooling were measured in some of the plate-castings.

The plates were sectioned into bars parallel to the bottom end; the bars were heat- treated to T6 condition (538 °C for 5 hours, quenched into water at 72 °C, and aged, within 2 hours after quenching, at 160 °C for 4 hours). Then the bars were machined into fatigue-specimens and polished to a surface finish of 16 micro-inches. The fatigue-specimens were made by Westmoreland Mechanical Testing & Research, Inc., Youngstown, PA. The specimens had a testing section of 10.2 mm in length and a diameter of 5.08 mm. The testing section transitioned to the unthreaded ends by a generous radius and were designed to fit into the gripping mechanisms of the testing instrument.

Axial Fatigue Testing and Microstructural Characterization

High-cycle fatigue was done with a closed-loop servo-hydraulic testing instrument (MTS 810), in load-control at 15 Hz. The maximum/minimum stresses were set to 175/17.5 MPa ($R = 0.1$). The yield strength of the subject alloy is in the range of 189 MPa to 207 MPa [17]. Image-analysis software (Image-Pro) was used to characterize the porosity; pore size is defined as the maximum pore diameter, which is the length of the longest chord passing through the pore centroid. Dendrite arm spacing (DAS) was measured by the line-intercept method, and the images were also used to ascertain whether Sr-modification was effective.

Results and Discussion

Porosity and Secondary Dendrite Arm Spacings

The reader is reminded that the plates were top poured into molds without gating systems. Although the atmosphere in the furnace was either a vacuum (1.3×10^{-4} atm) or a gas mixture

Table 1: Features of the four groups of plate castings.

Group	Characteristics of the cast alloy	Plate	Casting Pressure, atm
1	A356 without flux	4	1.3×10^{-4} (vacuum)
	"	12	1
	"	13	10
	"	15	20
2	A356 + Sr without flux	5	1.3×10^{-4} (vacuum)
	"	8	1
	"	9	10
	"	10	20
3	A356 with flux	7	1.3×10^{-4} (vacuum)
	"	11	1
	"	14	10
	"	16	20
4	A356 + Sr with flux	6	1.3×10^{-4} (vacuum)
	"	17	1
	"	18	10
	"	19	20

of Ar and H_2, even small O_2 potentials would have oxidized the melts during pouring. The porosity in the plate-castings, therefore, likely resulted from the presence of oxide-bifilms. Indeed, motivation for using flux in the mold-cavities was to capture the bifilms. Another point is that in the upper parts of the plates the DAS is rather large because the molds were preheated to above the liquidus and backed with sand surrounded by an insulating blanket. From cooling curves in the instrumented molds, it was seen that heat transfer from plate-to-plate did not yield consistent solidification cooling rates as a function of distance from the chill. Therefore, results from plate-to-plate are compared at the same DAS rather than at the same distance from the chill.

Percent porosities versus DAS for Group 1 (no flux) are presented as Fig. 1. The maximum DAS in the plates cast under the high over-pressures is about 120 μm; in the other two (vacuum and 1 atm) the maximum DAS is about 185 μm. With arm spacings in the range of 60 to 130 μm, there is a modest decrease in the percent porosity with 10 and 20 atm. The comparison cannot be made at larger arm spacings. The two "outliers" with the relatively high percent porosity in the 1 atm - plate can be explained by the sporadic formation of oxide-bifilms during top pouring. Bifilms are also expected in the pressurized plates, but of course the pores were not allowed to expand.

Percent porosities in Group 2 are presented as Fig. 2. Group 2 is similar to Group 1 except that Group 2 is Sr-modified. It appears that oxide-bifilms are more prevalent, probably exacerbated by Sr, as seen by the exceptionally high amount of porosity in the plates solidified under 1 atm. The amounts of porosity in Groups 1 and 2 are otherwise comparable. To save publication space, the porosities in the plates cast in molds with the flux are not shown. Again the most porosity is present in the plates solidified under 1 atm, and the presence of Sr increases the percent of porosity. For the pressurized plates and DAS > 50 μm, the percent porosity in Groups 3 and 4 (with flux) is about the same as in Groups 1 and 2 (no flux). With DAS < 50 μm, however, very low percents of porosity resulted in pressurized plate-castings.

In fatigue, maximum pore sizes are more important than are the average percentages of

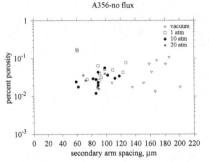

Figure 1: Percent of porosity in samples taken from the plate-castings of Group 1 (no flux).

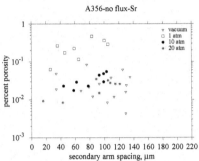

Figure 2: Percent of porosity in samples taken from the plate-castings of Group 2 (no flux and Sr-modified).

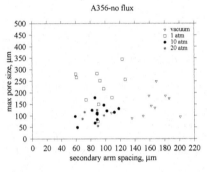

Figure 3: Maximum pore sizes in samples taken from the plate-castings of Group 1 (no flux).

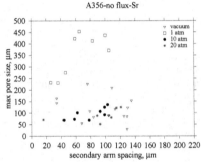

Figure 4: Maximum pore sizes in samples taken from the plate-castings of Group 2 (no flux and Sr-modified).

porosity. The maximum pore sizes for Groups 1 and 2 are shown in Figs. 3 and 4, respectively. Again, in order to save publication space, the maximum pore sizes in Groups 3 and 4 are not shown. The efficacy of solidifying under over-pressures of 10 and 20 atm is very apparent in Figs. 3 and 4. The worst are those that solidified under 1 atm. These results clearly indicate the importance of minimizing the formation of oxide-bifilms when filling molds for aluminum castings (viz., as in bottom-filling processes used industrially). Large pores can also be minimized by applying over-pressure during solidification (e.g., the "pressurized lost foam" process [5]). In the unmodified alloys, the use of flux does not significantly reduce the maximum pore size. In the Sr-modified alloys, the flux reduces the maximum pore size in samples when DAS < about 50 μm.

Sr-Modification

Without modification the faceted Si within the eutectic constituents had the expected high aspect ratios. Many of the Si-particles are needle-like, especially when DAS is > 120 μm.

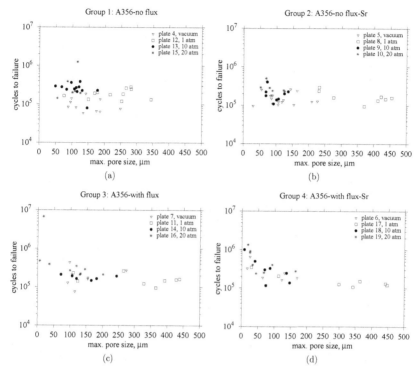

Figure 5: Effect of the maximum pore size on the fatigue life of the four groups: (a) Group 1 – no flux; (b) Group 2 – no flux and Sr-modified; (c) Group 3 – with flux; and (d) Group 4 – with flux and Sr-modified.

The castings that contained Sr had the usual well-rounded Si-particles. Plate 6 with Sr, however (melted and cast under vacuum), did not have the modified structure, and Plate 5, also with Sr and melted and cast under vacuum, was partially modified.

Effect of Porosity on Fatigue Life

Figures 5(a)-5(d) are plots of fatigue life versus the maximum pore size for Groups 1-4, respectively. Many of the maximum pore sizes are considerably larger than those reported by Frueh et al. [4], who restricted measurements of the pores to plate-positions where the cooling rate during solidification was 0.5 K s^{-1} and the DAS was about 55 μm. Figures 5(a)-5(d) show that the greater fatigue lives are achieved when the maximum pore size is < about 100 μm, which agrees with Zhang et al. [6-8]. The pressurized-plates typically have smaller maximum pore sizes and better fatigue lives.

The role of flux requires explanation. Specifically, during sectioning the plates cast in the fluxed molds, visible pores were found. These pores were filled with solidified flux that pulverized by the action of the band saw. During casting, the flux did not completely separate

279

from the molten alloy by floatation. Hence, before heat treating and machining all bars from the fluxed castings were radiographed. Only those bars with no visible pores or with no pores detected in the radiographs were processed further. With this proviso, the use of flux in the molds increased fatigue life when fatigue lives are about 10^5 cycles or less in "no flux" samples. Otherwise the use of flux appears marginally beneficial at best; compare Groups 1 and 3. With Sr present, the use of flux appears inconsequential; compare groups 2 and 4. Perhaps, the major conclusion to be drawn from Fig. 5 is that fatigue lives are improved when maximum pore sizes are less than about 100 μm and the best fatigue lives are achieved where the maximum pore size is less than about 50 μm in Sr-modified allow with flux added to the mold.

Effect of DAS on Fatigue Life

The same fatigue lives of Figures 5(a)–(d) are re-plotted in Figures 6(a)–(d) as fatigue life versus DAS. Fatigue life increases as the DAS decreases for each group. This probably reflects that pore sizes decrease as the cooling rate during solidification increases for a given pressure. Also within each group, the fatigue life increases as the pressure increases. The benefit of flux is noticeable in Group 4, in which both flux and Sr-modification were employed. By reducing maximum pore sizes with high pressures and presumably eliminating oxide-bifilms with the flux, the benefit of Sr-modification can be seen when the DAS is less than about 50 μm.

Conclusions

(i) Pressures of 10 and 20 atm mitigated microporosity in the castings.

(ii) Fatigue life increased as the maximum pore size decreased, which resulted from the over-pressures of 10 and 20 atm and/or a decrease in the secondary dendrite arm spacing.

(iii) The use of flux in the mold and Sr-modification, especially when employed simultaneously, significantly increase fatigue life when the DAS is less than about 50 μm. This conclusion, however, is based on a population from which samples with entrapped flux were eliminated before fatigue testing.

Acknowledgements

The plates were cast at Sandia National Laboratories, Albuquerque, NM by C. Frueh and M.E. Miszkiel. They have our special thanks. Q. Wang of GM Powertrain graciously arranged to make the fatigue test-specimens. Heat-treating was done by K. Smith at The University of Arizona, and microstructures were by L.C. Ortiz at CIMAV. The first author (M.A.N.) is grateful to CIMAV for his sabbatical leave and to The University of Arizona for a financial supplement. D.R.P. and M.A.N. are also especially grateful to P.K. Sung (now at Howmet in Whitehall, MI) for his guidance during the early stages of the fatigue testing.

References

1. J. Campbell, *Castings* (Oxford, UK: Butterworth-Heinemann, 1991), 1-26 & 261-263.

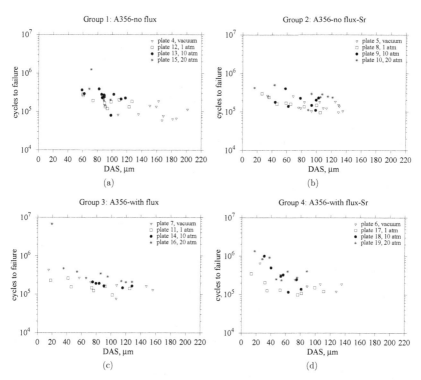

Figure 6: Effect of secondary dendrite arm spacing (DAS) on the fatigue life of the four groups: (a) Group 1 – no flux; (b) Group 2 – no flux and Sr-modified; (c) Group 3 – with flux; and (d) Group 4 – with flux and Sr-modified.

2. J.A. Eady, and D.M. Smith, "The Effect of Porosity on the Tensile Properties of Aluminum Castings," *Mater. Forum,* 10 (1986), 217-223.

3. C.H. Caceres, and B.I. Selling, "Casting Defects and the Tensile Properties of an Al-Si-Mg Alloy," *Mat. Sci. Eng. A,* A220 (1996), 109-116.

4. C. Frueh et al., "Microporosity in A356.2 Aluminum Alloy Cast under Pressure," *Advances in Aluminum Casting Technology II,* ed. M. Tiryakioğlu and J. Campbell (Materials Park, OH: ASM International, 2002), 99-105.

5. A.T. Spada, " 'Core Competency' Focus Includes Lost Foam at Mercury Marine," *Modern Castings,* 91 (5) (2001), 27-30.

6. W. Chen et al., "Microstructure Dependence of Fatigue Life for A356.2," *Automotive Alloys II,* ed. S.K. Das (Warrendale, PA: TMS, 1998), 99-113.

7. B. Zhang, D.R. Poirier and W. Chen, "Microstructural Effects on High Cycle Fatigue-Crack Initiation in A356.2 Casting Alloy," *Metall. Mat. Trans. A,* 30A (1999), 2659-2666.

8. B. Zhang, W. Chen and D.R. Poirier, "Effect of Solidification Cooling Rate during Solidification on the Fatigue Life of A356.2-T6 Aluminum Alloy," *Fatigue Fract. Engng. Mater. Struct.,* 23 (2000), 417-423.

9. B. Closset and J.E. Gruzleski, "Structure and Properties of Hypoeutectic Al-Si-Mg Alloys Modified with Pure Strontium," *Metall. Trans. A,* 13A (1982), 945-288.

10. N. Fat-Halla, "Structural Modifications of Al-Si Eutectic Alloy by Sr and Its Effect on Tensile and Fracture Characteristics," *J. Mat. Sci.,* 24 (1989), 2488-2492.

11. S. Shivkumar, L. Wang and D. Apelian, "Molten Metal Processing of Advanced Cast Aluminum Alloys," *JOM,* 43 (1991), 26-32.

12. W. Schneider and F.J. Feikus, "Heat Treatment of Aluminum Casting Alloys for Vacuum Die Casting," *Light Metal Age,* 56 (10) (1998), 12-37.

13. N. Fat-Halla, M. Hafiz and M. Abdulkhalek, "Effect of Microstructure on the Mechanical Properties and Fracture of Commercial Hypoeutectic Al-Si Alloy Modified with Na, Sb, and Sr, *J. Mater. Sci.,* 34 (1999), 3555-3564.

14. F.T. Lee, J.F. Mayor and F.H. Samuel, "Effect of Silicon Particles on the Fatigue Crack Growth Characteristics of Al - 12 wt. pct. Si-035 Wt. Pct. Mg- 0-0.02 Wt. Pct. Sr Casting Alloys," *Metall. Mat. Trans. A,* 26A (1995), 1553-1570.

15. B. Zhang et al., "Effects of Strontium Modification and Hydrogen Content on the Fatigue Behavior of A356.2 Aluminum Alloy," *Trans. AFS,* 108 (2000), 383-389.

16. B. Zhang, D.R. Poirier and W. Chen, "Effects of Hipping and Strontium Modification on the Fatigue Behavior of A356.2 Aluminum Alloy," *Trans. AFS,* 110 (2002), 393-405.

17. B. Zhang, "Fatigue Behavior in an Aluminum Casting Alloy (A356.2): Effects of Some Defects, SDA, Hipping and Strontium Modification," (Ph.D. dissertation, The University of Arizona, 2002), 166.

EFFECTS OF CASTING PROCESS PARAMETERS ON POROSITY AND MECHANICAL PROPERTIES OF HIGH PRESSURE DIE CAST ADC12Z ALLOY

YAN Yan-fu[1], XIONG Shou-mei[1*], LIU Bai-cheng[1], Mei Li[2] and John Allison[2]
*Corresponding author: smxiong@tsinghua.edu.cn

[1]Department of Mechanical Engineering, Tsinghua University, Beijing 100084, P.R. China
[2]Ford Research & Advanced Engineering, Ford Motor Company, Dearborn, MI, USA 48121

Keywords: High pressure die casting, Casting process parameters; Porosity; Mechanical property, ADC12Z alloy

Abstract

High pressure die casting experiments were systematically conducted using a 650t cold chamber machine to study the influence of operating conditions on porosity and mechanical properties of a step-shape casting using ADC12Z alloy. The step-shape casting had five steps each with a different thickness. The results showed that process parameters had differing effects on density and porosity within the die cast part at different thicknesses. At the same thickness the mechanical properties were greatly influenced by operating conditions and closely related to the porosity of the cast parts. Under the same die casting conditions, the density and mechanical properties of cast part decreased dramatically with an increase in thickness. Furthermore, the relationship between mechanical properties, porosity and microstructure of the die cast parts was discussed.

Introduction

High pressure die casting process is a near-net shape manufacturing process in which molten metal is injected into a metal mold at high speed and allowed to solidify under high pressure [1]. The main disadvantage of this process is the formation of gas porosities as a consequence of the high speed injection of the molten metal into the die cavity[2-5]. The presence of gas porosities in die castings is harmful as the mechanical properties and pressure tightness are adversely affected. Therefore, the applications of aluminum die castings are normally limited to non-structural components that do not require heat treatment.

Studies on die casting process to improve the die casting properties have been carried out for many years [6-8]. Recently, some researchers have been focusing on the study of the influence of process parameters on die casting defects[6-13]. This project was designed to systematically investigate the influences of die casting process parameters on the quality of die castings. In this paper, the effects of the casting pressure, the slow and fast shot speeds, as well as the biscuit thickness on porosity and mechanical properties of ADC12Z die castings were discussed.

Experimental procedure

Alloy and sample preparation. The casting alloy used in this study is ADC12Z aluminum alloy. Its chemical compositions are shown in Table 1. In order to study the influences of the casting pressure, the slow shot speed, the fast shot speed and the biscuit thickness, combinations of these process parameters at different levels were carefully selected to conduct the experiments. The baseline process parameters are shown in Table 2. Five levels of each of these casting parameters, as shown in Table 3, were tested separately while other die casting process parameters were kept constant.

Table 1. Chemical compositions of ADC12Z

Alloy	Cu	Si	Fe	Mg	Ni	Al
ADC12Z	2.3	10.6	0.9	0.18	0.21	Balance

Table 2 Baseline die casting process parameters of ADC12Z alloy

Parameter	Value
Injection temperature of metals (°C)	680
Die equilibrium temperature (°C)	160
Casting pressure (MPa)	66.7
Slow shot speed (m/s)	0.2
Fast shot speed (m/s)	2
Fast shot start position (mm)	240
Pressurization position (mm)	290
Biscuit thickness (mm)	20
Pressurization time (ms)	40

Table 3 Levels of die casting process parameters

Parameters	Level 1	Level 2	Level 3	Level 4	Level 5
Casting pressure (MPa)	24	39	44	55	66.7
Slow shot speed (m/s)	0.1	0.2	0.3	0.4	0.5
Fast shot speed (m/s)	0.7	1	2	3	4
Biscuit thicknesses (mm)	15	20	25	30	35

The experiments were conducted on a 650t cold chamber die casting machine using a step-shape casting. The geometry of the casting is shown in Fig.1. All experiments were performed consecutively under thermal equilibrium condition, and for each set of operating condition, five castings were taken for the testing purpose. Five customized tensile specimens with different thickness were taken from each of the five steps of the test casting. The geometry and dimension of the tensile specimen are shown in Fig.2 (thickness of the specimen is the as-cast thickness of each step from the casting).

Porosity measurement. Porosity was determined via density measurement. Specimen mass was measured in both air and distilled water. The actual density (ρ) of each specimen was determined by equation (1) based on ASTM Standard D3800.

$$\rho = \left(\frac{m_s}{m_s - m_0} \right) \times \rho_{H_2O} \qquad (1)$$

where m_s and m_0 are the mass of the specimen in air and in distilled water respectively. ρ and ρ_{H_2O} are the density of the specimen and the distilled water.

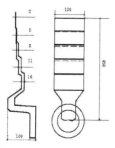

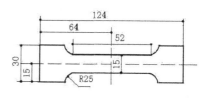

Fig.1.Geometry of the step-shape casting

Fig.2. Geometry and dimensions of the tensile specimen

Mechanical properties measurement. Tensile strength and yield strength were measured on a CC-5510 electronic tensile test machine at a nominal strain rate of 1mm/min. For each set of experimental conditions, five specimens were tested to get an average value.

Microstructural analysis. Microstructural samples were taken from the step with a thickness of 14mm. Mounting and polishing of the samples were conducted following the standard metallographic procedures. A Buehler Optical Image Analyzer 2002 system was used to analyze the porosity distribution of the specimens. Ten microstructural images were taken from each sample. The area percentage of porosities, the average minimum distance between porosities and the average diameter of porosities were evaluated using the image analysis software.

Results and Discussions

Effects of die casting process parameters on porosities
Porosities were evaluated based on the measured densities of the test specimens. The effects of the casting pressure, the biscuit thickness, the slow shot speed and the fast shot speed on density are shown in Fig.3. It can be seen that the density of the test specimen increases as the specimen thickness decreases for all test conditions. Fig 3a shows the effect of the casting pressure on density. It can be seen that the density of the die casting increases with the increase of the casting pressure and this effect becomes more prominent for thicker steps.

The biscuit thickness (Fig. 3b) has a similar effect on the density as the casting pressure since the biscuit thickness strongly affects the pressure that can be transferred to the casting. With the increase of the biscuit thickness, the density of the casting increases and the biscuit thickness has a greater effect on the thicker steps than on the thinner ones.

285

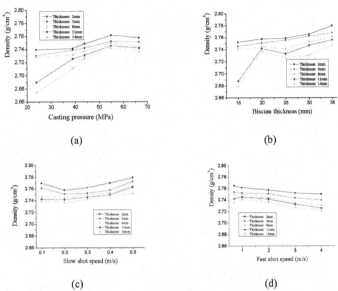

Fig.3 Influences of die casting process parameters on densities

(a) casting pressure   (b) biscuit thickness   (c) slow shot speed   (d) fast shot speed

Compared to casting pressure and biscuit thickness, the slow shot (Fig. 3c) and the fast shot speed (Fig.3d) have a relatively small effect on the density of the casting. The slow shot speed has different influence on the density of the casting with different thickness. The densities of the specimens with thickness of 8mm, 11mm and 14mm increase as the slow shot speed increases, while densities of specimens with thickness of 2mm and 5mm decrease first and then gradually rise as the slow shot speed increases. However, with the increase of the fast shot speed, the densities decrease at all thicknesses, which is related to the air entrapment as a consequence of the high speed injection of the molten metal into the die cavity.

In order to study the porosity distribution, samples were taken from the 14mm step of the castings. The area percentage of porosity, the average distance between porosities and the average diameter of porosities were analyzed, and the influences of the casting pressure, the biscuit thickness, the slow shot speed and the fast shot speed on these microstructural features are shown in Fig. 4. It can be seen that the casting pressure has the most significant effect on porosity. As the casting pressure increases, the area percentage of porosity decreases (Fig. 5), while the average diameter and the average distance between porosities increase. With the increase of the biscuit thickness, the area percentage of porosity decreases and the average distance between porosities increases. The slow shot speed and the fast shot speed do not have obvious influences on the average diameter or the average distance of porosities, however, increasing the fast shot speed increases the area percentage of porosity, while increasing the slow shot speed decreases the area percentage of porosity.

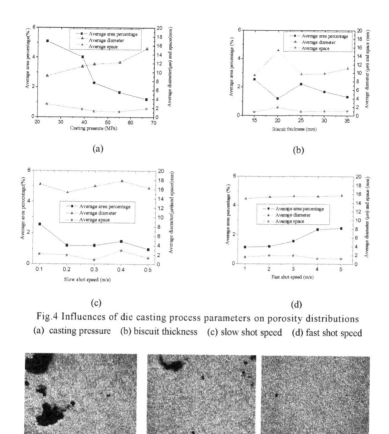

(a)   (b)

(c)   (d)

Fig.4 Influences of die casting process parameters on porosity distributions
(a) casting pressure (b) biscuit thickness (c) slow shot speed (d) fast shot speed

(a) 24MPa   (b) 44MPa   (c) 55MPa

Fig. 5 Porosities of the 14mm step of the castings under different casting pressures

Effects of the die casting process parameters on mechanical properties.

Tensile strength. The influences of the casting pressure, the biscuit thickness, the slow shot speed and the fast shot speed on the tensile strength of the test specimens are shown in Fig.6. The effects of the die casting process parameters on tensile strength follow the same trend as the effects of the parameters on density. Under the same operating conditions, the tensile strength increases as the thickness of the test specimen decreases. With the increase of the casting pressure and the biscuit thickness, the tensile strength increases, and the influence of the casting pressure and the biscuit thickness becomes more prominent when the casting thickness is greater than 8 mm. It can also be seen from Fig. 6 that with the increase of the slow shot speed, the tensile strength increases in the casting steps of 8mm, 11 mm and 14mm. There might be a critical slow shot speed corresponding to the minimum tensile strength in the casting steps of

287

2mm and 5mm. As for the fast shot speed, the tensile strength decreases as the fast shot speed increases.

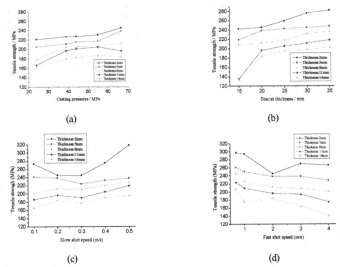

(a)   (b)

(c)   (d)

Fig.6 Influences of die casting process parameters on tensile strength

(a) casting pressure　(b) biscuit thickness　(c) slow shot speed　(d) fast shot speed

Since the tensile strength of die castings is closely related to their density, a relationship between tensile strength and density was established as shown in Fig. 7, and can be expressed as:

$$\sigma_b = -575.2 + 4.12 * \rho^{5.2} \tag{2}$$

where σ_b is the tensile strength of the casting and ρ is the density of the casting.

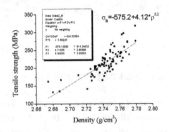

Fig. 7 Tensile strength vs. density of the test specimens

Elongation. The effects of the casting pressure, the biscuit thickness, the slow shot speed and the fast shot speed on the elongation of the test specimens are shown in Fig.8. It can be seen that under the same operating conditions, the elongation of the test specimen increases as the casting step thickness decreases. However, the effects of the die casting process parameters on elongations are insignificant. This indicates that the elongation is mainly determined by the grain size of the casting, and the porosity does not have strong influence on the elongation.

288

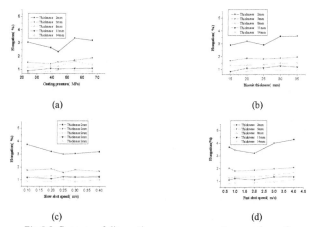

(a)   (b)

(c)   (d)

Fig.8 Influences of die casting process parameters on elongation

(a) casting pressure   (b) biscuit thickness   (c) slow shot speed   (d) fast shot speed

Yield strength. The influences of the die casting process parameters on the yield strength of the test specimens are shown in Fig. 9. It can be seen that the yield strength of the 2mm step is lower than those of the other steps and there is no significant influence of process parameters on the yield strength of the 2mm step. Increasing casting pressure increases the yield strength of the 5mm and 14mm casting step, but only change the yield strengths of the other steps slightly. The influence of biscuit thickness on the yield strength is small except in the 14mm casting step, where yield strength increases with biscuit thickness. Both the slow and fast shot speed have small effect on the yield strength.

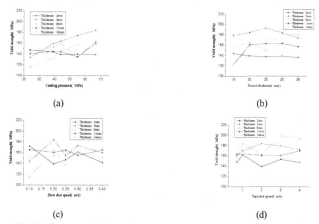

(a)   (b)

(c)   (d)

Fig.9 Influences of die casting process parameters on yield strength

(a) casting pressure   (b) biscuit thickness   (c)slow shot speed   (d) fast shot speed

Conclusions

(1) The density of die castings increases as the casting thickness decreases under all the test conditions. Increasing casting pressure and biscuit thickness increases the density of the die casting, and this effect becomes more significant when the casting thickness is larger. The density of the casting decreases with the increase of the fast shot speed.

(2) Microstructual analysis of the porosity distribution of the 14mm casting step shows that the effect of the die casting process parameters on the area percentage of porosity is identical to that on the casting density.

(3) The tensile strength of die castings increases with the decrease of the casting thickness. The influence of the die casting process parameters on the tensile strength follows the same trend as their influence on the casting density. A relationship between the tensile strength and casting density was established.

(4) The elongation of die castings increases with the decrease of the casting thickness and die casting process parameters do not have significant effect on the elongation.

(5) The effects of the casting thickness and the die casting process parameters on yield strength are not clear enough and further investigations are needed.

Acknowledgements

This research work was financially supported by the Ford Motor Company (0509A67) and by Tsinghua-TOYO R&D Center of Magnesium and Aluminum Alloys Processing Technology which was jointly established by Tsinghua University and TOYO Machinery & Metal Co., Ltd.

References

1. Metals Handbook. Cast, Vol, 9th edition, ASM Ohio: International, Metals Park (1998), 294-295.

2. M.R. Ghomashchi, A. Vikhrov, J. Mater. Process.Technol.101(2000), 429-436.

3. J.R. Morton, J. Barlow. The Foundrymen (1994), 23-28.

4. M.A.Savas, S.Altintas, Mater.Sci.Eng.A, 137(1993), 227-231.

5. F. Weinberg, Proceedings of an International Conference Organized by the Application Metallurgy and Metals technology on Solidification Technology in the Foundry and Cast House(1980), 131-136.

6. Bob R. Powell, Vadim Rezhets, Michael P. Balogh and Richard A. Woaldo. JOM (2002), 34-38.

7. V.D. Tsoukala. Int. J. Cast Metals Res., 15 (2003), 581-558.

8. Haavard T. Gjestland et al. Die Cast Engineer, 7(2004), 56-65.

9. Wenhui Liu et al. The 1st International Light Metals Technology Conference, Brisbane, Australia (2003), 43-147.

10. Gerry G. Wang, Bryan Froese and Per Bakke. Aluminum Technology 2003, ed. H.I.Kaplan (TMS,2003), 65-69.

11. Zheng Liu et al. Metall, 54(3)(2000), 122-125.

12. H. Mao,J. Brevick et al. SAE Technical Paper Series (1999), 27-33.

13. A.L. Bowles, J.R. Griffiths and C.J. Davidson. Aluminium Technology 2001, New Orleans, 2001, (TMS,2001), 161-168.

Shape Casting: The 2nd International Symposium *Edited by Paul N. Crepeau, Murat Tiryakioğlu and John Campbell*
TMS (The Minerals, Metals & Materials Society), 2007

THE EFFECT OF CASTING IN A HELIUM ATMOSPHERE ON THE COOLING RATE AND TENSILE PROPERTIES OF RESIN-BONDED SAND CASTINGS

Jean-Christophe Gebelin and W. D. Griffiths

IRC in Materials Processing, Department of Metallurgy and Materials,
School of Engineering, The University of Birmingham;
Edgbaston; Birmingham, B15 2TT, United Kingdom

Keywords: Solidification, resin-bonded sand casting, helium, aluminium

Abstract

The work presented here was aimed at examining the effect of replacing air by helium during casting of aluminium alloys in a resin-bonded sand casting process. Helium has a thermal conductivity about 4 to 5 times greater than that of air, and would be expected to increase cooling rates and improve mechanical properties of castings. Plates of different thicknesses (5, 10, 20 and 50 mm) were cast in a Sr-modified A356 alloy, in air and in a >95vol.% helium atmosphere. Cooling rates were enhanced by the use of helium by 59% to 84% over the solidification range. The mechanical properties were improved in the as-cast condition by up to 16% for the yield stress, by up to 14 % for the maximum stress, and by up to 5% for the maximum strain.

Introduction

As-cast mechanical properties in castings are greatly dependant on the cooling rate experienced during solidification. Increasing the cooling rate, increases the mechanical properties. The parameters controlling the rate of heat transfer from a casting are the mould or die material, and the interface between the casting and the mould. In the work presented here, a method to increase the rate of cooling by replacing the interfacial gas by helium has been used. Helium has a thermal conductivity about 4 to 5 times greater than that of air, and may be expected to have a noticeable impact on the rate of cooling of the casting. Doutre [1] showed that in the case of low pressure die casting, the use of helium in the die prior to casting can improve productivity from 23 to 29%.

In a sand mould the principal resistance to heat transfer from the casting is the sand mould itself, with the interface playing a minor role. From considerations of the thickness of a layer of gas trapped between the sand mould and the casting the interfacial heat transfer coefficient is probably around 770 $Wm^{-2}K^{-1}$ (determined by consideration of the thermal resistance of a layer of air with a depth of about 100 μm, about the diameter of a grain of sand). Applying He to the mould cavity should result in an interfacial heat transfer coefficient of 3670 $Wm^{-2}K^{-1}$; i.e., the heat transfer from the casting to the mould surface is increased by a factor of about 5.

The heat transfer within the sand mould is less well understood. Heat transfer occurs by conduction through the sand grains, via their points of contact, but also by conduction to, and within, the atmosphere between them[2]. By increasing the He content of the mould atmosphere, a significant increase in the overall thermal conductivity of the sand mould would be achieved. However, it is not known what fraction of the overall heat transfer through the mould is carried through the sand grains, and how much through the gas in the intergranular pores.

The present paper presents the results obtained from casting aluminium alloy plates in air and helium in a resin-bounded sand mould, and the influence of the interfacial gas on cooling rate, microstructure and tensile properties.

Experimental procedure

Plates of thicknesses of 5, 10, 20 and 50 mm were cast in both air and in a chamber containing a He atmosphere greater than 95 vol.%, to determine the effect of increased cooling rate on microstructure and mechanical properties. In the first stage of the experimental work the running and feeding systems of the plates were designed so as to avoid the introduction of defects associated with mould filling, and to avoid solidification shrinkage that would detract from the mechanical properties. To achieve this, commercial simulation packages MAGMAsoft® and Flow-3D® were used to design a running system that gave quiescent filling, without missrun, and that minimized shrinkage. The level of shrinkage porosity was assessed using MAGMAsoft® predictions. The final running system design used has been shown in Figure 1. It consisted of a tapered downsprue, followed by two ceramic foam filters, then a thin runner bar of 5 mm height and two feeders placed on both sides of the plate. At the end of the runner there was a dross trap to retain the metal damaged during the initial filling of the downsprue and the priming of the filters.

A pattern was produced and used to make resin-bonded sand moulds, and the mould filling models were validated by examining the filling of the mould for a 5 mm plate using real-time X-ray radiography, which confirmed that quiescent filling occurred. The microstructure of the casting was also examined and was found to be free of any porosity.

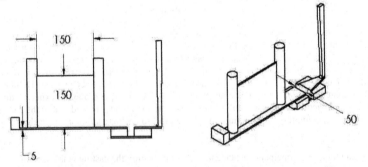

Figure 1. Final running system developed for the 5mm plate. The two filters used were 20 ppi ceramic foam filters.

An air-tight steel chamber was built to contain an electric resistance-heated furnace of 10 kg capacity, (of Al). All castings, whether cast in air or in He, were made inside the chamber following the same procedure. The A356 alloy charge was initially melted in air, followed by an addition of 0.03%Ti (to bring it to 0.18%) to act as a grain refiner, and an addition of 0.03% Sr for modification. The chamber was then closed and partially evacuated to 0.5 atmospheres and held for one hour to bring about some removal of hydrogen from the melt (the only means of degassing). For casting in air the chamber pressure would be returned to 1 atmosphere pressure and the mould cast. For casting in He the chamber would be backfilled with He, and then re-evacuated and backfilled with He several times, until the atmosphere reached the desired concentration of, typically, 95 to 96 vol.% He. As the concentration of He and the temperature increases so too does the thermal conductivity. This is shown in Figure 2 which shows the calculated thermal conductivity for air-He mixtures at room temperature (20°C) and the expected metal-mould interface temperature of 500°C.

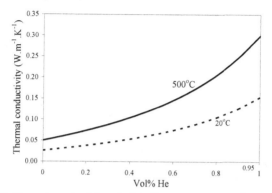

Figure 2. Thermal conductivity of air-helium mixtures at 20°C and 500°C calculated from Lindsay-Bromley equation [3].

The alloy used was a commercial A356 alloy and its composition is given in Table 1. Castings were poured at 740°C, except the 50 mm thick plate, which was cast at 720°C.

Table 1: Composition of the A356 alloy used

Element	Si	Fe	Cu	Mn	Mg	Cr	Ni	Zn	Pb	Sn	Ti	Sr	Al
Weight%	6.99	0.07	0.01	0.01	0.38	0.01	0.01	0.01	0.01	0.01	0.18	0.03	Bal

Some moulds contained thermocouples placed in the centre of the plate in order to record cooling curves. These castings were also used to examine the microstructure of the casting, also at the centre of the plate, and to determine their secondary dendrite arm spacing.

For each casting thickness and atmosphere two castings were made in order to determine and compare the mechanical properties in the as-cast and in the heat-treated T6 condition. The heat treatment for the T6 condition was as follows: solution heat treatment for 12 hours at 540°C, quench in water at 65°C, aging for 5 hours at 150°C, followed by cooling in air.

Tensile testing was carried out on a computer controlled Zwick tensile testing machine equipped with an extensometer, at a displacement rate of 1 mm.min⁻¹. The tensile specimens were taken from the centre of the cast plates, and so that the material characterized was from the middle plane of the plates. The tensile specimens from the 5 mm plates had a gauge length of 27 mm and a diameter in the gauge length of 3.8 mm. The specimens from the 10, 20 and 50 mm plates had a gauge length of 37 mm and a diameter in the gauge length of 6.75 mm.

The samples used for determination of the secondary dendrite arm spacing were also taken from the central plane of the plate, with the samples ground and polished using a 0.25 μm alumina suspension. Optical microscopy was used in conjunction with AxioVision image analysis software to measure the Secondary Dendrite Arm Spacing (SDAS). The number of measurements made for each plate was in the range of 15-20. Figure 3 shows an example of a micrograph with SDAS measurements.

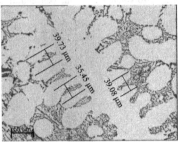

Figure 3. An example of SDAS measurement made on a 5mm plate cast in a helium atmosphere

Results

Table 2 and Figure 4 gives the cooling rates measured over the solidification range for the different plates in the different atmospheres.

Table 2: Average cooling rates between 603 and 555°C for the different plates.

Plate thickness	atmosphere	Cooling rate (K.s^{-1})	Cooling rate improvement (%)
5	He	1.03	71
	air	0.60	
10	He	0.38	84
	air	0.27	
20	He	0.12	59
	air	0.076	
50	He	0.031	79
	air	0.017	

These results showed that the use of a He atmosphere increased the cooling rate, with the improvement in the cooling rate being in the range of 59 to 84%.

Table 3 and Figure 5 shows the measured values of the secondary dendrite arm spacing for the different thickness plates cast in the different atmospheres.

Table 3: Secondary dendrite arm spacing measured in the different plates

Plate thickness		5 mm	10 mm	20 mm	50 mm
Helium	average (µm)	30	47	58	104
	std. dev. (µm)	5	9	11	18
Air	average (µm)	39	47	68	134
	std. dev. (µm)	6	12	10	21

Figure 5 plots the secondary dendrite arm spacing (SDAS) versus plate thickness for the plates cast in both air and helium atmospheres. The error bars for each plate are for plus or minus one standard deviation. As expected, for a given thickness, plates cast in helium gave smaller SDAS than when cast in air. The cooling rates obtained from the different plates spanned nearly two

294

orders of magnitude (1.7×10^{-2} to 1 K.s^{-1}), giving SDAS values of between 30 and 134 μm. The maximum relative change of SDAS observed for a given thickness was for the thinnest plates, 5 mm, with a change from 39 μm in air to 30 μm in helium atmosphere, a decrease in SDAS of 23%. The 10 mm plate, surprisingly, gave no change in SDAS despite a large change in cooling rate (84% increase in cooling rate for helium compared to air). This may be explained by the method used to measure the SDAS as, in the 10 mm plate sample, it was very difficult to find dendrites that were in the observed plane, so most of the measurements were made from dendrites that were not contained or were not parallel to the observed plane, perhaps resulting in a greater error in the measured values.

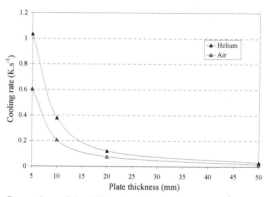

Figure 4. Comparison of the different cooling rates.

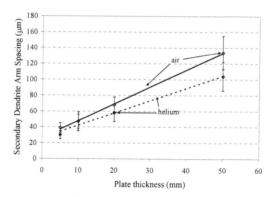

Figure 5. Evolution of the mean Secondary Dendrite Arm Spacing (SDAS) with plate thickness for the different plates cast in different atmospheres.

Table 4 summarises the mechanical property results from all of the plates, cast with the different thicknesses and atmospheres and these results are also shown in Figures 6 to 8. Figure 6 shows the evolution of the yield stress at 0.1% plastic deformation for the different plate thicknesses, with error bars showing +/- one standard deviation. It appeared that the as-cast plates (diamonds) cast in a helium atmosphere (filled markers) had a slightly higher yield stress than the air-cast plates (hollow markers), from 9% for the 10 mm plate to 17% for the 20 mm plate. It also showed that as the thickness increased the yield stress decreased. After heat treatment, the difference was less apparent, with some air cast plates having higher yield stresses than the equivalent helium cast plate. Looking at the error bars plotted in Fig.6 it appears that the significance of the differences associated with the air and helium atmospheres was low. Table 4 also shows the change in the yield stress at 0.2% plastic deformation for the different plate thicknesses. Values for the heat treated 50 mm plates were not given as most of the tensile test bars did not reach 0.2% plastic deformation. Observation of the tensile test bars machined from the 20 and 50 mm plates revealed some porosity within the test bars. The morphology of the pores suggested that they were due to hydrogen porosity rather than shrinkage and the decomposition of the resin in the mould during casting was possibly the origin of the gas content.

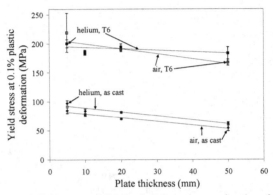

Figure 6: Evolution of Yield stress at 0.1% plastic deformation as a function of plate thickness for the different atmospheres in as-cast and T6 conditions.

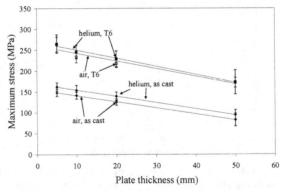

Figure 7: Maximum stress versus plate thicknesses.

Figure 7 shows the evolution of the maximum stress with plate thickness in the as-cast condition, with error bars of +/- one standard deviation. The graph shows that the plates cast in a helium atmosphere exhibited a higher maximum stress than when cast in air. The improvement of the maximum stress by using helium ranged from 9% for the 10 mm plate to 15% for the 20 mm plate, in the as-cast condition. After heat treatment, the improvement of the maximum stress by

296

casting in a helium atmosphere ranged from only 1% for the 50 mm plate to 5% for the 10 mm plate. However, the error bars plotted shown that these improvements were not statistically significant.

Table 4: Tensile tests results

Plate thickness	5				10			
Heat treatment	as-cast		T6		as-cast		T6	
Atmosphere	air	helium	air	helium	air	helium	air	helium
E (MPa)	75000	75247	89733	83651	61058	81109	85338	93781
$\sigma_{y0.1}$ (MPa)	85.0	97.6	219.4	200.7	77.0	83.9	184.1	185.2
$\sigma_{y0.2}$ (MPa)	93.1	107.5	242.3	212.3	85.3	94.9	198.5	199.3
Maximum stress (MPa)	147	162	263	265	141	154	234	246
Final plastic strain (%)	2.83	2.24	1.14	2.53	2.59	2.42	1.23	1.8
Final total strain (%)	2.63	2.46	0.81	2.86	2.82	2.62	1.51	2.07
Plate thickness	20				50			
Heat treatment	as-cast		T6		as-cast		T6	
Atmosphere	air	helium	air	helium	air	helium	air	helium
E (MPa)	55242	61611	85617	81175	67948	53052	96578	70946
$\sigma_{y0.1}$ (MPa)	70.1	81.7	189.9	194.8	54.2	62.6	167.7	183.5
$\sigma_{y0.2}$ (MPa)	78.5	90.6	204.6	208.8	67.0	70.6	-	-
Maximum stress (MPa)	125	139	218	229	82	94	172	173
Final plastic strain (%)	1.78	1.86	0.41	0.7	0.78	0.77	0.14	0.15
Final total strain (%)	2.01	2.09	0.67	0.98	0.9	0.95	0.32	0.39

Figure 8 shows the maximum strain versus plate thickness, with error bars of +/- one standard deviation. It appears that the final strain decreased with plate thickness. In the as-cast condition, the two thickest helium cast plates exhibited higher final strain than when cast in air, whereas the two thinnest had a lower final strain. After heat treatment, all the plates cast in helium possessed a higher final strain than the air-cast plates. Here again, the error bars show that the differences in the mean values were not significant.

Discussion

This preliminary work showed that the use of a helium atmosphere in a resin-bonded sand casting process produced increased cooling rates, a shorter solidification time and enhanced mechanical properties compared to casting in air. The improvement in the average cooling rate during solidification was between 60 and 80%, and that led to an improvement in the mean values of the mechanical properties in the as-cast condition, of between 9 or 17% for the yield stress at 0.1% plastic deformation, and between 9 to 15% for the maximum stress. Nevertheless the differences recorded were not significant. Also, the mechanical properties after heat treatment did not show a substantial improvement, probably due to the fact that the aim of the heat treatment is to create a required microstructure, which may have "erased" the thermal history of the material therefore reducing the effect of the initial solidification. For reference, ASM handbooks[4] give for sand cast A356 a tensile strength of 227 MPa, a yield stress at 0.2% of 165 MPa and a maximum strain of 3.5% at 24°C. This was achieved in these experiments but any enhancement in the mechanical properties with the application of He was showed mostly in the as-cast alloy, and was diminished by the T6 heat treatment. In many of the castings, particularly in the larger section thicknesses the properties were probably strongly affected by porosity. While the improvements in mechanical properties were slight, the application of He produced a 60 to 80% increase in cooling rate in the sand moulds which could be exploited to increase productivity.

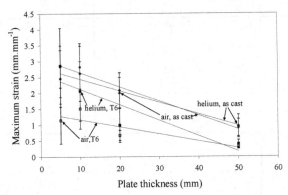

Figure 8: Final strain versus plate thickness.

Summary

1. The use of a He atmosphere in the resin-bonded sand casting process resulted in an increase in the cooling rate, compared to casting in air, of from 60 to 80%.
2. The use of a He atmosphere caused a decrease in SDAS by up to 23%.
3. Slight improvements in mechanical properties of as-cast plates were associated with the use of a He atmosphere, but these improvements had low statistical significance.
4. A T6 heat treatment further reduced the slight improvements in mechanical properties found.

Acknowledgements

The authors would like to thank EPSRC for its financial support with grant no. EP/C514718/1. The technical help of Mr. Adrian Caden throughout the project is also gratefully acknowledged.

References

[1] D. A. Doutre, *The influence of helium injection on the cooling rate and productivity of the permanent mold casting process* (5th AFS International Conference on Permanent Mold Casting of Aluminum, Milwaukee, WI, USA, 23-24 Oct 2000), 96-105
[2] K. Kubo and R. D. Pehlke, *Heat and moisture transfer in sand molds containing water* (Metall. Trans. B, 1986, vol. 17B), 903-911.
[3] A.L. Lindsay and L.A. Bromley, *Thermal conductivity of gas mixtures* (Ind. Engng. Chem., Volume 42, 1950), 1508 -1511
[4] ASM International handbook committee, ASM handbooks, Volume 2: Properties and selection--nonferrous alloys and special-purpose materials (American Society for Metals, 1999).

DEVELOPMENT OF RHEO-DIECASTING (RDC) PROCESS FOR PRODUCTION OF HIGH INTEGRITY COMPONENTS

Z. Fan, S. Ji, X. Fang, G. Liu, J. Patel and A. Das

BCAST (Brunel Centre for Advanced Solidification Technology),
Brunel University, Uxbridge, Middlesex, UB8 3PH, UK

Keywords: Semisolid processing, Rheo Diecasting, Microstructure, Mechanical Property

Abstract

Rheo-diecasting (RDC) is an innovative one-step semisolid metal (SSM) processing technique to manufacture near net shape components of high integrity directly from liquid alloys. The RDC process innovatively adapts the well-established high shear dispersive mixing action of a twin-screw mechanism for *in situ* creation of SSM slurry with fine and spherical solid followed by direct shaping of the SSM slurry into near-net shape components using cold chamber high pressure diecasting (HPDC) process. Based on the results achieved so far, RDC shows several advantages over conventional HPDC process such as, finer and more uniform microstructure, close to zero porosity, more tolerant to inclusions, ability to process a wider range of alloys, and improved mechanical properties. In this paper, we report the rheo-diecasting process and the scientific understandings behind this technological development. We also present the microstructures and mechanical properties of various Al- and Mg-alloys processed by RDC

Introduction

Since the discovery of thixotropic and pseudoplastic behavior of solid-liquid slurries at MIT, semisolid processing of metallic materials has experienced 3 decades of intense research and development efforts [1]. The increasing demand for energy efficiency, environmental impact and stronger yet lighter materials have led to a tremendous potential for semisolid processing of light alloys in the future, especially in the automotive sector where the commitment to reduce CO_2 emission has prompted a steady increase of Al and Mg usage. At present, high pressure die casting (HPDC) is the preferred fabrication route for Mg and Al components in the transportation sector due to high production efficiency, low production cost and near net shape component production. However, HPDC components suffer from a substantial amount of porosity introduced due to turbulent mould filling [2] severely limiting their usability in demanding applications and post casting property improvement through heat treatment due to blistering from sub-surface pores. Semisolid processing addresses such limitations through laminar mould filling of solid-liquid slurry of higher viscosity compared to liquid metal. In addition, the non-dendritic and refined microstructure reduces segregation and improves strength, while the low heat content of the slurry improves the life of processing equipment. The initial development of semisolid technologies through partial melting and shaping of specially prepared thixotropic feedstock (also known as thixoprocessing) has been severely plagued by the high operating cost and lack of in-house recycling. Thereafter, slurry-on-demand (rheoprocessing) routes are being developed that utilizes in house slurry generation directly from the melt and subsequent processing. Rheoprocessing holds the key to successful development and acceptance of semisolid technologies and worldwide initiatives have started to develop efficient slurry generation techniques and integration to component shaping [1]. BCAST has been involved in developing novel semisolid technologies based on the outcome of fundamental research on solidification under high intensity melt shearing. This paper presents an overview of the newly developed rheo diecasting (RDC) technology for the casting of high integrity components including the process description, physical mechanism of microstructure formation, and the microstructure and mechanical properties of a range of alloy systems produced through RDC.

The RDC process

The single-step RDC process, as shown in Fig.1, consists of two basic functional units. The unique twin-screw slurry generator uses high intensity shear and dispersive mixing action for *in-situ* creation of semisolid slurry with fine and spherical solid particles followed by direct shaping of the slurry into near-net shape components using an existing cold chamber HPDC process. A detailed description of the RDC process can be found elsewhere [3].

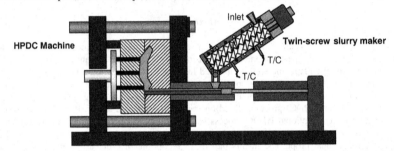

Fig. 1 The RDC equipment consisting of a mobile twin-screw slurry-making device connected to a conventional cold chamber diecasting machine.

The slurry maker consists of a pair of specially designed, fully intermeshing and self-wiping screws co-rotating inside a closely matched barrel. The barrel contains a heating and cooling arrangement with a precise temperature control of ±1 °C. The melt fed into the slurry maker experience high intensity shear and turbulence while being transported forward through a positive displacement action. The enormous amount of ever-changing contact area between the melt and the slurry maker provides fast and efficient heat transfer converting the melt into semisolid slurry of requisite solid fraction. For Mg-alloys a protective environment is used in the melting furnace and the slurry maker. Flow inside the slurry maker is characterized by high shear rate, turbulence, and cyclic variation of shear rate. Both shear rate and degree of turbulence is proportional to the screw rotation speed and are functions of the screw profile and arrangement.

During RDC, a predetermined dose of melt is continuously cooled to the semisolid temperature while being intensely sheared in the slurry maker. Solid fraction is controlled by setting the barrel temperature. The slurry maker works in a batch manner providing semisolid slurry every 20-30s, which is then transferred to the shot chamber of the HPDC machine for component shaping. Both the slurry maker and the HPDC work in parallel rather than sequentially and at the end of a HPDC cycle a new dose of slurry is ready for subsequent component shaping.

Solidification in the twin-screw device

Since the early rheological work, it has been realized that globular solid particles can ensure sufficient slurry fluidity even at high (>20) vol.% solid [4]. This is because spheres can move past each other fairly easily than say dendrites, and fewer points of contact between spherical particles compared to irregular ones in an agglomerate promotes faster deagglomeration under applied shear. To generate such slurry one has to promote nucleation (or grain multiplication)

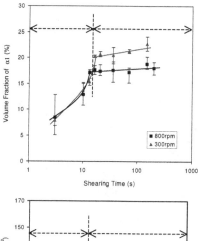

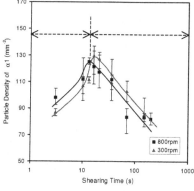

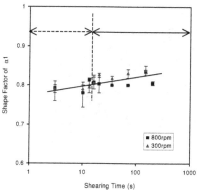

Fig. 2 Evolution of (a) solid fraction, (b) particle density and (c) shape factor for AZ91D alloy as a function of shearing time in the twin-screw device.

and subsequent non-dendritic growth of the solid, which is the basis for twin-screw slurry maker. By employing high intensity shear and turbulent mixing, uniform temperature and composition is created in the entire melt along with effective dispersion of natural nucleants. Also, efficient heat transfer ensures effective removal of latent heat of solidification minimizing recalescence. Under these conditions, most of the nuclei formed below the liquidus can survive leading to an effective increase in nuclei density in the melt. Fig. 2 presents the volume fraction, number density and shape factor of solid formed inside the twin-screw device as a function of shearing time for AZ91D alloy.

Fig. 2a shows that the melt is continuously cooled to the semisolid temperature within 15 sec with a corresponding rise in the solid fraction in the twin-screw device. Following this, the slurry is sheared isothermally. Fig. 2b indicates a continuous increase in the particle density during the continuous cooling while remaining highly spherical (high shape factor) from the very beginning as shown in Fig. 2c. During the isothermal shearing, shape factor increases nominally but particle density falls due to Ostwald ripening. The steady increase in particle density during cooling is indicative of continuous nucleation as non-dendritic particles from the beginning discard any grain multiplication through dendrite fragmentation. Accordingly, the increase in solid fraction is contributed by fresh nucleation rather than growth in the twin-screw device. Continuous nucleation with high nuclei survival leads to a slurry with high density of refined particles. It is also necessary to ensure compact growth morphology of the solid. Constitutional undercooling from the rejected solute at the interface promotes growth instability leading to dendritic morphology. Such instability can occur beyond a critical cooling rate even when the initial particle density is moderately high. Employing computer simulation it has been found that convection can modify the growth morphology by altering the solutal diffusion layer as shown schematically in Fig. 3.

Dendrite formation is observed without convection, while particle rotation under laminar flow promotes a rosette morphology. Turbulent flow, by destroying the solute diffusion layer, promotes compact growth morphology. Turbulence inside the twin-screw device thereby ensures stable compact growth at all solid fractions and cooling rates. The combination of high effective nucleation, limited growth, stabilization of compact morphology and dispersive mixing is utilized to produce an ideal semisolid slurry with fine spherical particles uniformly dispersed in the liquid matrix for subsequent component shaping.

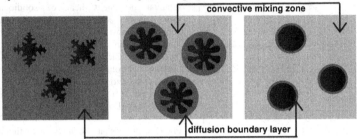

Fig. 3 Effect of the nature of fluid flow on the solute diffusion boundary layer at the solid-liquid interface, and consequently, the growth morphology.

Microstructural features of RDC samples

RDC has so far been used to process a wide variety of Al and Mg based cast and wrought alloys and the microstructural features observed are very similar independent of the alloy system. Fig. 4 presents the microstructural features across the entire cross section of RDC processed AZ91D alloy bar of 6 mm diameter. The spherical and fine light areas represent the primary solid of the semisolid slurry uniformly distributed in the matrix of remnant liquid that has solidified into extremely fine structure. Microstructural uniformity across the entire cross section clearly indicates that no liquid-solid segregation occurs during RDC. The total porosity observed is less than 0.3% illustrating the benefits of semisolid processing over HPDC. Fig 5 presents detailed microstructural features from RDC processed A357 cast alloy and 2014 wrought Al-alloy.

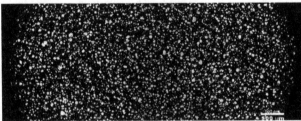

Fig. 4 Optical micrograph illustrating the microstructure across the cross section of a ϕ 6 mm AZ91D alloy bar processed through RDC at 593 °C.

The large white particles are the solid formed inside the slurry maker, henceforth called primary solid. The fine white solid (α-Al) formed from the sheared remnant liquid of the slurry inside the die cavity (during subsequent diecasting) is termed secondary solid. Both the primary and secondary solid are highly globular and around 50 μm and 10 μm in size, respectively, irrespective of the alloy system. The primary solid is very stable and maintains its globular morphology during further solidification inside the die. The sheared remnant liquid solidifies in a

302

fine non-dendritic structure resulting from the temperature and compositional uniformity (due to prior shearing), no superheat and fast heat extraction during diecasting. The processed Mg and Al based wrought alloys showed reduced hot tearing susceptibility from a combination of reduced temperature, heat content and solidification shrinkage. Also, unlike the interdendritic cracking observed in normal cast conditions, RDC samples showed no prominent cracking indicative of better melt feeding during the last stages of solidification. The efficient dispersion ability of RDC was also successfully utilized in processing immiscible alloys and particulate reinforced metal matrix composites, some representative microstructures are shown in Fig. 6.

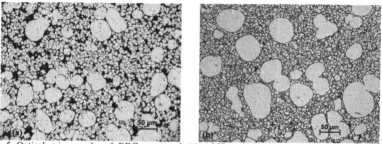

Fig. 5 Optical micrographs of RDC processed (a) A357 and (b) 2014 Al-alloys illustrating the morphology and size of primary α-Al solid formed inside twin-screw device and during diecasting.

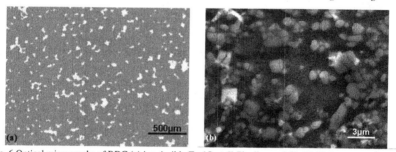

Fig. 6 Optical micrographs of RDC (a) immiscible Zn-15 wt.% Pb alloy and (b) Al-Al$_2$O$_3$ composite.

Impurity tolerance of RDC

One potential benefit of rheo- over thixo-processing is the opportunity for in-house scrap recycling. It is, however, essential for a process to be tolerant to impurity pickup in terms of both changes in the processability arising out of changing alloy chemistry and minimizing the harmful effect of impurities on the mechanical properties of processed components. Successful processing of wide range of alloys has been pointed out in the preceding section. Due to the fine size and globular shape of the solid particles, changes in the viscosity of the slurry are minor until moderately high solid fraction. Consequently, RDC has a stable and wide processing window to tolerate moderate changes in the alloy chemistry. The dispersive mixing and high shear was found effective in breaking up and distributing oxide inclusions reducing their harmful effect. One major detrimental impurity for Al alloys is Fe, normally picked up from the use of industrial steel tooling and through recycling. Due to the low solid solubility of Fe in Al, low symmetry Fe-bearing intermetallics crystallize in the form of long needles and plates and substantially reduce the ductility of cast components. For commercial Al-Si alloys, the Fe content is limited below 0.7 wt.% [5] and keeping the Fe content low economically is one of the concerns for recycling. Fig. 7 compares the tensile elongation obtained from RDC and HPDC samples as a function of total Fe content (extra Fe added through Al-Fe master alloy) for LM24 (Al-9.3Si-3.2Cu-1.58Zn-0.3Mn in wt.%) and LM25 (Al-10.3Si-0.3Cu-0.16Mn) alloys. Both RDC and HPDC samples suffered from rapid reduction of elongation as the Fe content increased. However, RDC showed higher Fe tolerance compared to HPDC (0.7 and 0.35 wt.%, respectively, for LM24 and LM25) maintaining similar level of elongation. For a specific Fe level, RDC specimens showed higher elongation compared to HPDC. The drastic reduction in elongation in HPDC samples is contributed by long needles and plates of βAlFeSi and compact but coarse αAlFeMnSi phases. The refinement of these Fe-bearing intermetallic particles under shear reduced the detrimental effect on ductility, improving the Fe tolerance of RDC.

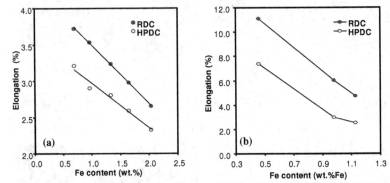

Fig. 7 Tensile elongation as a function of Fe content in HPDC and RDC (a) LM24 and (b) LM25 alloys.

Mechanical properties of RDC samples compared to other processes

The microstructural refinement observed in RDC is expected to improve the mechanical property of components. However, the major impact on mechanical property, especially elongation, is likely to result from reduction in porosity level over HPDC. Table 1 summarises the mechanical properties of Mg-alloys processed using different methods and various heat-treated condition.

Table 1. Mechanical properties of RDC Mg-alloys in different heat treated condition compared against reported values in the literature.

Alloy	Process	Mechanical property			Ref.
		YS (MPa)	UTS (MPa)	Elongation (%)	
AZ61	RDC	126	253	12.3	
	HPDC	146	212	3.3	[6]
	Thixocasting	134	223	3.6	[7]
	Thixomoulding	—	150-241	3-5	[8]
	NRC	—	230	5.5	[9]
AZ91D	RDC	145	248	7.4	
	RDC + T4	91	230	10.7	
	RDC + T5	133	236	6.2	
	RDC + T6	134	255	6.4	
	RDC + Tx	125	259	9.0	
	HPDC	128.7	194.8	6.7	[10]
AM50	Thixocasting	108 ± 6	200 ± 3	7 ± 0.7	[11]
	RDC	120	240	19.5	

304

Comparing the mechanical property data in Table 1, it is evident that RDC shows improvement in tensile strength and elongation values compared to HPDC and other SSM techniques resulting from the high integrity of samples. For AZ91D, solution treatment improved the elongation but age hardening deteriorated the elongation without improving the tensile strength. It was realised that discontinuous precipitation of $Mg_{17}Al_{12}$ phase hinders strengthening while continuous grain boundary network of the phase reduces ductility. A heat treatment, Tx, was developed to partially dissolve the grain boundary intermetallic network by short holding near the solvus that improved the ductility of as-cast components without sacrificing strength. Similar property enhancement is observed for Al-alloys. Fig. 8 presents the tensile properties of A357 alloy processed through different semisolid routes. Both the tensile strength and elongation obtained in RDC compares favourably against other established semisolid processes.

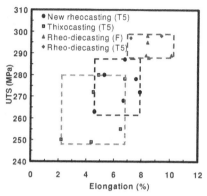

Fig. 8 Tensile properties of RDC A357 alloy compared against the properties obtained in other processes. New rheocasting and thixocasting data are from Ref. [12].

Component production trials

After successful development, the RDC process is now being explored for component production, at the same time being marketed commercially by Zyomax Ltd., Uxbridge, UK. Components with varying section thickness ranging from 2-6 mm have been successfully produced with high integrity and good surface finish. Fig. 9a illustrates the overall uniformity and soundness of the microstructures obtained from various parts of a cast component including runners. A consortium has been formed with the material supplier, component manufacturer and end user, under the financial support of Department of Trade and Industry (DTI), UK and the private sector, to scale up the process to industrial scale. Pilot scale production trial of automotive components is currently underway such as the Mg-alloy seat frame shown in Fig. 9b.

(a) (b)

Fig. 9 Industrial scale trial components produced using RDC showing (a) microstructure from various parts of an LM24 component and (b) a Mg-alloy seat frame for Jaguar XJ and J-series.

305

Summary

Rheo Diecasting (RDC) is a new semisolid casting technology that integrates the novel twin-screw slurry making with existing cold chamber diecasting operation. After successful development, the process is now undergoing component production trials. The unique characteristics and benefits of RDC over HPDC can be summarized as follows:

- High effective nucleation and compact growth inside the slurry maker under shear and in the die following shearing contributes to a fine and uniform microstructure with highly globular solid particles.
- Significantly reduced porosity (< 0.3%) and hot tearing resulting from laminar mould filling and lower operating temperature, heat content and solidification shrinkage. This makes RDC components amenable to post casting heat treatment.
- High process stability and wide processing window, capability to process a wide range of alloys ranging from cast and wrought alloys, large and short freezing range alloys, immiscible alloys and metal-matrix composites.
- High tolerance to impurity and oxide inclusions due to the refinement and dispersion of impurities reducing the harmful influence on mechanical properties. In particular, RDC is found to be more tolerant to Fe contamination of Al alloys compared to HPDC.
- Improvement of mechanical properties, especially tensile elongation, compared to HPDC and other semisolid processing technologies for a wide variety of alloy systems.
- Additional processing benefits result from increased die life due to lower operating temperature, high energy efficiency, lower scrap rate, high material yield and high productivity leading to lower overall component production cost.

Acknowledgements

The financial and material support from EPSRC, DTI, Magnesium Elektron Ltd. and Ford Motor Co., UK is gratefully acknowledged.

References

[1] A. Das, Z. Fan, in Z. Xiao Guo (Ed.), 'The deformation and processing of structural materials', Woodhead Publ. Ltd., Cambridge, 2005, p. 252.

[2] A. Balasundaram, A.M. Gokhale, in: J. Hryn (Ed.), Mg Technology, TMS, Warrendale, PA, 2001, p. 155.

[3] Z. Fan, S. Ji, G. Liu, Mater. Sci. Forum 488-489 (2005) 405.

[4] M.C. Flemings, Metall. Mater. Trans. A, 22A (1991) 957.

[5] Al and Al-alloys, ASM speciality handbook, J.R. Davis (Ed.), ASM International, 1993.

[6] C. Pitsaris, T. Abbott, C.H.J. Davies, G. Savage, in: K.U. Kainer (Ed.), Proc. 6th Int. Conf. on Magnesium Technology and Their Applications, Wiley-VCH, Weinheim, 2003, p. 694.

[7] F. Czerwinski, A. Zielinska-Lipiec, P.J. Pinet, J. Overbeeke, Acta Mater. 49 (2001) 1225.

[8] J. Aguilar, T. Grimming, A Bührig-Polaczek, in: K.U. Kainer (Ed.), Proc. 6th Int. Conf. on Magnesium Technology and Their Applications, Wiley-VCH, Weinheim, 2003, p. 767.

[9] H. Kaufmann, P.J. Uggowitzer, in: K.U. Kainer (Ed.), Magnesium Alloy and Their Application, Wiley-VCH, Verlay GmbH, Weinheim, 2000, p. 533.

[10] R.M. Wang, A. Elizer, E.M. Gutman, Mater.Sci. Eng. A355 (2003) 201

[11] Z. Koren, H. Rosenson, E.M. Gutman, YaB Unigovski, A. Eliezer, J. Light Metals, 2 (2002) 81.

[12] P. Giordano, G.L. Chiarmetta, in: Y. Tsutsui et al. (Eds.), Proc. 7th Int. Conf. on Semisolid Metal Processing, Tsukuba, 2002, p. 665

AUTHOR INDEX

SUBJECT INDEX